CLIMATE JUSTICE

UNIVERSITY OF CALGARY
LCR Publishing Services

CLIMATE JUSTICE AND

Participatory Research

BUILDING CLIMATE-RESILIENT COMMONS

EDITED BY

Patricia E. Perkins

LCR Publishing Services
An imprint of University of Calgary Press
2500 University Drive NW
Calgary, Alberta
Canada T2N 1N4
press.ucalgary.ca

This book's Open Access publication, and the research on which it is based, were supported by the Canadian Queen Elizabeth Scholars program (QES-AS), with financial support from the International Development Research Council (IDRC) and the Social Sciences and Humanities Research Council of Canada (SSHRC).

LIBRARY AND ARCHIVES CANADA CATALOGUING IN PUBLICATION

Title: Climate justice and participatory research : building climate-resilient commons / edited by Patricia E. Perkins.
Names: Perkins, Patricia E., editor.
Description: Includes bibliographical references and index.
Identifiers: Canadiana (print) 20230199895 | Canadiana (ebook) 20230199992 | ISBN 9781773854663 (hardcover) | ISBN 9781773854076 (softcover) | ISBN 9781773854106 (EPUB) | ISBN 9781773854090 (PDF) | ISBN 9781773854083 (Open Access PDF)
Subjects: LCSH: Climate justice—International cooperation. | LCSH: Participant observation. | LCSH: Commons. | LCSH: Resilience (Ecology) | LCSH: Climatic changes—Social aspects.
Classification: LCC GE220 .C55 2023 | DDC 363.7/0526—dc23

The University of Calgary Press acknowledges the support of the Government of Alberta through the Alberta Media Fund for our publications. We acknowledge the financial support of the Government of Canada. We acknowledge the financial support of the Canada Council for the Arts for our publishing program.

Copyediting by Tania Therien
Cover image: Colourbox 49263254
Cover design, page design, and typesetting by Melina Cusano

Contents

Part IV: Collective Resilience for Climate Justice

List of Maps

Rhys Davies

Introduction: Participatory Research, Knowledge and Livelihood Commons Build Community-Based Climate Resilience

Patricia E. (Ellie) Perkins

The authors of this book live and work in many territories across Africa and the Americas, where settler colonialism over more than five hundred years has violently dispossessed original peoples, vastly enriched colonizers and fuelled capitalist globalization that externalizes environmental costs, feeds on inequities, and is now endangering the planet. When we acknowledge the original peoples, we recognize our responsibility to keep working for justice for those affected by colonial violence, including climate change. Canada (as it is now called) is the homeland of more than fifty First Nations; they include the Anishinaabe, Haudenosaunee, Huron-Wendat, and Abenaki in what are now called Ontario and Quebec, and the Niitsitapi, Nêhiyawêwin, Denesuline and other Dene peoples, Tsuut'ina, and Nakoda in what is now called Alberta. In coastal Brazil, the Tupi-Guarani. In north-central Chile, the Mapuche. In Africa too, many original peoples preserve their governance traditions, languages, and cultures. Colonial diasporas exacerbated by climate catastrophe continue to uproot and disperse Indigenous peoples. Climate justice centres their struggles, their cultures' wisdom, and their claims to land, water, and political agency.

This book brings together a variety of articulations of the What, Why, and How of climate justice, through the voices of motivated and energetic scholar-activists who are building alliances across Latin America, Africa, and Canada. Exemplifying socio-ecological transformation through equitable public engagement, each chapter describes processes that are underway in many settings to build the post-fossil-fuel energy transition and transform socio-economies from situations of vulnerability to collective well-being.

"Climate justice" names the vision of removing climate-related inequities, both within countries/regions and worldwide. The poorest and most marginalized, who are least responsible for the consumption and emissions that cause climate change, are the first and hardest impacted, and also those least able to protect themselves due to socially perpetuated inequities including racism, gender-based inequities and violence, poverty, and discrimination. Climate justice is simultaneously a movement, an academic field, an organizing principle, and a political demand. Climate catastrophe throws into stark relief the extreme inequities that colonialism creates and capitalism relies on, which are life-threatening for growing numbers of people worldwide: building climate justice is a matter of life and death for millions. Where to start, in untangling the many interrelated challenges posed by climate justice intentions? This book offers ideas and inspiration.

Grounded in our varied experiences as researchers, climate activists, community educators and teachers, we show how participatory research, knowledge and livelihood commons can help to build community-based climate resilience. "Commons" are resources vital for survival to which many people have access (such as farmland, water, aquifers, forests, fisheries, and other ecosystem-based productive areas).[1] Rights to use the resources, and responsibilities for caring for and maintaining them, are mediated by community-organized and enforced commons governance systems that prevent "open access" (the "tragedy of the commons" where outsiders, disregarding local needs and practices, swoop in to "enclose," claim and seize what they opportunistically see as individually profitable). In times of ecological and social crisis such as those driven by climate chaos, war, or societal breakdown, the governance systems that maintain commons can be as protective of local well-being as the commonly held resources themselves (Fournier, 2013; Farooqui et al., 2021; Berkes, 2017; Bollier, 2014; Burger et al., 2001).

Collective action, commonly held assets, mutual aid, environmental protection by those who know the area best, and partnerships with allies (including academics) from outside the area, are some of the themes that emerge from the first-person stories told in this book. Seventeen chapter authors from Canada and eight Latin American and African countries describe their research related to climate justice and commons, carried out in partnership with local communities and civil society organizations. Some of the stories show the negative effects of climate-related actions that run roughshod over local communities' interests and well-being—for example, REDD+ (reducing emissions from deforestation and forest degradation) carbon-sequestration projects in Mozambique that further impoverish small farmers, or organic food initiatives in Brazil that cause producers to depend on niche upper-class markets in big cities rather than feeding the local community. Participatory research itself can present challenges, as when technologies are intimidating for community participants with less access or familiarity, or when long distances prevent easy communication.

To learn from such grassroots perspectives, you have to talk with the people directly affected. The authors share their diverse stories of commitment to participatory research as a means to further climate justice—what works and what doesn't.

Our focus on commons and collective resilience-building continues a long tradition of mutual aid, which has always been the source of protections for the most vulnerable in times of chaos and deprivation (Hossein& Kinyanjui, 2022). It extends "commons-building" to include "commons-reclamation" in its descriptions of how this works in specific Latin American and African contexts. The stories told here make current climate justice processes and activism richly understandable in relation to each other.

For example, in South Africa, collective land ownership systems dating back to pre-apartheid times provide a model for collective water rights and activism by smallholder farmers—including recourse to the courts—to reclaim the means to produce their livelihoods. In Brazil, quilombola community members in towns that began as refuges for self-liberated slaves still hold land and work together collectively, using mutual aid to minimize climate risks for all. In Chile, Indigenous fishers request official recognition of their expertise in marine conservation as a way of preserving wisdom about cross-species environmental protection in marine commons. And in Kenya, local community lawsuits against port and industry construction that destroys mangroves and fishers' livelihoods results in the project's funders pulling out: communities can successfully protect local commons.

The lead authors for all chapters in this book were part of an international project on Climate Justice, Commons Governance, and Ecological Economics (2018–2021) that helped to support their participatory climate-justice related research in their own countries, linking sixteen universities in Latin America, Africa and Canada.[2] It was funded by the Social Sciences and Humanities Research Council of Canada (SSHRC) and the International Development Research Centre (IDRC) through the Queen Elizabeth Scholars—Advanced Scholars (QES-AS) program. Despite the challenges of carrying out participatory research during the COVID-19 pandemic (which delayed some researchers' projects and forced modifications to others), the scholars communicated virtually, established regional and special-interest sub-networks, and peer-reviewed each other's chapters for this book. They are part of a growing global network of QES Scholars who share experience and commitment to participatory research approaches in their academic careers. Other contributors to the project and this book include Patricia Figueiredo Walker and Kathryn Wells, coordinators of the project who are also chapter authors; and Rhys Davies, who drew the maps and illustrations for this book. This book thus represents the hard work of a large group of much appreciated collaborators.

The book is organized into sections on "Knowledge Commons," "Food, Land, and Agricultural Commons," "Water and Fisheries Commons," and "Collective Resilience for Climate Justice." There are many overlaps and cross-connections across sections and chapters.

The first section focuses on knowledge commons: information and understandings about the world human beings inhabit, which includes knowledge produced in research processes as well as cultural and traditional knowledge collected and developed by humans over time in interactive relationships with territories and their other human and non-human inhabitants. Kathryn Wells discusses knowledge commons in the context of decolonization, ethics, and climate justice. Allan Iwama and co-authors focus on widening access to knowledge via citizen or community science, describing their experience with a Brazilian-Chilean example of knowledge-building and sharing. They identify both potentials and limitations of technological tools for geographic and ecosystem measurement.

In section two, on food, land, and agriculture, Ayansina Ayanlade and co-authors describe their community-based research with small farmers in two Nigerian ecological zones to assess climate impacts, farmers' options and local resilience strategies. Guy Donald Abassombe recounts his extensive knowledge exchange process with Cameroonian palm farmers in Ngwéi province. Kátia Carolino and Marcos Sorrentino discuss Brazilian land laws and how they constrain options for urban food production and community development, with particular reference to community gardens in São Paulo. Marcondes Coelho and co-authors describe their research on soils as a commons with *Quilombolo* community members in Brazil who have collective farming and commons traditions that predate the abolition of slavery. Aico Nogueira traces influences on Brazilian small farmers' agro-ecological choices as they juggle their proximity to conservation areas, changing government policies, urban food markets, and the energy transition.

Section three focuses on water and fisheries commons. Daniela Campolino and Lussandra Gianasi describe the chaotic, toxic impacts on watersheds when poorly maintained mine tailings dams collapse, and how information and education are key to building public capacity to respond—in Brazil, Canada, and elsewhere. Solomon Njenga discusses his work with a climate justice non-governmental organization (NGO) and local fisherfolk in Lamu, Kenya, where development of an oil export terminal is destroying mangroves and coral reefs. Camila Bañales-Seguel tells the story of her work with Mapuche community members to document climate impacts on the Queuco watershed in Chile while transmitting Indigenous knowledge via place-names (toponyms). Francisco Araos and co-authors show how they have built relationships with Indigenous partners to carry out community mapping that blends traditional knowledge and "scientific" understandings of ecosystem change. Patience Mukuyu and Mary Galvin show how water commons in South Africa's Inkomati watershed are being defended by Black farmers, relying on legal rights and pre-apartheid land governance frameworks. Ferrial Adam highlights women's leadership in democratizing water management in the Vaal watershed of South Africa, using community science approaches.

The last section includes examples of collective resilience for climate justice. Andries Motau's chapter explains how participatory engagement with community stakeholders (civil society organizations, unions, and workers) in the Mpumalanga coal mining region of South Africa is helping to develop a detailed and nuanced understanding of Just Transitions-related tensions in that coal-intensive region, from a community-based perspective. Chrislain Eric Kenfack describes his faith-based work with Indigenous environmental activists in Canada and Brazil. Natacha Bruna and Boaventura Monjane's chapter details how climate change mitigation policies such as REDD+ are harming Mozambican farmers, while community-based interventions led by a climate justice NGO in another area are building collaborative agro-eco-logical futures. Patricia Walker overviews the strength and potential of youth movements for intergenerational climate justice that benefits everyone.

The methodologies, conclusions, and climate justice implications of all these chapters are diverse, situation-specific, and best explained by the authors themselves. Together, in conversation with each other, they provide important inspiration, motivation and guidance about how academic-community alliances can be developed, and the promise of such alliances to advance equitable socio-ecological transformation in the face of climate chaos. The authors demonstrate the potentials and importance of participatory engagement to address climate-related inequities, laying the foundation for a fairer post-fossil fuel future.

Figure 0.1 represents the multi-directional interactions and reinforcing feedbacks that the stories in this book describe in relation to knowledge production and sharing across the boundaries of academia, class, race, gender, and the global geographic divides that are heightened by capitalism and colonialism such as rich and poor, Global North and South (Minority and Majority Worlds). When participatory research and action-oriented research are based in partnerships leading to ongoing relationships of trust between academics and community groups, published research more knowledgeably reflects community worldviews and priorities, which then become available and may enter policy discourse, while also increasing the voice, agency, confidence and organization of traditionally marginalized climate justice experts with lived experience of climate emergencies and priorities regarding what should be addressed first.

At the same time, local communities may gain access to academic information sources, broader-scale allies, and political networking opportunities that help strengthen their demands for land, water, and livelihood security. This also increases political pressure for equity across society, since climate change (and pandemics) demonstrate that human and ecosystem futures are closely interrelated. Skills for successful commons governance and reclamation may be highlighted and resuscitated. Indigenous experts may be recognized and sought out. Movements for food sovereignty, inclusive watersheds and water access, just energy transitions, citizen science/community science

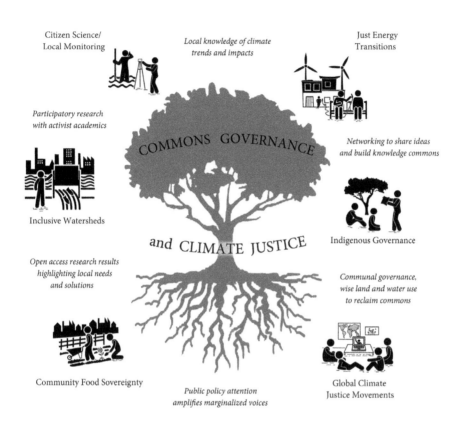

Citizen Science/
Local Monitoring

*Local knowledge of climate
trends and impacts*

Just Energy
Transitions

*Participatory research
with activist academics*

COMMONS GOVERNANCE

*Networking to share ideas
and build knowledge commons*

Inclusive Watersheds

and CLIMATE JUSTICE

Indigenous Governance

*Open access research results
highlighting local needs
and solutions*

*Communal governance,
wise land and water use
to reclaim commons*

Community Food Sovereignty

*Public policy attention
amplifies marginalized voices*

Global Climate
Justice Movements

Fig. 01 Synergistic aspects of community agency for climate justice are shown in this diagram. **Credit:** Rhys Davies.

and local environmental monitoring and care may become seen as interrelated, synergistic means of mitigating climate catastrophe.

These processes are not new ideas; they are receiving growing attention worldwide. What the chapters in this book contribute are specific, detailed accounts of how these processes emerge in action.

NOTES

1   Human cultures, languages, and shared knowledge are also commons: available to all (subject to protective rules) and necessary for survival.

2   The universities represented in this project included: York University, Toronto, Canada; University of Johannesburg, Johannesburg, South Africa; University of Cape Town, Cape Town, South Africa; University of KwaZulu-Natal, Centre for Civil Society, Durban, South Africa; Eduardo Mondlane University, Maputo, Mozambique; Technical University of Mozambique, Maputo, Mozambique; University of Nairobi, Institute for Climate Change and Adaptation, Nairobi, Kenya; University of Yaoundé I, Yaoundé, Cameroon; Obafemi Awolowo University, Ile-Ife, Nigeria; University of Concepción, Concepción, Chile; University of Los Lagos, Osorno, Chile; University of São Paulo, Luis de Queiroz College of Agriculture, Piracicaba, Brazil; Federal Rural University of Rio de Janeiro, Seropédica, RJ, Brazil; Federal University of Minas Gerais, Belo Horizonte, Brazil; McGill University, Montreal, Canada; and the University of Alberta, Edmonton, Canada.

Reference List

Berkes, F. (2017). Environmental governance for the Anthropocene? Social-ecological systems, resilience, and collaborative learning. *Sustainability, 9*(7), 1232.

Bollier, D. (2014). *The commons as a template for transformation.* Great Transformation Initiative Essay. https://greattransition.org/publication/the-commons-as-a-template-for-transformation

Burger, J., Ostrom, E., Norgaard, R.B., Policansky, D., & Goldstein, B.D. (Eds.). (2001). *Protecting the commons: a framework for resource management in the Americas.* Island Press.

Farooqui, U., Ratner, B., Rao, J., Chaturvedi, R., & Priyadarshini, P. (2021, November 4). Systems change for the commons. *Stanford Social Innovation Review.* https://ssir.org/articles/entry/systems_change_for_the_commons

Fournier, V. (2013/4). Commoning: On the social organisation of the commons. *M@n@gement, 16,* 433–453. https://doi.org/10.3917/mana.164.0433

Hossein, C., & Kinyanjui, M.N. (2022). Indigeneity, politicized consciousness, and lived experience in community economies. In C. Hossein and P.J. Christabell, *Community economies in the Global South: Case studies of rotating savings and credit associations and economic cooperation* (pp. 223–238). Oxford University Press.

PART I

Knowledge Commons

Putting Ethos into Practice: Climate Justice Research in the Global Knowledge Commons

Kathryn Wells[1]

Introduction: Human Knowledge Is a Commons

Climate change is a product of colonial globalization, which has also made global communication possible with unprecedented ease. Human knowledge has become a global commons; knowledge produced in one place influences people across the world (Hess & Ostrom, 2007; Levine, 2007). The global knowledge commons includes a vast array of research, stories, history, and traditions—understandings of everything that is shared, such as oceans and watersheds, the atmosphere, seeds, soil, ecosystems, etc., and for people, cultures, histories, languages, and ontologies (Mazé, Domenech, & Goldringer, 2021; Perkins, 2019, p. 184). These include all the ways in which people connect to one another and the world.

This chapter considers the ethical implications of knowledge commons and how an ethos—a distinguishing set of beliefs, spirit, or character of a person, group, or culture—might emerge to help address the injustices inherent in the current knowledge commons, including those driven by climate chaos (Joranson, 2013; Puckett et al., 2012; Kranich, 2007). An ethos that amplifies historically marginalized perspectives through a decolonial and transformative lens emphasizes moving away from coloniality and towards more inclusive climate justice. Knowledge co-production through participatory research is one way to begin to shift the power dynamics of institutional and community climate research. This has wider applications and implications for the shared commons, such as knowledge acquisition and dissemination, and for governance in general.

The Global Knowledge Commons Is Not Open Access

Knowledge commons, and its implications for governance, are increasingly discussed in environmental and climate justice spaces (Henscher et al., 2020; Janssen, 2022). The knowledge we have and share through the commons, who it is available to, and who has the privilege of understanding the changes happening to the ecosystems and environments we live in, are not equally shared and validated. There is a system of power embedded within institutionalized knowledge acquisition and dissemination pathways, particularly in the academy. The vast majority of published research on climate justice comes from the "Global North." As a result, Western-colonial assumptions, validation, and publication systems are imposed upon the "Global South," who are disproportionately impacted by the climate crisis. The mechanisms at work in systems of knowledge production and publication are colonial, resulting in research from colonial places and perspectives being seen and validated as more valuable. When it comes to discussions of climate justice, Western or Global North and Global South are used in much of the literature to distinguish between those who benefit from capitalist exploitation and those who suffer from it, from a global perspective.

As researchers, we need to be critical of from where, from whom, and for whom, and how knowledge is being produced and shared (Sultana, 2019). Vital, relevant knowledge does not necessarily follow the regimented hierarchies of Western academic institutions. Rather, knowledge is shared in a wide variety of ways that have not been legitimized by the colonial institutions we privilege and prioritize in knowledge production.

Capitalist globalization has created circumstances where much of the published research on climate justice is in English. Languages themselves are commons that help us share knowledge, and being able to communicate in various ways allows us to further dismantle the coloniality involved in communicating about climate justice research. Being more inclusive to different styles of communication also necessitates a conversation about various worldviews and ethical systems. Global divisions include not only economic beneficiaries, the exploited, and those most afflicted by climate chaos, but also geographical and geopolitical divides. Those who inhabit extraction and fossil-fuel sacrifice zones are the most marginalized, exploited, and poor in all geographic locations.

Climate Justice Depends Upon Open Access to Knowledge

For environmental and climate justice, wide-reaching transformations to social and economic systems are required to avoid irreparable damage to the Earth's climate systems (Krause, 2018; IPCC, 2022). Within these discussions, besides attention to government policies and hand-wringing about why they have been so ineffectual thus far, much of the focus is on two ways of addressing deep system changes through i) degrowth, and ii) just transitions. These discussions often do not address the underlying capitalist and colonial roots that built society as we know it in the Anthropocene—the current geological age where humans are the dominant influence on climate and the environment. The deep-seated inequities of current governance systems will not necessarily be addressed through an energy transition that substitutes renewable forms of energy for fossil fuels (Temper et al., 2020). Another approach, iii) just transformation, recognizes the need for more than marginal political shifts and understands that transformations often occur in response to crises and disasters.

Just transformations involve changes in both political structures and social relations. Transformative change addresses "the growing economic and political power of elites, and patterns of stratification related to class, gender, ethnicity, religion or location that can lock people into disadvantage" (Krause, 2018, p. 511). Sustainability transformation goals are "grounded in universal and rights-based policy approaches; revers[ing] normative hierarchies within integrated policy frameworks; re-embed[ing] economic policies and activities in social and environmental norms; and foster[ing] truly participatory decision-making approaches" (Krause, 2018, p. 511). This requires inclusive empowerment for active and ongoing participation by all members of society in order to "consider how deeper social, economic and political structures create and reinforce vulnerability and hence are part of the problem" (Newell et al., 2021, p. 7). In this sense, climate justice activism focuses largely on "the global dynamics of rights and responsibilities, mostly taking nation-states, and to a lesser extent, corporations, as the focal point" (Newell et al., 2021, p. 6), and for a better understanding of "how inequalities in global decision-making interact with and mirror local power dynamics of exclusion" (pp. 6–7). This call acknowledges the wide reach and impact that global climate decisions have on the planet, via the creation and adoption of new technologies. Democratization of governance, and therefore of access to knowledge, is a crucial part of such just transformations.

Decolonizing Knowledge Access Requires Political Activism

A few examples serve to demonstrate how transformational change in governance systems and institutions, when it happens at all, is usually very slow, and only takes place in response to political pressure from constituencies.

The Environmental Justice Movement (EJM) emerged in the USA in the 1960s in response to toxic waste sites and hazardous facilities' being sited in or near low-income residential areas where racialized people lived, and suffered terrible environmental and health impacts. It is no coincidence that the movement emerged "in the wake of the civil rights movement and was shaped by African-American (predominantly women's) resistance in the South" (Opperman, 2019, pp. 59–60). The EJM movement is inextricably linked to environmental racism, "the differential distribution of environmental burdens according to race, perpetuated by the exclusion of people of color from environmental decision-making" (Opperman, 2019, p. 58). The framing of "environmental justice" within this movement, although recognized as an intersectional way of approaching climate activism, also left the emphasis and importance of racial and economic justice out of focus, thus eliding the central role of white supremacy and capitalism in determining environmental injustice across the globe (Opperman, 2019, pp. 60–61). As the EJM gained momentum in the 1980s and 1990s (Mohai, 2018), it remained mostly a Western endeavour (Reed & George, 2011), defined through a lens of Western (i.e., colonial) thinking (Álvarez & Coolsaet, 2020, p. 50) in its history and practice.

Environmental justice and climate justice are terms that have been used interchangeably in some instances. Yet, there is a particular history that informs the use of these ideas. Environmental justice emphasizes the intersection of social inequalities with environmental harm. Using ecological justice, which centres the relationship of humans to the nonhuman world (Opperman, 2019, p. 62), the EJM politicized the meaning of "environment" while simultaneously framing an opposition of social justice versus "an ecological valorization of the more-than-human" (Opperman, 2019, p. 62). As a result, the EJM has been critiqued for discounting the socio-cultural inequalities that are tied to race, space, and place in favour of boosting human intervention in ecological crises, making the movement appear to be motivated by white colonial-settler saviourism, while omitting any blame on capitalistic structures (Gonzalez, 2020; Pulido, 2016; Sperber, 2003; Dorsey, 2001, p. 69). In more recent years, the term "climate justice" has gained popularity in an effort to become a more inclusive and rights-based way of expressing the need for more than environmental justice—including global climate and commons justice as well. Many similar critiques remain relevant to those of the EJM in that climate justice often uses universalist philosophy and is deeply rooted in Western-colonial or "Northern" ideology (Newell et al., 2021, p. 2). The distinction between environmental and climate justice, although sometimes arbitrary, allows us to begin to see how the framing of environmental and climate movements serves to perpetuate Western colonial ideology (Kojola & Pellow, 2020; Cock & Fig, 2012; Arthur, 2017; Whyte, 2020) within climate research and what is validated in these spaces.

Many of the current climate solutions being put forward by governments serve to perpetuate and entrench pre-existing structural inequalities that drive climate crises (Deranger et al., 2022, p. 54). These power dynamics continue to privilege certain knowledges and discount others. Decolonial movements work to maintain "sustainable ecological practices, communal wealth-sharing, and institutions that preserve long-term quality of life" (Perkins, 2019, p. 187). This is in stark contrast to capitalist-coloniality (Álvarez & Coolsaet, 2020, p. 52), which is contingent upon industrial resource extraction of the land and human exploitation while simultaneously utilizing human exceptionalism and hetero-patriarchy to dismiss the ties between women and nature (Perkins, 2019, p. 188). This produces escalating inequities between those who are the cause of the climate crisis, primarily those benefiting from resource extraction industries, and those most negatively impacted by it. In a conference panel discussion on whether Canadian federal climate policy includes Indigenous Peoples and their rights that resulted in a report by Indigenous Climate Action (ICA, 2021), climate justice activist

Ariel Deranger noted, "Indigenous Peoples and our rights, knowledge, and climate leadership were mentioned again and again in both (the 2016 and 2020 federal climate) plans, and yet we were structurally excluded from the decision-making tables where these plans were made" (Deranger et al., 2022, p. 53). This in turn perpetuates the reproduction of bureaucratic structural inequalities that are driving the climate crisis.

Global participatory research networks and information sharing to build the global knowledge commons—open-access, freely available research results on current and future conditions, technologies, and options—are a way to work toward just transformations that conserve and protect the ecological commons on which all life depends.[2]

Climate Justice Aspects of the Global Knowledge Commons Are Emergent

Along with (and often in conjunction with) participatory community-based research, there are many processes underway that further the development of, and open access to, global knowledge commons. These include:

- Recognizing non-Western and Indigenous knowledges by disrupting and unsettling time-space distinctions as part of the commons: Examination of "commons" shifts focus away from the human connection with, reliance on, and domination of nature. Commons discourse tends to focus on collective action, voluntary associations, and collaborations by questioning governance systems and building participatory processes with interest in shared values and ethical responsibility (Perkins, 2019, p. 185). Climate crises need to be understood by listening to those who are experiencing them first hand. In many cases that means those who do not have a voice in the global knowledge commons. Indigenous peoples across the globe are knowledgeable about how to adapt and survive the changes that are happening, but the scientific methods of research are limiting the ways in which governments and colonial societies address these issues. Indigeneity is foundational for knowledge co-production. The only way to decolonize is to disrupt and undo the colonial frameworks we are accustomed to; dismantling the structures of capitalistic hyper-individualism, patriarchy, heteronormativity, extractivism, and systems of white supremacy among other oppressions (Deranger et al., 2022, p. 67).

- Focusing on cognitive justice to explore and trace interactions among inequities: Cognitive justice is a concept that examines whose knowledge is seen as valid, who creates and disseminates knowledge, and who participates in authorizing and holding accountability for knowledge production (Newell et al., 2021, p. 9). In this sense, those in the global North are usually validated as more "objective" and universalistic assumptions about individualism within nation-states are seen as correct and just (Newell et al., 2021, pp. 6–7). Taking a different approach and adopting pluralistic, bottom-up, decolonial, and community-oriented methods of knowledge creation and dissemination, implies difficulties in gaining validation from the established systems of power. One aim of coloniality, in the context of Environmental Justice movements, is to anchor oppression in psychological structures in order to disempower through internalized oppression and affix marginalized people and groups to certain immovable spaces within movements (Álvarez & Coolsaet, 2020, pp. 62–63). The ways in which academic research has been historically and often continues to be done, is a microcosm with the same underpinnings. As researchers we must continuously challenge ourselves both through understanding of cognitive justice and the broader systemic oppressions, in order to address these concepts in our work.

- Unsettling human exceptionalism: There is an inherent focus on human experiences and needs in climate justice research. While human development and capitalist globalization cause the climate crisis, an emphasis on human survival above all other species and commodification of nature for human gain is called human exceptionalism (Newell et al., 2021, p. 7). Indigenous environmentalism, a key aspect of decolonization, rejects "colonialism, extractivism and dispossession in the current distribution and accumulation of wealth between nations, classes and social groups" (Newell et al., 2021, p. 7) in favour of a pluralistic way of understanding and pursuing justice by ascribing value to all living things. Decolonization disperses human exceptionalism to focus on transforming the systems and practices into more complex and nuanced ways of approaching social, political and climate justice as intersectional movements.

- Prioritizing process-oriented participatory knowledge creation, co-production, and sharing: Knowledge co-production is linked to citizen or community science, interactive and creative research, co-design, and participatory research, among other methods (Newell et al., 2021, p. 9; Norström et al., 2020, p. 183). Participatory research for climate justice involves the participation of those who are knowledgeable in varying ways, and also seeks out perspectives that are hidden and/or formerly undervalued. This approach can be linked to decolonial ways of knowledge co-production, allowing for various perspectives to be seen as valid in the face of power structures. Decolonization demands detachment from the false concept of scientific neutrality; participatory research demands active participation of knowledge-holders from communities who are feeling the climate crisis first hand, who can help to reimagine meanings and lead climate justice movements (Álvarez & Coolsaet, 2020, p. 63; Deranger et al., 2022, p. 70). There are many ways of doing and knowing, so when we approach research with the goal of pluralistic knowledge co-production in mind, this necessarily means bringing together academics from various disciplines with many others, such as local communities, Indigenous communities, government, civil society, beneficiaries of the status quo, etc. However, such processes require a range of skills and types of knowledge and expertise to address the power dynamics, activate change and generate knowledge (Norström et al., 2020, p. 186) if they are not to lead to less engagement and simply reproduce knowledge hierarchies where certain kinds of knowledge and expertise are seen as more legitimate than others (Norström et al., 2020, p. 186). Without addressing power imbalances directly, the quality of engagement and process outcomes suffers (Álvarez & Coolsaet, 2020, pp. 59–60; Norström et al., 2020, p. 186).

- Protecting commons spaces: The physical spaces in which people live and move around are of importance, because (and to remind us that) we are not disconnected from the environment where we live. Commons governance relies on self-organized social systems and networks that are outside of political governance systems.

- Sharing knowledge with and for all: The traditional ways of scientific knowledge production and dissemination are siloed, do not value Indigenous knowledge (Deranger et al., 2022, p. 60), and instead focus on extracting data for supplemental use in Western science (Arsenault, et al., 2019, p. 122). There are many ways in which knowledge production can become more inclusive and decolonial: community-based approaches to research, which includes external accountability strategies; providing accessible capacity-building resources for communities to develop their own plans, assessments, and standards when conducting climate research; or participating in, documenting, and supporting the growing Indigenous guardian movement that trains Indigenous scientists as community monitors (Arsenault, et al., 2019, p. 128). Examples of how Indigenous and traditional communities are maintaining commons governance and knowledge commons include the Quilombos in Brazil where former slaves created small settlements of liberation, maintaining harmonious relationships with the land, in the face of systemic oppression; the Indigenous water protectors, land defenders, and pipeline fighters in Canada who are protecting their inherent rights and sovereignty of the land against government and private-sector oppression; and community gardens, often found in urban areas, that bolster community food sovereignty for neighbourhoods in food deserts. Other ways to facilitate knowledge sharing through commons include community radio and social media; open-access information sharing and making innovative technologies available; co-operative institutions that utilize and facilitate networks for and by community members; equal access to education and government processes to allow for social and political participation; integrating accessibility and different ways of learning, such as storytelling and language translation into design and dissemination; and many more ways of sharing knowledge in the commons. One key element of knowledge sharing is relationship building, which include trust, consent, accountability, and reciprocity (Whyte, 2020, p. 1). Without this, access to resources will not be utilized. Decolonization of these concepts is the responsibility of powerholders, in many cases white colonial-settlers, like myself, who need to take responsibility and bolster relational reciprocity with Indigenous and marginalized communities.

Climate justice links the historical ways in which colonialism and coloniality harm nature and at the same time harm the most marginalized in society. Bridging the gap between academic pursuit of knowledge and communities who know the most about their own environments is crucial for climate justice transformations. Participatory research is one way to facilitate this shift. By prioritizing decolonial methods of knowledge creation and dissemination, researchers can move toward a more just way of participating in both academic pursuits and inclusive holistic community supports that reverse the dangerous impacts of the climate crisis.

Another important form of power relations is the position of researchers themselves. We have a responsibility to be sensitive to the "importance of local autonomies and self-recognition in overcoming injustices" (Álvarez & Coolsaet, 2020, p. 60). Being careful, humble, transparent, and taking time to discuss and share when approaching participatory research is of utmost importance. If we approach this by shifting and diffusing power to research participants, while making efforts to learn how and actively try to decolonize both ourselves and the systems of which we are a part, through steps that prioritize participation, we can start to build robust shared decolonial knowledge commons.

When we approach climate justice research through an ethos of decolonial and transformative justice, we begin to unravel the systems of power established by coloniality and global capitalism that are responsible for climate catastrophe.

1   I want to acknowledge my subject position as a researcher. I am always learning and as such I strive to be everchanging, growing, and improving. As a white settler living on Turtle Island [1], I want to acknowledge the many Indigenous Peoples whose land was stolen and who continue to be oppressed by the colonialist state in which I reside. I carry a specific privilege that has afforded me my education, among other privileges, and that contributes to how I move around in the world more freely than others. My approach to this work comes from a deep learning and deeper unlearning that I have and will continue to experience throughout my life-course. This learning has been informed and heavily influenced by many BIPOC [2] women and Queer [3] folks to whom I owe a great deal of gratitude. The idea that there is one ultimately clear way to say, write, or express a point is in and of itself a colonial idea. It is therefore necessary to carefully unpack the "what, why, for whom, by whom, and how" of any decolonial endeavour (Sultana, 2019, p. 40).

[1] Turtle Island refers to the continent of North America for some Indigenous Peoples. https://www.thecanadianencyclopedia.ca/en/article/turtle-island

[2] BIPOC is an acronym for Black, Indigenous, People of Colour.

[3] Queer has been adopted as an umbrella term by some people who identify within the 2SLGBTQIA+ community and refers to themselves or the community as a whole.

2   Alarmingly, there are indications that the concept of "knowledge commons" is already being subverted and warped into a form of advertising for climate entrepreneurs (Luiken & Shah, 2022; Sperfeld et al., 2021).

Reference List

Álvarez, L., & Coolsaet, B. (2020). Decolonizing environmental justice studies: A Latin American perspective. *Capitalism Nature Socialism*, *31*(2), 50–69. https://doi.org/10.1080/10455752.2018.1558272

Arsenault, R., Bourassa, C., Diver, S., McGregor, D., & Witham, A. (2019). Including Indigenous knowledge systems in environmental assessments: Restructuring the process. *Global Environmental Politics*, *19*(3), 120–132. https://doi.org/10.1162/glep_a_00519

Arthur, K.K.L. (2017). *Frontlines of crisis, forefront of change: Climate justice as an intervention into (neo)colonial climate action narratives and practices* [Unpublished master's thesis]. Massachusetts Institute of Technology. http://hdl.handle.net/1721.1/111292

Cock, J., & Fig, D. (2012). From colonial to community-based conservation: Environmental justice and the national parks of South Africa. *Society in Transition*, *31*(1), 22–35.

Deranger, E.T., Sinclair, R., Gray, B., McGregor, D., & Gobby, J. (2022). Decolonizing Climate Research and Policy: Making space to tell our own stories, in our own ways. *Community Development Journal*, *57*(1), 52–73. https://doi.org/10.1093/cdj/bsab050

Dorsey, J.W. (2001). The presence of African American men in the environmental movement (or lack thereof). *Journal of African American Men, 6*, 63–83.

Gonzalez, Carmen G. (2020). Racial capitalism, climate justice, and climate displacement. *Oñati Socio-Legal Series, Climate Justice in the Anthropocene, 11*(1), 108–147.

Henscher, M., Kish, K., Farley, J., Quilley, S., & Zywert, K. (2020). Open knowledge commons versus privatized gain in a fractured information ecology: lessons from COVID-19 for the future of sustainability. *Global Sustainability, 3*, E26. https://doi.org/10.1017/sus.2020.21

Hess, C., & Ostrom, E. (2007). *Understanding knowledge as a commons: From theory to practice*. MIT Press.

ICA (Indigenous Climate Action). (2021). *Decolonizing climate policy in Canada: Report from phase one*. Indigenous Climate Action. https://www.indigenousclimateaction.com/publications

IPCC (Intergovernmental Panel on Climate Change). (2022). *Sixth assessment report*. https://www.ipcc.ch/assessment-report/ar6/

Janssen, M.A. (2022, July 28). A perspective on the future of studying the commons. *International Journal of the Commons*. https://www.thecommonsjournal.org/articles/10.5334/ijc.1207/print/

Joranson, K. (2013). Indigenous knowledge and the knowledge commons. *International Information & Library Review, 40*(1), 64–72. https://doi.org/10.1080/10572317.2008.10762763

Kojola, E., & Pellow, D.N. (2020). New directions in environmental justice studies: Examining the state and violence. *Environmental Politics, 30*(1–2), 100–118.

Kranich, N. (2007). Countering enclosure: Reclaiming the knowledge commons. In C. Hess & E. Ostrom (Eds.), *Understanding knowledge as a commons: From theory to practice* (pp. 85–122). MIT Press.

Krause, D. (2018). Transformative approaches to address climate change and achieve climate justice. In T. Jafry (Ed.), *Routledge Handbook of Climate Justice*. Routledge.

Levine, P. (2007). Collective action, civic engagement, and the knowledge commons. In C. Hess & E. Ostrom (Eds.), *Understanding knowledge as a commons: From theory to practice* (pp. 247–276). MIT Press.

Luiken, M., & Shah, A. (2022, April 22). *Policy brief: Scaling climate goals through the use of technical experts & digital technical knowledge commons*. T7 Task Force Climate and Environment. https://www.think7.org/wp-content/uploads/2022/04/Climate_Scaling-Climate-Goals-through-the-use-of-technical-experts-Digital-technical-knowledge-commons_Luken_Shah.pdf

Mazé, A., Donenech, A.C., & Goldringer, I. (2021). Commoning the seeds: Alternative models of collective action and open innovation within French peasant seed groups for recreating local knowledge commons. *Agriculture and Human Values, 38*, 541–559.

Mohai, P. (2018). Environmental justice and the Flint water crisis. *Michigan Sociological Review, 32*, 1–41.

Newell, P., Srivastava, S., Naess, L.O., Torres Contreras, G.A., & Price, R. (2021). Toward transformative climate justice: An emerging research agenda. *WIREs Climate Change, 12*(6), e733. https://doi.org/10.1002/wcc.733

Norström, A.V., Cvitanovic, C., Löf, M.F., West, S., Wyborn, C., Balvanera, P., Bednarek, A.T., Bennett, E.M., Biggs, R., de Bremond, A., Campbell, B.M., Canadell, J.G., Carpenter, S. R., Folke, C., Fulton, E.A., Gaffney, O., Gelcich, S., Jouffray, J.-B., Leach, M., ... Österblom, H. (2020). Principles for knowledge co-production in sustainability research. *Nature Sustainability, 3*(3), 182–190. https://doi.org/10.1038/s41893-019-0448-2

Opperman, R. (2019). A permanent struggle against an omnipresent death: Revisiting environmental racism with Frantz Fanon. *Critical Philosophy of Race, 7*(1), 57–80.

Perkins, P.E. (Ellie). (2019). Climate justice, commons, and degrowth. *Ecological Economics, 160*, 183–190. https://doi.org/10.1016/j.ecolecon.2019.02.005

Puckett, A., Fine, E., Hufford, M. Kingssolver, A., & Taylor, B. (2012). Who knows? Who tells? Creating a knowledge commons. In S. Fisher and B.E. Smith (Eds.), *Transforming places: Lessons in movement-building from Appalachia* (pp. 239–251). University of Illinois Press. https://www.researchgate.net/profile/Betsy-Taylor-2/publication/287185981_Who_knows_Who_tells_Creating_a_knowledge_commons/links/5ad3c15f458515c60f53b781/Who-knows-Who-tells-Creating-a-knowledge-commons.pdf

Pulido, L. (2016). Flint, environmental racism, and racial capitalism. *Capitalism Nature Socialism, 27*(3), 1–16. https://doi.org/10.1080/10455752.2016.1213013

Reed, M., & George, C. (2011). Where in the world is environmental justice? *Progress in Human Geography, 35*(6), 835–842.

Sperber, I. (2003). Alienation in the environmental movement: Regressive tendencies in the struggle for environmental justice. *Capitalism Nature Socialism, 14*(3), 1–43. https://doi.org/10.1080/10455750308565532

Sperfeld, F., Kovac, S., Dolinga, S., Nagel, L., & Rodriguez, H. (Eds.). (2021). *Co-benefits knowledge commons: Renewable energy, employment opportunities, and skill requirements*. IASS (Institute für transformative Nachhaltigkeitsforschung). https://publications.iass-potsdam.de/pubman/faces/ViewItemFullPage.jsp?itemId=item_6001370_6&view=EXPORT

Sultana, F. (2019). Decolonizing development education and the pursuit of social justice. *Human Geography, 12*(3), 31–46. https://doi.org/10.1177/194277861901200305

Temper, L., Avila, S., Del Bene, D., Gobby, J., Kosoy, N., Le Billon, P., Martinez-Alier, J., Perkins, P., Roy, B., Scheidel, A., & Walter, M. (2020). Movements shaping climate futures: a systematic mapping of protests against fossil fuel and low-carbon energy projects. *Environmental Research Letters, 15*, 123004.

Whyte, K. (2020). Too late for Indigenous climate justice: Ecological and relational tipping points. *WIREs Climate Change, 11*(1), e603. https://doi.org/10.1002/wcc.603

2

Integrating Citizen Science Observations in Climate Mapping: Lessons from Coastal-Zone Geovisualization in Chilean Patagonia and the Brazilian Southeast

Allan Yu Iwama, Francisco Brañas, David Núñez,
Daniela Collao, Ramin Soleymani-Fard, Carla Lanyon,
Adrien Tofighi-Niaki, Petra Benyei, Lara da Silva,
Rafael Pereira, Francisco Ther-Ríos, and Sarita Albagli

Introduction

The effects of climate change have been observed on a global scale, especially in coastal areas. The Intergovernmental Panel on Climate Change (IPCC, 2022) emphasized the increase in frequency and magnitude of extreme events such as storms, floods, and heat waves, while studies have guided efforts to create climate-forecast models at different scales of analysis to identify the risks of these threats and to support mitigation and adaptation strategies.

Climate-prediction models are fundamental for representing possible impact trajectories on a regional scale. On the other hand, several studies also point to the need for climate change research to include local observations to contextualize the causes of impacts, in addition to expanding the scale of observations at the community level (David-Chavez & Gavin, 2018; Iwama et al., 2021; Reyes-García & Benyei, 2019; Savo et al., 2016).

Studies have shown that scientists are increasingly recognizing the relevance of Traditional and Local Knowledge (TLK) in providing observations for adaptation to the effects of climate change (García-del-Amo et al., 2020; Hill et al., 2020; Iwama et al., 2021; Nakashima et al., 2018; Reyes-García et al., 2019; Tengö et al., 2017) and global environmental change (Berkes, 2009; Fernández-Llamazares & Virtanen, 2020; Merçon et al., 2019). These studies identify the importance of participatory approaches in including local observations and understanding their appropriate contexts, scales, and ways to assess people's engagement in risk governance (Iwama et al., 2021); such participatory approaches can reduce impacts and risks overall.

Similarly, citizen science initiatives all over the world—understood as those scientific activities in which non-professional scientists participate (Kullenberg & Kasperowski, 2016; Bonney et al., 2014; Sauermann et al., 2020)—involve citizens and local communities in data-gathering processes for knowledge production, monitoring strategies, and cooperative governance.

Working with different knowledge systems (e.g., scientific knowledge and TLK) requires scientists and participants involved in citizen science initiatives to engage one another in more flexible, reflective, and diverse ways, because different kinds of knowledge and worldviews often clash. These efforts toward collaboration often show how the dialogue spaces created amidst such epistemological tensions can in turn create new narratives and ways of producing science that is more appropriate to the local context (Merçon et al., 2019; Tengö et al., 2021).

TLK can collaborate with scientific knowledge to co-produce new knowledge for disaster risk reduction (DRR) and climate justice. Where TLK interacts with scientific knowledge through citizen science monitoring programs or other participatory methods, local perceptions and knowledge of climate change and disaster risks can supplement poor baselines with data that scientists and communities would otherwise lack access to.

In this chapter we show how using action research—experiential, reflexive knowledge production leading to transformative change—and participatory geographic information systems (GIS) to record local observations on sea-level rise, floods, landslides, and coastal erosion produces socio-cultural responses to natural hazards and climate/environmental impacts. We held capacity-building workshops with local communities to produce their own maps using social cartography (collective mapping of socially important relationships, histories, and features based on the community's views about socio-environmental hazards, livelihoods, and natural resources), Quantum GIS, and Google Earth (for data geovisualization), and conducted interviews using mobile GPS (global positioning system) applications—e.g., Mobile Topographer, QField for QGIS, and Google Maps—to record the geographic coordinates of the information provided in the interviews.

Through our work with Mapuche Indigenous and artisanal fishers in Chiloé archipelago (Chile) (see Map 1), and with traditional communities of artisanal fishers and *Quilombolas* on Brazil's southeast coast (see Map 2), we promoted and helped build capacity to use community-based data-management tools and systems. We found that local communities had difficulty handling the geovisualization tools, even with capacity-building support. Moreover, poor access to the internet kept them from accessing the platform and using the interactive maps, which reinforces the importance of using the social cartography approach to identify critical points and using local knowledge to build escape routes in the event of a natural disaster. Based on our results, we discuss how, despite the increased use of digital platforms and social technologies to facilitate dialogue between TLK and scientific knowledge for climate change adaptation and DRR, citizen science initiatives need to move forward with a focus on long-term participation processes.

Currently, in both Brazil and Chile and throughout the countries of Latin America and the Caribbean, initiatives such as spatial data infrastructures (SDIs) are being developed at the national level.[1] The SDIs are available for data visualization by local communities, but our research shows that these infrastructures have been little—if ever—used at the local community level, and usually their use is limited to researchers and managers. Taking a prototype of such global platforms from the Local Indicators of Climate Change Impacts Observation Network (LICCION), and initiatives to improve government programs at the national level, such as the Cemaden-Educação[2] programme in Brazil, we consider how centralized SDIs might support local initiatives. In this sense, we discuss how and why it is necessary to expand the discussion of the barriers and opportunities for integrating traditional and scientific knowledge in long-term citizen science initiatives on climate change and DRR.

This chapter presents two case studies, one in Brazil and the other in Chile, where citizen science approaches were developed with participatory methodologies such as action research and social cartography. Community work groups were formed at the Federal University of Rio de Janeiro in Brazil and at the University of Los Lagos in Osorno, Chile.[3] Our initiative proposed to develop a quantitative and qualitative approach for vulnerability analysis and adaptation to climate change, focusing on communities living with elevated climate-related disaster risks. The research seeks to advance scientific knowledge production based on citizen science, integrated with technical-scientific risk mapping in coastal zones in the south of Chile and the southeast of Brazil.

We recommend strengthening trust with local communities for citizen science initiatives. We also recommend raising funds to guarantee adequate infrastructure and continuous training for monitoring climate change at the local level for those communities eager for intra- and inter-institutional partnerships (Alonso-Yanez et al., 2019; David-Chavez & Gavin, 2018; Reyes-García et al., 2019). Several studies have pointed out that the question of sustainability is one of the biggest continuity challenges for citizen science initiatives—along with different perspectives/epistemologies, data sovereignty, and citizen engagement (Arriagada et al., 2018; Iwama et al., 2021; Lam et al., 2020; Reyes-García et al., 2022).

Study Areas

The geographical areas where we worked were in Chiloé Province (Chile) and the northern coast of São Paulo State (Brazil) (see Map 1, Map 2). The action research was carried out at two different times and places: 2017–2018 for the Brazilian case study (the cities of Caraguatatuba (often shortened to just Caraguá), Ubatuba in Sao Paulo State, and Paraty in Rio de Janeiro State; and 2019–2020 for the Chilean case study in the cities of Maullín in Llanquihue province and Dalcahue and Quellón on Chiloé Island (Map 2)).

Off Chile's Pacific coast lies the Sea of Chiloé, with coastal plains from Chiloé Island (41ºS) to the extreme south of Chile; the coast is made up of fjords, estuaries, channels, and gulfs (Fariña et al., 2008; Avaria & Barra, 2009). The marine currents (such as the cold Humboldt Current) that bathe Chile's coastal zone largely define its ecological characteristics and high biological productivity, essential for the country's fishing industry. The occurrence of periodic weather fluctuations and climatic phenomena such as El Niño and La Niña temporarily modify some of these conditions, complicating

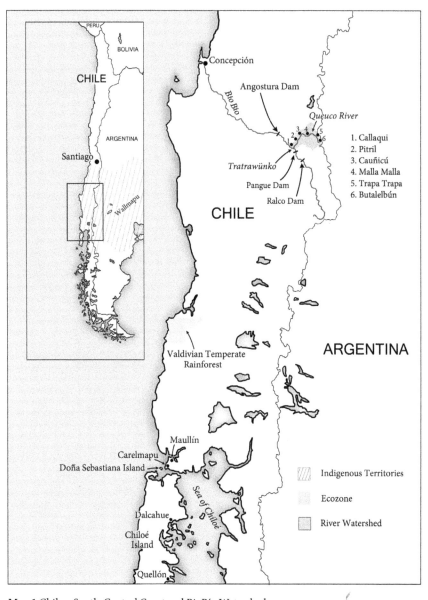

Map 1 Chile—South-Central Coast and BioBío Watershed

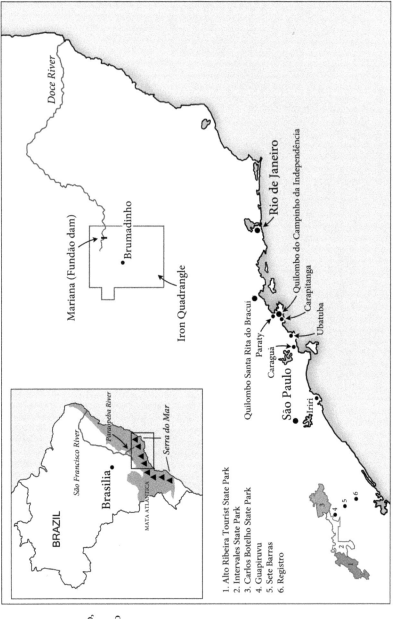

Map 2—Brazil—Areas of Minas Gerais, São Paulo, and Rio de Janeiro States

Doce River

Mariana (Fundão dam)

Brumadinho

Iron Quadrangle

Rio de Janeiro

Quilombo do Campinho da Independência

Carapitanga

Ubatuba

Quilombo Santa Rita do Bracuí

Paraty

Caraguá

São Paulo

Iriri

1. Alto Ribeira Tourist State Park
2. Intervales State Park
3. Carlos Botelho State Park
4. Guapiruvu
5. Sete Barras
6. Registro

BRAZIL

Brasília

São Francisco River

Paraíba River

Serra do Mar

MATA ATLÂNTICA

the ecological dynamics in Chilean seas (Camus, 2001). In Brazil, the Serra do Mar along the coastline intensifies atmospheric flows coming from the sea to produce high rainfall (Nunes & Calbete 2000; Scofield et al., 2014). In periods of intense and prolonged rains, landslides and floods are frequent (Tavares & Mendonça 2017; Koga-Vicente & Nunes, 2011).

The communities in both study areas have developed important productive activities associated with exploitation of natural resources (for example, in Brazil with the oil and gas industry, and in Chile with salmon farming and artisanal fishing). They are also both tourist areas, with strong pressure on infrastructure services in the summer months. The study areas, Chiloé and the north coast of São Paulo State, were selected because residents in both territories experience problems related to disasters, almost daily. The social and cultural contexts are different, however, and this may demonstrate different climate change adaptation strategies.

Both areas have recurring problems, such as disaster risks related to tidal waves and floods. In Chile, in particular, the threat of earthquakes and tsunamis is added. The history of Chile records dozens of destructive tsunamis, with earthquakes named Huara (2005), Aysén (2007), Tocopilla (2007), Cobquecura (2010–event 27F), Iquique (2014), and Illapel (2015) being especially important. Earthquakes accompanied by tsunamis have caused national catastrophes with hundreds of victims and great economic damage. In addition, the effects of the El Niño phenomenon, associated with other threats such as floods and droughts, affect ecosystem dynamics, enhancing the effects of red tides (harmful algal blooms), and causing immeasurable economic, environmental, and social impacts for fishing communities.

Methods

Using social cartography, this work engaged six local community groups in the collective production of their own maps of social risk, evacuation routes, and adaptation strategies at the local level––maps of flooding, sea-level rise, coastal erosion, tsunamis, and droughts. The maps drawn up in participatory processes were digitalized using open-source QGIS mapping software, and compared to data produced by scientific and technical institutions. In addition, interviews were conducted with long-time residents. Figure 2.1 outlines the CoAdapta methodology adapted from previous work in Brazil (Albagli & Iwama, 2022).

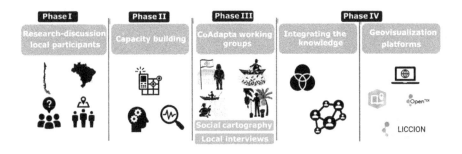

Fig. 2.1 The CoAdapta approach, using participatory citizen science.

We built on a methodology developed earlier in the CoAdapta | Litoral project for linking citizen science with GIS and SDI mapping (Figure 2.1). In each country, three local CoAdapta groups were created: in Brazil, groups from Iriri/Onça (Ubatuba), Juqueriquere (Caraguatatuba), and from Carapitanga (Paraty); and in Chile, groups from Maullín (Llanquihue), Dalcahue, and Quellón (Chiloé Island).

Following discussions about the purpose and value of this research for local participants (Phase I), the groups were trained in global positioning systems (GPS), and GPS devices coupled to phone applications allowed participants to collect local observations on climate change impacts (Phase II). Each group co-constructed semi-structured questionnaires with questions on how climate change is perceived to be happening, what impacts people observe, and how they respond with adaptation strategies based on local knowledge. Using social cartography, the groups carried out participatory mapping, showing the places where impacts occur and the responding local adaptation strategies, which include escape routes in case of disasters (Phase III).

The results on impact sites and adaptation strategies were organized into a prototype CoAdapta platform to design custom maps using the Story Map tool, an application from the ESRI company that uses ArcGIS software (a common GIS computer program). These maps enhance digital storytelling about observations on climate change and adaptation. Other tools used were OpenTEK and Oblo, digital platforms implemented by the LICCI[4] (local indicators of climate change impacts) and LICCI(ON)[5] (LICCI observation network) projects. LICCI is a project aimed at bringing Indigenous and local knowledge into climate change research, funded by the European

Research Council. Through cutting-edge science, the LICCI project strives to deepen understanding of perceived climate change impacts, include TLK in policy-making processes, and influence international climate change negotiations. LICCION was created to help share the climate research of the LICCI project. CoAdapta data were adapted to the LICCI climate impact classification protocol to geovisualize data on both platforms. Both LICCI and LICCION projects were coordinated by the Institute of Environmental Science and Technology of the Autonomous University of Barcelona, in collaboration with the CoAdapta | Litoral project. As shown by all these connections, global networks, and partnerships among research institutes and universities, there is no shortage of funding or academic interest in expanding sources of data on climate change to address its risks and impacts.

This chapter explains our participatory research process and presents preliminary results on how the local coastal communities in Brazil and Chile are responding to the impacts of climate change (Figure 2.1, Phase IV). We also discuss the potential roles, opportunities, and challenges for co-building a geovisualization platform in appropriate, culturally sensitive language accessible to traditional communities that do not frequently use such technologies or the internet.

Between Phase III and Phase IV, there are possibilities of combining high-tech methods with participatory, lower-tech methods and testing the efficacy, effectiveness, and value of this approach. We experimented with several high-tech mapping platforms to explore this.

Story Map Platform

The CoAdapta platform used the Story Map tool, a web-based application linked to ArcGIS, to build stories with custom maps that inform and inspire local observations about climate change impacts and adaptation. Storytelling helps people comprehend and navigate the climate crisis together, building agency through shared understandings (Ellis & Gladwin, 2022). Maps are an integral part of storytelling about climate change impacts, offering narratives a stronger sense of place, illustrating spatial relationships, and adding visualizations of local data.

With a map-maker, we created custom maps to enhance the participants' digital storytelling. We added text, photos, and videos to their existing web maps built in ArcGIS, and web scenes to create an interactive narrative that is easy to publish and share.

OpenTEK and Oblo Platforms

LICCION is based on LICCI research methodology (Reyes-García et al., 2020), which proposes a classification of climate change impacts through indicators, built from qualitative place-based observation of climate change impacts. The indicators are grouped into four main systems: climatological, physical, biological, and socio-cultural/economic. LICCION is grounded in a co-production process where local actors have access to first-draft LICCI indicators that are transformed according to their local realities, interests, and concerns, and are able to modify the classification to develop surveys that best meet their needs and can be integrated into the Oblo platform.

Oblo is a free and open-source technology designed by the Institute of Environmental Science and Technology of the Autonomous University of Barcelona. It allows anyone to create online platforms to document and visualize geolocalized data. The first platform the LICCI research team built with Oblo is OpenTEK, a citizen science tool designed to encourage participation in climate change research by allowing anyone in the world to document and classify observations on local climate change impacts. This tool is currently being extended in collaboration with non-governmental organizations (NGOs) and researchers to provide more relevant biocultural options and functionalities for community-based data collection.[6]

The adaptability of the Oblo technology and LICCI methodology allows communities and organisations to determine what dataset can be included and to develop multiple local-level platforms for collecting policy-relevant or context-specific data.

In this research, we aimed to test the flexibility of the Oblo platform and LICCI using qualitative and quantitative data collected through the CoAdapta process. This process provided qualitative data from surveys, and spatial data collected and created through Phases I, II, III, IV of the process (Figure 2.1). To organise data, firstly we compared the LICCI indicators with CoAdapta survey questions and participatory mapping, in order to recognize common elements and classify CoAdapta information according to the LICCI indicators. The data were organised under the natural disaster umbrella selecting different LICCI indicators of each system and creating new ones.

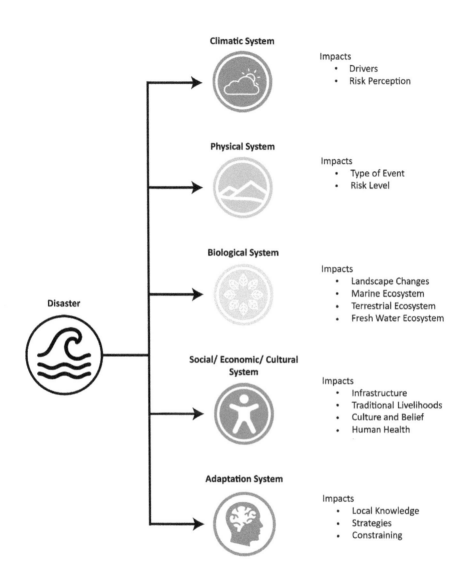

Fig. 2.2 CoAdapta data organization, adapted from LICCI tree.

Each CoAdapta interview represented a data-entry point in the platform, complemented with qualitative spatial information; some indicators had georeferenced points, for example flooding areas, security areas, etc. The data was organized into five systems, followed by their indicators (see Figure 2.2).

Geovisualization of Local Observation Data on Climate Change Impacts and Adaptation in Different Platforms

We are working with data collected from the CoAdapta | Litoral project from 2017 to 2020 using three digital platforms: Story Map, OpenTEK, and Oblo.

In Story Map, data collected from interviews, social cartography data, and audiovisual materials are made available on the platform in an interactive system.[7] It is an intuitive platform, easy to use and easy to interact with. However, it is also a paid platform, sold by the ESRI company, limiting access to those who can pay for the use and design of the maps created (Figure 2.3 and 2.4).

Looking to provide greater accessibility and replicability in how data is visualized, the CoAdapta | Litoral and LICCION projects sought to show how the free, open-source Oblo platform could adapt to the context of previously co-designed projects at the community level. Based on revisions/adaptations of classification systems from LICCI protocols on climate change impact observations, and questions related to the development of the CoAdapta project, an initial Oblo prototype was created.[8]

Thus, data mapped from social cartography were gradually transferred to the university-built Oblo platform in order to give visibility to the data co-built with traditional and local communities in the coastal CoAdapta and LICCION project, following a co-production process (Figure 2.5 and 2.6). Both projects involved local actors and their knowledge to inform climate change impacts at local levels. Despite the similarities in process, different frameworks or lenses were used to understand climate change, creating challenges for integrating both projects.

Barriers and Opportunities of Map Visualising Platforms

Our conclusions about this effort to combine TLK and digitized scientific mapping methods for use by communities are mixed. One challenge relates to the design and purpose of the mapping software and platforms. CoAdapta has been using a risk-based approach and LICCION an impact-based approach,

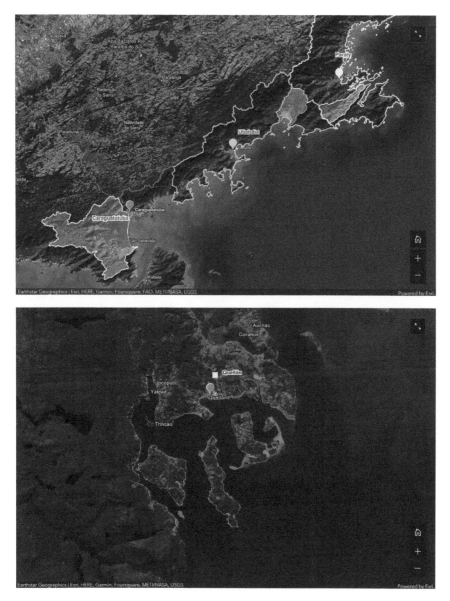

Figs. 2.3 Brazilian sites and **2.4** Chilean sites. Co-Adapta data in Story Map platform for southeast Brazil (Fig 2.3) and for Chiloé Island, Chile (Fig. 2.4).

Fig. 2.5 CoAdapta data presented in the OpenTEK platfom for southeast Brazil.

Fig. 2.6 Oblo platform for southern Chile. The Oblo and OpenTEK platforms directly share local citizen science observations, recorded in Portuguese or Spanish. All data is conveyed in the language in which it was generated use such technologies on the internet."

making it difficult to fit CoAdapta data into the Oblo platform. Converting data into the proper data format was also challenging. For instance, spatial information was in shapefiles, which is typical for geoprocessing data. Using this information as Oblo data required extra steps to transform it to meet Oblo requirements, designed initially to geospatialize data from JSON (JavaScript Object Notation), a standard text-based format for representing structured data based on JavaScript object syntax.

A more fundamental problem relates to how the technologies were used during the process of participatory research. Researchers responsible for organizing CoAdapta information into the LICCION platform did not participate in the data collection; therefore, interpretation and the limited knowledge regarding each social-ecological system might have led to a loss of information. Consequently, we highly recommend that researchers be involved in the knowledge co-production process from the beginning of the project in order to adjust the survey and interviews to create the data following the LICCION tree, or to re-organize the data together with local participants and organizations, during the fieldwork period of engagement in the communities.

Finally, the LICCION project has several protocols related to the CARE (collective benefit, authority to control, responsibility, ethics) principles of Indigenous data governance, which were developed by the Global Indigenous Data Alliance,[9] and relate to Indigenous data sovereignty and ethical data management practices. CoAdapta also has a set of principles around protecting sensitive cultural or traditional knowledge. It is important to spend the time and resources to understand these goals, and the differences between these protocols and other similar measures, and to develop appropriate principles in each community context—perhaps in collaboration with local organizations and community members—so as to ensure that data collection and analysis meet each project's standards for ethics and community respect.

We found that some community members were less comfortable than others with the technological aspects of mapping that we tried to use, despite our efforts to familiarize everyone with the software, platforms, and methodologies of this kind of mapping. Since climate change impacts have grave implications for traditional communities who hold important and relevant knowledge and understandings, and government decision-makers usually rely on high-tech data analysis methods, we see this process of continuing to relate TLK and high-tech science as a long-term participatory priority.

With Indigenous and civil society partners, LICCION has extended Oblo and developed distinct community-centred domains and surveys to not only facilitate the documentation of observations on local climate change impacts but also to allow community-specific livelihoods, perspectives, and protocols to be considered and integrated. The purpose of these new domains and surveys is to enable customary and community-led research and evidence-building on climate change while upholding Indigenous data sovereignty principles and values.

Final Remarks

Citizen science initiatives using digital platforms have been widely used at various scales and levels in many local communities. Our chapter presented our reflections on the processes of collecting climate observations in traditional coastal communities in Chile and Brazil, which were co-designed in prototypes of digital geovisualization platforms that allowed for quick visualisation in vector data format (points, polygons, and lines), as well as the formats of photographic records, text, and videos.

The use of data from Indigenous and traditional communities has raised concerns about data sovereignty in projects such as CoAdapta, LICCI, and LICCION, underscoring the need for traditional communities to use technologies that presuppose principles of open access, easy access, and cognitive flexibility (Albagli & Iwama, 2022; Reyes-García et al., 2022; Serret et al., 2019).

Citizen science initiatives with traditional communities often have a methodological design that seeks to build bonds of trust via local working groups and establish research relationships at the level of collaboration or co-design. In this sense, it is also important to emphasize the principles of the right to research (Appadurai, 2006), guaranteeing community participants their legitimate citizens' rights.

Our results demonstrate the importance of seeking open and freely accessible methodologies for visualizing and sharing data produced at the community level, together with protocols that guarantee free decisions about what types of information the community wants to share, so that the purpose of knowledge production is situated in the local context.

NOTES

1   See, for example, SDI development in Chile at https://www.ide.cl/ and in Brazil at https://inde.gov.br/.

2   See http://www2.cemaden.gov.br/cemaden-educacao/.

3   The Brazilian team was part of the CoAdapta | Litoral project, carried out from 2017 to 2021 and funded by the Rio de Janeiro State Foundation for Research Support at the Brazilian Science and Technology Information Institute, Federal University of Rio de Janeiro. For researchers at the University of Los Lagos in Chile, funding support came through a Queen Elizabeth II Diamond Jubilee Scholarships Advanced Scholars Program (QES-AS) project based at York University (Canada).

4   LICCI project—https://licci.eu/.

5   LICCION project—https://licci.eu/liccion/.

6   See https://oblo.network/. Features and technical documentation can be found here: https://oblo-cit-sci.github.io/.

7   See https://www.coadaptalitoral.net/mapping-platform.html.

8   See https://oblo.network/domain?d=coadapta.

9   See https://www.gida-global.org/care.

Reference List

Albagli, S., & Iwama, A.Y. (2022). Citizen science and the right to research: Building local knowledge of climate change impacts. *Humanit. Soc. Sci. Commun.*, *9*, 39. https://doi.org/10.1057/s41599–022–01040–8

Alonso-Yanez, G., House-Peters, L., Garcia-Cartagena, M., Bonelli, S., Lorenzo-Arana, I., & Ohira, M. (2019). Mobilizing transdisciplinary collaborations: collective reflections on *de* centering academia in knowledge production. *Glob. Sustain.*, *2*, e5. https://doi.org/10.1017/sus.2019.2

Appadurai, A. (2006). The right to research. *Glob. Soc. Educ.*, *4*, 167–177. https://doi.org/10.1080/14767720600750696

Arriagada, R.A., Aldunce, P., Blanco, G., Ibarra, C., Moraga, P., Nahuelhual, L., O'Ryan, R., Urquiza, A., & Gallardo, L. (2018). Climate change governance in the anthropocene: Emergence of polycentrismin Chile. *Elem Sci Anth*, *6*, 68. https://doi.org/10.1525/elementa.329

Avaria, C.C., & Barra, C.A. (2009). La Gestión del Litoral Chileno: Un diagnóstico. Universidad Católica de Chile, Instituto de Geografía. CYTED Ibermar Red Iberoamericana—Manejo Costero Integrado. https://patrimonioceanico.cl/wp-content/uploads/2021/06/RedIBERMAR.-2009.-LA-GESTIO%CC%81N-DEL-LITORAL-CHILENO-UN-DIAGNO%CC%81STICO-.-CYTED-IBERMAR-..pdf

Berkes, F. (2009). Indigenous ways of knowing and the study of environmental change. *J. R. Soc. N.Z.*, *39*, 151–156. https://doi.org/10.1080/03014220909510568

Bonney, R., Shirk, J.L., Phillips, T.B., Wiggins, A., et al. (2014). Next steps for citizen science. *Science, 343*(6178), 1436–1437. https://doi.org/10.1126/science.1251554

Camus, P.A. (2001). Biogeografía marina de Chile continental/Marine biogeography of continental Chile. *Revista Chilena de historia natural, 74*(3), 587–617. http://dx.doi.org/10.4067/S0716-078X2001000300008

David-Chavez, D.M., & Gavin, M.C. (2018). A global assessment of Indigenous community engagement in climate research. *Environ. Res. Lett., 13*, 123005. https://doi.org/10.1088/1748–9326/aaf300

Ellis, N., & Gladwin, D. (2022, July 19). How climate storytelling helps people navigate complexity and find solutions. *The Conversation.* https://theconversation.com/how-climate-storytelling-helps-people-navigate-complexity-and-find-solutions-185354

Fariña, J.M., Palma, A.T., & Ojeda, F.P. (2008). Subtidal kelp-associated communities off the temperate Chilean coast. In T. McClanahan & G.M. Branch (Eds.), *Food webs and the dynamics of marine reefs* (pp. 79–102). Oxford University Press.

Fernández-Llamazares, Á., & Virtanen, P.K. (2020). Game masters and Amazonian Indigenous views on sustainability. *Curr. Opin. Environ. Sustain., 43*, 21–27. https://doi.org/10.1016/j.cosust.2020.01.004

García-del-Amo, D., Mortyn, P.G., & Reyes-García, V. (2020). Including Indigenous and local knowledge in climate research: An assessment of the opinion of Spanish climate change researchers. *Clim. Change.* https://doi.org/10.1007/s10584–019–02628-x

Hill, R., Adem, Ç., Alangui, W.V., Molnár, Z., Aumeeruddy-Thomas, Y., Bridgewater, P., Tengö, M., Thaman, R., Adou Yao, C.Y., Berkes, F., Carino, J., Carneiro da Cunha, M., Diaw, M.C., Díaz, S., Figueroa, V.E., Fisher, J., Hardison, P., Ichikawa, K., Kariuki, P., ... Xue, D. (2020). Working with Indigenous, local and scientific knowledge in assessments of nature and nature's linkages with people. *Curr. Opin. Environ. Sustain., 43*, 8–20. https://doi.org/10.1016/j.cosust.2019.12.006

IPCC (Intergovernmental Panel on Climate Change). (2022). *Climate change 2022: Impacts, adaptation, and vulnerability: Working Group II contribution to the 6th Assessment Report of the IPCC.* https://report.ipcc.ch/ar6/wg2/IPCC_AR6_WGII_FullReport.pdf

Iwama, A.Y., Araos, F., Anbleyth-Evans, J., Marchezini, V., Ruiz-Luna, A., Ther-Ríos, F., Bacigalupe, G., & Perkins, P.E. (2021). Multiple knowledge systems and participatory actions in slow-onset effects of climate change: Insights and perspectives in Latin America and the Caribbean. *Curr. Opin. Environ. Sustain., 50*, 31–42. https://doi.org/10.1016/j.cosust.2021.01.010

Koga-Vicente, A., & Nunes, L.H. (2011). Impactos socioambientais associados à precipitação em municípios do Litoral Paulista. *Geografia, 36*(3), 571–588).

Kullenberg, C., & Kasperowski, D. (2016). What is citizen science?—A scientometric metaanalysis. *PLoS ONE, 11*(1), e0147152https://doi.org/10.1371/journal.pone.0147152

Lam, D.P.M., Hinz, E., Lang, D.J., Tengö, M., von Wehrden, H., & Martín-López, B. (2020). Indigenous and local knowledge in sustainability transformations research: A literature review. *Ecol. Soc.*, *25*, art3. https://doi.org/10.5751/ES-11305-250103

Merçon, J., Vetter, S., Tengö, M., Cocks, M., Balvanera, P., Rosell, J.A., & Ayala-Orozco, B. (2019). From local landscapes to international policy: Contributions of the biocultural paradigm to global sustainability. *Glob. Sustain.*, *2*, e7. https://doi.org/10.1017/sus.2019.4

Nunes, L.H., & Calbete, N.D. (2000). Variabilidade pluviométrica no Vale do Paraíba Paulista. *Congresso Brasileiro de Meteorología*, *11*, 3987–3984. http://mtc-m16b.sid.inpe.br/col/sid.inpe.br/iris@1915/2005/03.15.18.27/doc/Nunes_Variabilidade%20pluviometrica.pdf

Nakashima, D., Rubis, J.T., & Krupnik, I. (2018). *Indigenous knowledge for climate change assessment and adaptation*. UNESCO Publishing.

Reyes-García, V., & Benyei, P. (2019). Indigenous knowledge for conservation. *Nat. Sustain.*, *2*, 657–658. https://doi.org/10.1038/s41893-019-0341-z

Reyes-García, V., García-del-Amo, D., Benyei, P., Fernández-Llamazares, Á., Gravani, K., Junqueira, A.B., Labeyrie, V., Li, X., Matias, D.M., McAlvay, A., Mortyn, P.G., Porcuna-Ferrer, A., Schlingmann, A., & Soleymani-Fard, R. (2019). A collaborative approach to bring insights from local observations of climate change impacts into global climate change research. *Curr. Opin. Environ. Sustain.*, *39*, 1–8. https://doi.org/10.1016/j.cosust.2019.04.007

Reyes-Garcia, V., Fernández-Llamazares, A., García-del-Amo, D., & Cabeza, M. (2020). Operationalizing local ecological knowledge in climate change research: Challenges and opportunities of citizen science. In M. Welch-Devine, A. Sourdril, & B. Burke (Eds.), *Changing climate, changing worlds* (pp. 183–197). (Springer). https://doi.org/10.1007/978-3-030-37312-2_9

Reyes-García, V., Tofighi-Niaki, A., Austin, B.J., Benyei, P., Danielsen, F., Fernández-Llamazares, Á., Sharma, A., Soleymani-Fard, R., & Tengö, M. (2022). Data sovereignty in community-based environmental monitoring: Toward equitable environmental data governance. *BioScience*, *72*(8), 714–717. https://doi.org/10.1093/biosci/biac048

Sauermann, H., Vohland, K., Antoniou, V., Balázs, B., Göbel, C., Karatzas, K., Mooney, P., Perelló, J., Ponti, M., Samson, R., & Winter, S. (2020). Citizen science and sustainability transitions. *Research Policy*, *49*(5). https://doi.org/10.1016/j.respol.2020.103978

Savo, V., Lepofsky, D., Benner, J.P., Kohfeld, K.E., Bailey, J., & Lertzman, K. (2016). Observations of climate change among subsistence-oriented communities around the world. *Nat. Clim. Change*, *6*, 462–473. https://doi.org/10.1038/nclimate2958

Scofield, G.B., De Angelis, C.F., & de Sousa Júnior, W.C. (2014). Estudo das tendências do total de precipitação e do número de dias para eventos extremos no Litoral Norte, S.P. *Geografia*, *39*(1), 109–124. https://www.periodicos.rc.biblioteca.unesp.br/index.php/ageteo/article/view/9310/6742

Serret, H., Deguines, N., Jang, Y., Lois, G., & Julliard, R. (2019). Data quality and participant engagement in citizen science: Comparing two approaches for monitoring pollinators in France and South Korea. *Citiz. Sci. Theory Pract.*, *4*, 22. https://doi.org/10.5334/cstp.200

Tavares, R., & Mendonça, F. (2017). Ritmo climático e ritmo social: Pluviosidade e deslizamentos de terra na Serra do Mar (Ubatuba/SP). In F. de Assis Mendonça (Ed.), *Riscos climáticos: Vulnerabilidades e resiliência associados* (pp. 13–22). Paco Editorial.

Tengö, M., Austin, B.J., Danielsen, F., & Fernández-Llamazares, Á. (2021). Creating synergies between citizen science and Indigenous and local knowledge. *BioScience*, *71*(5), 503–518. https://doi.org/10.1093/biosci/biab023

Tengö, M., Hill, R., Malmer, P., Raymond, C.M., Spierenburg, M., Danielsen, F., Elmqvist, T., & Folke, C. (2017). Weaving knowledge systems in IPBES, CBD and beyond—lessons learned for sustainability. *Curr. Opin. Environ. Sustain.*, *26–27*, 17–25. https://doi.org/10.1016/j.cosust.2016.12.005

PART II

Food, Land, and Agricultural Commons

3

Enhancing Local Sensitivities to Climate Change Impacts and Adaptation Capacities of Smallholder Farmers: Community-Based Participatory Research

Ayansina Ayanlade, Abimbola Oluwaranti, Oluwatoyin S. Ayanlade, Margaret O. Jegede, Lemlem F. Weldemariam, Adefunke F.O. Ayinde, Adewale M. Olayiwola, and Moses O. Olawole

Introduction: The Importance of Local Responses to Climate Change

Climate change and extreme weather events have led to multi-vulnerabilities worldwide, particularly in many African countries where recent severe droughts, floods, and intra-seasonal dry spells have impacted smallholder farming productivity. This chapter examines crop and livestock smallholder farmers' sensitivities to climate change and their adaptive strategies at the local level in southwestern Nigeria. Using participatory research methods, we investigated local indicators of climate change impacts and adaptation options being adopted by hundreds of rural farmers. We carried out our three-month study with the support of farmers' organizations in two major farm communities. We found that nearly 97 per cent of farmers have experienced delays in rainy-season onset and changes in times for the ending of the rains during the growing seasons, and have noticed low yields of some crops in recent years compared to the average over the past thirty years. We were able to derive locally relevant climate change assessment indicators in the study sites, which included such factors as rural farmers' awareness of climate change, its impacts, and specific adaptation measures. This research has improved our understanding of how climate change affects smallholder farmers and their socio-economic systems through the documentation and analysis of local knowledge and perceived effects of climate change. The participatory research process has also raised awareness, understanding, and agency in the local communities.

Climate change is a global crisis which, out of necessity, is being addressed in diverse and innovative ways at local levels worldwide. While there is vast general evidence of climate change, its impacts, and the many ways in which it affects agriculture, sensitive local adjustments are a vital means of adaptation in African countries. Continent-based climate assessments show that Africa is positioned to experience significant climatic changes, as extreme drying and warming occur in most African regions, with regional variations (Ayanlade et al., 2020b; Boko et al., 2007; Dunning et al., 2018). But "bottom-up" assessments from the perspective of local people, focused on understanding the interactions among multiple types of climate change risks and impacts, the sensitivities of different agrarian rural communities, and particularly gender impacts, have been relatively scarce and limited. Multi-hazards resulting from extreme weather events due to climate change have affected nearly six billion people (due mainly to water scarcity) and caused over eighty million casualties globally; most affected people are seniors, children, and women who live in rural agrarian communities (Lal, 2004; Rippke et al., 2016; Woolf et al., 2010). In Africa, extreme weather events have led to recent severe droughts (Ayanlade et al., 2018b; Ogunrinde et al., 2019), floods (Adelekan & Asiyanbi, 2016; Ahmadalipour et al., 2019; Lamond et al., 2019), and intra-seasonal dry spells (Fall et al., 2019; Han et al., 2019) which have great impacts on agricultural productivity, putting stresses on food security (Ayanlade et al., 2017; Lipper et al., 2014), water scarcity resources (Schilling et al., 2020; Shiru et al., 2019), and human health (Ayanlade et al., 2020a; Ayanlade & Radeny, 2020; Ayanlade et al., 2020b; Sergi et al., 2019). Some studies have shown that climate change represents the major challenge for future development, particularly in the drier parts of the continent, due to its increasing impact on crops and pastoral farming, ecosystem services, human health, and livelihoods (Adger et al., 2009; Ayanlade et al., 2018b; Mbow et al., 2014).

Year-to-year climate change impacts on crop production have resulted in severe agricultural losses in many African countries, which in some cases have led to unprecedented famines (Adejuwon, 2004; Kang et al., 2009). For example, a study by Adejuwon (2004) has reported inter-annual rainfall variability as a major factor affecting crop yields in Nigeria. Local farmers often do not have sufficient scientific information relating to the full range of causes and implications of climate change, and as such, their actions and adaptation strategies sometimes fall short of the desired result. The Intergovernmental Panel on Climate Change 6th Assessment Report (IPCC, 2021) shows that for adaptation to be effective, local knowledge is needed in conjunction with other forms of knowledge. The report further notes that economic poverty, political instability, and low productivity, which constitute important challenges in Africa, worsen and interact with the impacts of climate change. It is obvious that adaptation planning and implementation at the local level in Africa are essential for developing robust responses to climate change.

As an exploration of how this can be done, we set out to assess smallholder farmers' sensitivities to climate change and their adaptive approaches and rationales at the local level. Smallholder farmers, in this study, include small-scale farmers who own or control the land they farm but do not use mechanised equipment. They are typically operating under a small-scale agriculture system where they grow and commercialize their products alone or in local groups of neighbouring farmers. An important motivation of the study was to compare the perceptions of rural crop and livestock farmers to meteorological analyses in order to assess weather variability/changes and how rural farmers understand and view these changes. The study focused on two major research questions: the sensitivity of smallholder farmers to climate change and their adaptive capacity. As noted in the literature, adaptation and mitigation of anthropogenic climate change are significantly dependent on human sensitivity to its impacts and risks (Cox et al., 2018; Pecl et al., 2014; Trisos et al., 2022). We therefore tried to document the farmers' awareness of climate change, its impacts and specific adaptation measures, as a locally driven way of appraising the impacts of changes in rainfall and temperature during the rainy and dry seasons. The farmers' own assessments provide information about how their strategies help protect their livelihoods, minimize risk, and shape interactions among variables that (unlike weather fluctuations) are within the farmers' control. This, in turn, indicates considerations that policy-makers, extension agents, education institutions, and community organizers should consider to support adaptation and socio-economic welfare in regions affected by climate change.

This chapter describes our study's context, methodology, and outcomes. In the conclusion, we comment also on the relationship between methodology and results: how our study's relatively attentive, community-based, participatory approach made possible some of our nuanced findings.

Study Area, Goals, Research Partners, and Methods

We carried out our research in the southwestern part of Nigeria (Map 3), with local farmers' organizations located in the rural communities of Odemuyiwa and Ilora/Ilu-Aje, which are among the most populous smallholder farming communities in southwestern Nigeria.[1] Yoruba is spoken in the study area, and the population also includes some Hausa farmers and Fulani livestock farmers. Crops include cocoa, maize, cassava, vegetables, and other farm produce.

We had several reasons for selecting these two communities. Besides the fact that they are the major smallholder farming communities in the region and are accessible from our university's location in Ile-Ife, they are located in the two main agro-climatic zones of Nigeria: Ilora is located in the Guinea Savanna; Odemuyiwa is located in the Rain Forest agro-climatic zone. The Guinea Savanna zone is known for cereals and tuber crop farming while the Rain Forest zone is known for cash crops and tree crops such as cocoa, coffee and kola nut. Thus, working with smallholder farmers in these two communities allowed us to sample perceptions of climate change impacts and adaptation across different agro-climatic zones and different kinds of farming practices. The soils across the study area are rich and appropriate for the kinds of cultivation undertaken.

We explored climate change impacts and multi-risks using mixed methods and multi-disciplinary approaches to develop what we call a multiplying vulnerability index. This relied on both quantitative and qualitative information from questionnaires and in-depth focus group discussions (FGDs) with farmers. A set of semi-structured questions was used for in-depth interviews.

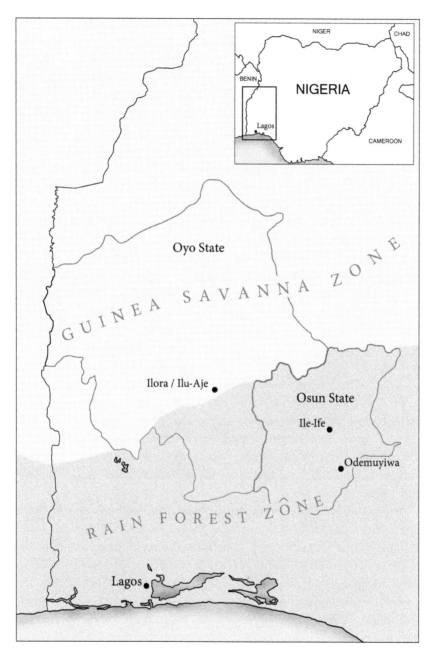

Map 3 Nigeria—Osun and Oyo States

Care was taken to purposely interview individuals who had been farming for periods longer than ten years, and who thus had experience with farm and weather conditions in the area over time. The questions were related to impacts of climate change, perceptions and responses to climate change impacts, adaptive capacity, and vulnerability. The FGDs facilitated information exchange on the farmers' perceptions, local indicators of climate change, its impacts, and their adaptation strategies. We also gathered information on the demographic characteristics of the farmers, their agricultural practices, means of climate change awareness, and other details. This was to understand the determinants of farmers' choices of adaptation methods and adaptive capacity through climate-smart agriculture or other initiatives.[2] As a way to increase our work's relevance, applicability, and potential usefulness for local people, we developed our research questions and approach in conjunction with the local farmers' organizations that were our partners in this research process: the Odemuyiwa Farmers' Association, Agbeloba Farmers Society of Ilu-Aje (Agbeloba means that as providers of food for the community, "farmers are the kings"), the Ilora Women's Farm Association, and the Ilora Smallholder Farmers Cooperative. The lead researcher, while introducing the project and requesting leaders' support and participation, also carried out "key informant" interviews with local chiefs, elders, and leaders. The farmers' organizations kindly introduced the lead researcher at their regular meetings, provided him with lists of member farmers, and appointed a liaison member to help him make appointments, accompany him, and introduce him to farmers to be interviewed each day. Interviews were mostly carried out in the fields, during the farmers' breaks. Women farmers were mostly interviewed on market days in their market stalls, in between their attending to customers.

Through collaborations with the farmers' organizations, we organized training workshops for farmers in each of the communities (Odemuyiwa and Ilora/Ilu-Aje) to discuss concepts, policies, and mechanisms relating to climate change impacts and adaption on agriculture and food security, linked to the livelihoods of smallholder farmers in the study sites (Figure 3.1). The invited participants included both livestock and crop farmers, representatives from local government authorities, and rural community leaders in the settlements where their primary occupations are farming. The workshops, FGDs, and interviews were led by a team of twelve researchers (six men, six women) made up of lecturers, graduate and undergraduate students from

Fig. 3.1 Workshop/training on climate change impacts held in Odemuyiwa village, Osun State, Nigeria.

the Department of Geography, Crop Production and Protection, and the African Institute for Science Policy and Innovation, Information Technology, of Obafemi Awolowo University in Ile-Ife, Nigeria. More than two hundred and fifty smallholder farmers and other guests participated in each workshop, roughly one-third of whom were women. The individual interviews with elders, leaders, and farmers numbered about forty-five in Odemuwiya (about fifteen of whom were women farmers) and about twenty-five in Ilora (ten of them women and four of them Hausa-speaking pastoralists from northern Nigeria who had moved to the area in recent years). The participants' ages ranged from thirty-five years to over sixty years. Many of the participants had at least fifteen years of farming experience and had been living in the communities over a long period of time; many were native to the communities.

The FGDs (Figure 3.2) were held either outside or in large churches in each community, and besides farmers, local youths were invited, along with a group of more than ninety undergraduate students from Obafemi Awolowo University

Fig. 3.2 Focus Group Discussion on climate change impacts, held in Ilu-Aje/ Ilora, Oyo State, Nigeria.

Fig. 3.3 University undergraduates visit Odemuyiwa and learn how climate variability affects agricultural productivity in the rural community.

who came by bus for the day, mingled and spoke with local youths and farmers, and learned about how to conduct participatory research. Among other presentations on climate change science and adaptation (Figure 3.3), university climate researchers brought a mobile weather station and showed interested farmers how it works to record and transmit rainfall and temperature data.

Other methods that research team members used to gather information included observation, investigation, measurements, field sketches, audio-video recording, photographs and GNSS (Global Navigation Satellite System) surveys. The workshops, with participants' permission, included digital audio and video recordings, and photographs (Figure 3.1), which were used to document and share farmers' sensitivities to climate change impacts, and their adaptation capacities, including from a gender perspective.

On the day of the team's arrival for the workshop that wound up our research in each community, the local participants were already assembled in a church hall facility. After introductions were made and participants' attendance was taken, an information session on climate change awareness was held. The main activities were based on FGDs in small groups, facilitated by members of the research team, who also made presentations on different climate change topics including the dynamics of climate as it affects crop yields; crop production, processing, and protection; and the technologies and incentives available to farmers. Team members joined in the Q+A discussions with all participants.

Farmers shared their suggestions regarding key areas researchers could explore to bring agricultural dividends to local farmers and facilitate their adaptation to climate change. At the end of the session, certificates were issued to participants, and cutlasses were given out too, as thanks for everyone's participation and to aid their farming activities. All participants shared a meal at the end of the workshop (Figures 3.4 and 3.5).

Participatory aspects of the research design included researchers' gradual introductions to community members, facilitated by local organizations and individuals; various opportunities for individual and group information-sharing in informal conversations, interviews, FGDs, and workshops; familiarization of farmers and local youth with climate communication terms, climate science, and potential adaptation measures; building political agency and stakeholder engagement through the farmers' organizations and local networks; respectful acknowledgement and sharing of local knowledge and innovations in facing climate change; special attention to the particular

Fig. 3.4 and **3.5** Workshop participants received cutlasses in appreciation of their sharing knowledge on climate change.

socio-economic contributions and multi-vulnerabilities of women farmers; and opening up possibilities for further ongoing communication and linkages between the villages and the university at both inter-personal and institutional levels.

Results and Discussion: Farmers' Climate Change Knowledge and Adaptation Strategies

In both the research design and the interpretation of results, we relied on Jost et al. (2014) and the framework for understanding local adaptive capacity (LAC) developed by Jones et al. (2019) as part of the Africa Climate Change Resilience Alliance (ACCRA) initiative. Within this framework, LAC for smallholder farmers is seen as depending on the context-specific interaction of governance institutions, social learning, trust, collective action, creativity / innovation, and the availability of assets; particularities of gender, politics, and power are also important. The LAC model groups these characteristics in five distinct but interrelated attributes of adaptive capacity. These include the asset base; institutions and entitlements; knowledge and information; innovation; and agile, forward-looking decision-making and governance.

Assets are very important for smallholder farmers, as the poorest are most vulnerable to the effects of climate change and broader developmental stresses. Lack of assets is likely to affect the ability of smallholder farmers to cope with the effects of climate change. Gender is a strong determinant of asset limitations for women farmers. Institutions and entitlements mediate access to and/or control of asset-based resources. Knowledge and information are required for a better understanding of the observed and possible future climate change impacts and their complexity, awareness of climate change adaptation options, ability to test options, and the ability to incorporate interventions. Decision-making and governance are key to supporting the capacities of smallholder farmers in coping with the consequences of climate change. Organising response options can help them better handle the effects of climate change along with other socio-economic and ecological pressures. This can be achieved through effective innovation that involves both scientifically and traditionally focused technology and innovation tailored to local projects that help smallholders respond to climate change impacts and risks. LAC theorists also note the importance of context and dialogue among community members in this regard (Jones, 2019; Jost et al., 2014). In our analysis of results from our research in the two communities, therefore, we consider all these factors as related to the adaptive capacity and practices of local farmers.

The outcomes of our research indicated that both crop and livestock farmers have certainly noticed changes in climate. During our participatory engagement with the farmers, the majority stated that they had perceived changes in times for the start and end of the rains during the growing seasons, and noticed that some crop yields are lower in recent years compared to the past thirty-year average. Livestock farmers are now finding it difficult to find water and green pastures during the prolonged dry spell. Nearly all the farmers perceived changes in the onset of rainfall. The level of awareness of local people about weather changes linked to global warming, and their sensitivities to this, are very strong.

Generally, the farmers perceive many changes in the climate system including the increase in the annual minimum temperature and reduction in the same in the coldest and hottest seasons, increase in the maximum temperatures in both seasons and reduction in the amount of rainfall in both seasons. Many farmers observed changes in the mean temperature, frequency of cold days, sunny days, and sunshine intensity and the intensity of heavy rainfall events was generally agreed to be on the increase. The frequency of warm days and changes in mean rainfall was agreed to be higher. The temperature is perceived to be on the increase only in the hottest season. Since rains are now delayed, cropping seasons are now shortened and planting dates are no longer fixed.

The farmers (both male and female) indicated that rainfall has been much more unreliable in recent years. The majority of the rural farmers perceived that "the climate is by far away from what we used to have in the past, the climate change has resulted into changes in the biophysical environment, poor yield of crops as a result of change associated with reduction in rainfall, attack of pests and diseases." They further stated that "some pests not known in years back are now prominent, while crops planted in the past are not as productive as they were, even tree crops are no longer sustainable."

The perceptions of farmers regarding changes in the intensity of rainfall and numbers of rainy days, heat changes/variability, and their sensitivities are generally consistent with climatic trend analysis. Previous studies have shown that sub-Saharan Africa is likely to be more vulnerable to climate change than other parts of the world, not only because the economy depends on rain-fed agriculture, but also due to the difficult challenges of poverty including food security, health challenges, and low levels of infrastructural and technological development (Ayanlade & Ojebisi, 2019; Ayanlade et al., 2018a; Bryan et al., 2018; Mogomotsi et al., 2020; Morton, 2007; Thornton et al., 2011). Drought and heavy rainfall-induced floods are projected to become more frequent and severe, thereby increasing pressure on freshwater resources. In particular, there is high confidence (Ayanlade et al., 2018b; Hillie & Hlophe, 2007) that risks associated with increases in drought frequencies and magnitudes are projected, even at an average global temperature increase of 1.5°C, for many African countries. What worsens the situation is that nearly 41 per cent of the population in Africa lives below the international poverty line of US$1.90 per day (World Bank, 2018).

Earlier studies have shown that because the majority of crop-farming activities in Nigeria are rain-fed, rainfall is the most important element of climate-related risk, a change of which will greatly affect both crop and live-stock farming in the country. Thus, crop and livestock farmers in Nigeria are likely to be severely affected because of their low levels of adaptive capacity to climate change/variability (Adejuwon, 2004; Boko et al., 2007; Liverpool-Tasie et al., 2019).

Farmers' perceptions of climate variability/change were based on the local climate parameters they identified, but it was apparent from all our information-gathering that these farmers are particularly vulnerable to climate change since the majority of them do not have enough assets or resources to cope with the situations they are experiencing. Many farmers who participated in our study made suggestions that facilitators should link them to policymakers who could intervene to help resolve their farming plight.

Some suggested that there is an urgent need to establish cooperatives headed by the local farmers with facilitators (perhaps from the university) as supporting members, who could provide advice on farming activities and help farmers gain access to agricultural loans and other incentives. In other words, the government could build the capacity of agricultural extension systems (Morton, 2007) and make available climate change education schemes (Ayanlade & Jegede, 2016), perhaps using communications technology innovations such as cell phone applications.

Our collaboration with local farmers' organizations allowed us to observe through participant observation the local mechanisms that exist in both communities to facilitate communication about problems farmers face: reliance on elders and leaders, development of social trust, information-sharing, and development of creative initiatives to advance collective well-being. For example, the farmers' repeated suggestions and questions related to funding options and support for their climate resilience projects showed that they are interested in organizing a collective community-based appeal and approach to climate adaptation to take advantage of any institutional opportunities. While the three-month time frame of our research may not have given us a thorough understanding of how local governance institutions work and how power is distributed or implemented in each community, we agree with the LAC framework that this is fundamentally important and has many gender implications. The majority of farmers stated that "governments need to help rural farmers, with financial and mechanical aids to cope during climate extreme events." Such aids are most needed by female farmers who are heads of households; the majority of them stated that they "do not have adequate financial aids and assistance like the male farmers in their communities." Consequently, "female farmers are less able to adapt to climate change than male farmers"; this was agreed on during the FGDs.

In Nigeria, like many other African countries, medium- and long-term adaptive measures have been identified in the national communications to the United Nations Framework Convention on Climate Change. The Ministry of the Environment in Nigeria, for example, has identified emergency measures for adaptation in the National Adaptation Programmes of Action (NAPAs), which focus on agriculture, food security, water resources management, and other sectors. Many of the measures have not yet been fully implemented, leaving many farmers without a sound understanding of the challenges facing agricultural production that result from climate change and the plans for facing them. Though campaigns towards behavioural and policy change as a result of climate change may be a long-term adaptation matter, farmers' awareness of climate impacts, such as the frequency of extreme weather events, needs to be addressed expeditiously (Ayanlade et al., 2022) in the rural areas of southwestern Nigeria, and throughout the country—a major gap that this study addresses. The findings of this study, identify a need to build farmers' capacity development programmes to assist them in coping with the changing climate.

Nonetheless, farmers are very resourceful and are using options open to them to increase their climate resilience. We were able to document a number of adaptation strategies that farmers are already adopting. These include the following:

- Shift planting dates to accommodate changes in onset of rains, even year to year.

- Irrigate using head-carried water from boreholes during droughts to keep plants alive (though this is very labour intensive and not practical at more than minimal scales).

- Plant new, water-intensive cash crops such as cucumber, watermelon, golden melon, carrot, and cassava species, at times of peak rainfall, to earn income to tide themselves through times of drought that may cause major rain-fed crops to fail.

- Minimize risk by sending family members to cities for work, who can remit funds to tide others through crop failures and drought.

- Seek funding for collective construction of new water wells and boreholes.

For all these adaptation strategies, we noted that several factors are related to specific farmers' ability to adapt and to determine their choice of adaptation methods. The principal factors determining their options are income, level of education, and years of farming experience. Farmers detailed that "lack of financial capital hinders the ability to purchase mechanized irrigation systems as adaptation methods during prolonged dry spells." Many of them claimed that they "have little income from smallholding farming and this is a major barrier to adopt some adaptation methods, especially those that are capital intensive." As per the LAC model, those with more assets and resources, knowledge and information, and entitlements are better able to innovate and take advantage of power differentials to adapt to climate change-driven shocks and trends.

With regard to differences between the Guinea Savanna and Rain Forest agro-climatic zones, we documented that rainfall variation and droughts are more frequent in the more northerly Guinea Savanna area, where there are fewer tree crops and thus more extreme dependence on annual rainfall and more pressure for irrigation. Through interviews with farmers, we also heard of incipient land conflicts between Hausa-speaking in-migrants who are animal herders and longer-term Yoruba-speaking residents whose crops are sometimes harmed by grazing animals. This reflects migration pressures on farmers living further north in Nigeria, also caused by climate change. Such conflicts have the potential to damage adaptation requisites noted in the LAC model such as social trust, effective governance institutions, entitlements, and asset distribution, reducing adaptation ability for some and likely increasing economic disparities within the farm communities.

Conclusion

In this study, we used participatory research methods to assess smallholder farmers' sensitivity to climate change and their adaptation strategies. Our goal was to improve understanding of how climate change affects smallholder farmers and their socio-economic systems through the documentation and analysis of local indicators and perceived effects of climate change. The results show that the majority of smallholder farmers are sensitive to climate change, as many of them are aware of changes in both rainfall and temperature in recent years. The majority of farmers claimed that recent changes in rainfall and temperature have significant impacts on the development and yield of many crops. They acknowledged the evidence of climate change in their rural communities. They had good understanding of changes in climate conditions in recent years, which they said include shortness of the duration of seasonal rainfall, and consistent intensification of temperatures during the daylight and sometimes also at night. They acknowledged that while some farm pests are no longer evident, others are still very much present and cause serious damage to their crops. They are changing their farm practices as best they can, given their differing vulnerabilities and options, to adapt to these changing farm conditions while attempting to preserve their socio-economic resilience.

This research adds to knowledge about participatory ways to assess multi-risks and multi-hazards resulting from climate change. Since there is high confidence that multi-risks of climate change are likely to aggravate poverty in Nigeria, where millions of people's livelihoods depend on rain-fed agriculture and the immediate natural environment, these approaches have great implications for well-being.

Our participatory fieldwork, which included a range of research methods, allowed us to explore climate change from the "grassroots" perspective of farmers, gathering detailed information about the many interacting factors identified in the LAC model as influencing climate adaptation abilities and overall social impacts of climate change. This in turn has enhanced the theoretical and scientific conclusions in the literature about the large impacts of climate change on smallholder agriculture in this particular context. Our experience with this participatory process, involving hundreds of local farmers and youth, researchers and university students, also gives us confidence that the research intervention itself has likely contributed to building social trust, innovation, and collective engagement—key factors in adaptation potential. As one senior chief told us at the final workshop, "I have never experienced such a thing before!" Several of the participants suggested, "we will appreciate it if this kind of programme can be organised for us again in this village, to help our farming activities under climate change."

We believe this kind of detailed, context-specific awareness needs to be replicated in other communities, since the ability to adapt at the local level is a prime determinant of community members' well-being.

Acknowledgements

This work was supported through a project on ecological economics, commons governance, and climate justice based at York University in Canada (qesclimatejustice.info.yorku.ca), with funding from the Canadian Queen Elizabeth II Diamond Jubilee Scholarships Advanced Scholars Program (QES-AS), supported by the Social Sciences and Humanities Research Council of Canada and the International Development Research Centre. This work was also funded by the Tertiary Education Trust Fund, TETFund NRF 2020 Nigeria (GrantAward—TETF/DR&D-CE/NRF2020/CC/17/VOL.1).

NOTES

1   All the authors contributed to this project in various ways: while the first author designed and coordinated the research, others assisted with focus groups and workshops, interviews with farmers, data analysis and writing, project logistics, language translation, and/or coordinating student participants' training as visitors while the research was underway.

2   We assessed our results using statistical models, and we report on those quantitative results more fully in other publications. The independent variables were the timing and duration of rainy and dry seasons, the adaptation methods currently employed by the farmers, and the length of time for which the respondent has been a farmer.

The dependent variables were crop yield, crop failure, and perceived variations in climate. These were used to calculate correlation and regression models. Data from the questionnaires and interviews were categorized based on different categories of farmers' perceptions of rainfall onset, amount, frequency and duration, intensity, variability/change, and cessation in the study area.

Reference List

Adejuwon, J. (2004). Crop yield response to climate variability in the Sudano-Sahelian ecological zones of Nigeria in southwestern Nigeria. *Messages from Dakar: Report of Second AIACC Regional Workshop for Africa and the Indian Ocean Islands, Dakar, Senegal*, 15–16.

Adelekan, I.O., & Asiyanbi, A.P. (2016). Flood risk perception in flood-affected communities in Lagos, Nigeria. *Natural Hazards, 80*, 445–469.

Adger, W.N., Dessai, S., Goulden, M., Hulme, M., Lorenzoni, I., Nelson, D.R., Naess, L.O., Wolf, J., & Wreford, A. (2009). Are there social limits to adaptation to climate change? *Climatic Change, 93*, 335–354.

Ahmadalipour, A., Moradkhani, H., Castelletti, A., & Magliocca, N. (2019). Future drought risk in Africa: Integrating vulnerability, climate change, and population growth. *Science of the Total Environment, 662*, 672–686.

Ayanlade, A., & Jegede, M.O. (2016). Climate change education and knowledge among Nigerian university graduates. *Weather, Climate, and Society, 8*, 465–473.

Ayanlade, A., Nwayor, I.J., Sergi, C., Ayanlade, O.S., Di Carlo, P., Jeje, O.D., & Jegede, M.O. (2020a). Early warning climate indices for malaria and meningitis in tropical ecological zones. *Scientific Reports, 10*, 1–13.

Ayanlade, A., & Ojebisi, S.M. (2019). Climate change impacts on cattle production: analysis of cattle herders' climate variability/change adaptation strategies in Nigeria. *Change and Adaptation in Socio-Ecological Systems, 5*, 12–23.

Ayanlade, A., & Radeny, M. (2020). COVID-19 and food security in Sub-Saharan Africa: Implications of lockdown during agricultural planting seasons. *npj Science of Food, 4*, 13.

Ayanlade, A., Radeny, M., & Morton, J.F. (2017). Comparing smallholder farmers' perception of climate change with meteorological data: A case study from southwestern Nigeria. *Weather and Climate Extremes, 15*, 24–33.

Ayanlade, A., Radeny, M., Morton, J.F., & Muchaba, T. (2018a). Drought characteristics in two agro-climatic zones in sub-Saharan Africa. *Climate Prediction S&T Digest*, 140.

Ayanlade, A., Radeny, M., Morton, J.F., & Muchaba, T. (2018b). Rainfall variability and drought characteristics in two agro-climatic zones: An assessment of climate change challenges in Africa. *Science of the Total Environment, 630*, 728–737.

Ayanlade, A., Sergi, C.M., Di Carlo, P., Ayanlade, O.S., & Agbalajobi, D.T. (2020b). When climate turns nasty, what are recent and future implications? Ecological and human health review of climate change impacts. *Current Climate Change Reports, 6*, 55–65.

Ayanlade, A., Oluwaranti, A., Ayanlade, O.S., Borderon, M., Sterly, H., Sakdapolrak, P., Jegede, M.O., Weldemariam, L.F., & Ayinde, A.F. (2022) Extreme climate events in sub-Saharan Africa: A call for improving agricultural technology transfer to enhance adaptive capacity. *Climate Services, 27*, 100311.

Boko, M., Niang, I., Nyong, A., Vogel, A., Githeko, A., Medany, M., Osman-Elasha, B., Tabo, R., & Yanda, P. (2007). Africa. In M.L. Parry, O.F. Canziani, J.P. Palutikof, P.J. van der Linden, & C.E. Hanson (Eds.), *Climate change 2007: Impacts, adaptation and vulnerability: Contribution of working group II to the fourth assessment report of the intergovernmental panel on climate change* (pp. 433–467). Cambridge University Press.

Bryan, E., Bernier, Q., Espinal, M., & Ringler, C. (2018). Making climate change adaptation programmes in sub-Saharan Africa more gender responsive: Insights from implementing organizations on the barriers and opportunities. *Climate and Development, 10*, 417–431.

Cox, P.M., Huntingford, C., & Williamson, M.S. (2018). Emergent constraint on equilibrium climate sensitivity from global temperature variability. *Nature, 553*(7688), 319–322.

Dunning, C.M., Black, E., & Allan, R.P. (2018). Later wet seasons with more intense rainfall over Africa under future climate change. *Journal of Climate, 31*, 9719–9738.

Fall, C.M.N., Lavaysse, C., Drame, M.S., Panthou, G., & Gaye, A.T. (2019). Wet and dry spells in Senegal: Evaluation of satellite-based and model re-analysis rainfall estimates. *Natural Hazards and Earth System Sciences Discussions*, 1–29.

Han, F., Cook, K.H., & Vizy, E.K. (2019). Changes in intense rainfall events and dry periods across Africa in the twenty-first century. *Climate Dynamics, 53*, 2757–2777.

Hillie, T. & Hlophe, M. (2007). Nanotechnology and the challenge of clean water. *Nature Nanotechnology, 2*, 663.

IPCC (Intergovernmental Panel on Climate Change). (2021). *Climate change 2021: The physical science basis: Contribution of Working Group I to the sixth assessment report of the intergovernmental panel on climate change* (V. Masson-Delmotte, P. Zhai, A. Pirani, S.L. Connors, C. Péan, S. Berger, N. Caud, Y. Chen, L. Goldfarb, M.I. Gomis, M. Huang, K. Leitzell, E. Lonnoy, J.B.R. Matthews, T.K. Maycock, T. Waterfield, O. Yelekçi, R. Yu, & B. Zhou, Eds.). Cambridge University Press.

Jones, L., Ludi, E., Jeans, H., & Barihaihi, M. (2019). Revisiting the local adaptive capacity framework: Learning from the implementation of a research and programming framework in Africa. *Climate and Development, 11*, 3–13.

Jost, C., Ferdous, N., & Spicer, T.D. (2014). *Gender and inclusion toolbox: Participatory research in climate change and agriculture.* CGIAR Research Program on Climate

Change, Agriculture and Food Security (CCAFS), World Agroforestry Centre (ICRAF), and CARE International.

Kang, Y., Khan, S., & Ma, X. (2009). Climate change impacts on crop yield, crop water productivity and food security—A review. *Progress in Natural Science, 19*, 1665–1674.

Lal, R. (2004). Soil carbon sequestration to mitigate climate change. *Geoderma, 123*, 1–22.

Lamond, J., Adekola, O., Adelekan, I., Eze, B., & Ujoh, F. (2019). Information for adaptation and response to flooding, multi-stakeholder perspectives in Nigeria. *Climate, 7*, 46.

Lipper, L., Thornton, P., Campbell, B.M., Baedeker, T., Braimoh, A., Bwalya, M., Caron, P., Cattaneo, A., Garrity, D., Henry, K., Hottle, R., Jackson, L., Jarvis, A., Kossam, F., Mann, W., McCarthy, N., Meybeck, A., Neufeldt, H., Remington, T., ... & Torquebiau, E.F. (2014). Climate-smart agriculture for food security. *Nature Climate Change, 4*, 1068–1072.

Liverpool-Tasie, L.S.O., Sanou, A., & Tambo, J.A. (2019). Climate change adaptation among poultry farmers: evidence from Nigeria. *Climatic Change, 157*, 527–544.

Mbow, C., Van Noordwijk, M., Luedeling, E., Neufeldt, H., Minang, P.A., & Kowero, G. (2014). Agroforestry solutions to address food security and climate change challenges in Africa. *Current Opinion in Environmental Sustainability, 6*, 61–67.

Mogomotsi, P.K., Sekelemani, A., & Mogomotsi, G.E. (2020). Climate change adaptation strategies of small-scale farmers in Ngamiland East, Botswana. *Climatic Change*, 1–20.

Morton, J.F. (2007). The impact of climate change on smallholder and subsistence agriculture. *Proceedings of the National Academy of Sciences, 104*, 19680–19685.

Ogunrinde, A., Oguntunde, P., Akinwumiju, A., & Fasinmirin, J. (2019). Analysis of recent changes in rainfall and drought indices in Nigeria, 1981–2015. *Hydrological Sciences Journal, 64*, 1755–1768.

Pecl, G.T., Ward, T.M., Doubleday, Z.A., Clarke, S., Day, J., Dixon, C., Frusher, S., Gibbs, P., Hobday, A.J., Hutchinson, N. & Jennings, S., (2014). Rapid assessment of fisheries species sensitivity to climate change. *Climatic Change, 127*(3), 505–520.

Rippke, U., Ramirez-Villegas, J., Jarvis, A., Vermeulen, S.J., Parker, L., Mer, F., Diekkrüger, B., Challinor, A.J., & Howden, M. (2016). Timescales of transformational climate change adaptation in sub-Saharan African agriculture. *Nature Climate Change, 6*, 605.

Schilling, J., Hertig, E., Tramblay, Y., & Scheffran, J. (2020). Climate change vulnerability, water resources and social implications in North Africa. *Regional Environmental Change, 20*, 1–12.

Sergi, C., Serra, N., Colomba, C., Ayanlade, A., & Di Carlo, P. (2019). Tuberculosis evolution and climate change: How much work is ahead? *Acta Tropica, 190*, 157–158.

Shiru, M.S., Shahid, S., Shiru, S., Chung, E.S., Alias, N., Ahmed, K., Dioha, E.C., Sa'adi, Z., Salman, S., Noor, M., Nashwan, M.S., Idlan, M.K., Khan, N., Momade, M.H., Houmsi, M.R., Iqbal, Z., Ishanch, Q., & Sediqi, M.N. (2019). Challenges in water resources of Lagos mega city of Nigeria in the context of climate change. *Journal of Water and Climate Change, 11*(4), 1067–1083.

Thornton, P.K., Jones, P.G., Ericksen, P.J., & Challinor, A.J. (2011). Agriculture and food systems in sub-Saharan Africa in a 4 C+ world. *Philosophical Transactions of the Royal Society A: Mathematical, Physical and Engineering Sciences, 369,* 117–136.

Trisos, C.H., Adelekan, I.O., Totin, E., Ayanlade, A., Efitre, J., Gemeda, A., Kalaba, K., Lennard, C., Masao, C., Mgaya, Y., Ngaruiya, G., Olago, D., Simpson, N.P., & Zakieldeen, A.S. (2022) Africa. In H.-O. Pörtner, D.C. Roberts, M. Tignor, E.S. Poloczanska, K. Mintenbeck, A. Alegría, M. Craig, S. Langsdorf, S. Löschke, V. Möller, A. Okem, & B. Rama (Eds.), *Climate change 2022: Impacts, adaptation, and vulnerability.* Cambridge University Press.

Woolf, D., Amonette, J.E., Street-Perrott, F.A., Lehmann, J., & Joseph, S. (2010). Sustainable biochar to mitigate global climate change. *Nature Communications, 1,* 56.

World Bank. (2018). *World bank open data.* World Bank Group.

4

The Oil Palm Sector in the Climate Crisis: Resilience and Social Justice in the Commune of Ngwéi (Littoral-Cameroon)

Guy Donald Abassombe, Mesmin Tchindjang, and Vadel Eneckdem Tsopgni

Introduction

Climate change is currently a central concern for both scientists and political decision-makers at the global level (Niang, 2009); it constitutes one of the many obstacles to human development (Brown & Crawford, 2008; Boko, 1988). The intrinsic injustice of global warming, which makes the poorest pay the consequences of the actions of the richest, is even more flagrant for peasants (Capocci et al., 2015; Development and Peace, 2015). This is because agriculture, which is one of the main levers of economic development, essentially depends on climatic conditions (Chanzy et al., 2015; Bélanger & Bootsma, 2004). Agriculture can be seen as both a victim of climate change, and also as one of its major causes (Baudouin, 2021). Deforestation for agriculture contributes to greenhouse gas (GHG) emissions, reduces habitat and biodiversity, and can reduce carbon sequestration, thus exacerbating climate change. On the other hand, climatic disturbances have a direct impact on agricultural production and yields (Boko et al., 2007; Mertz et al., 2009). This impact is particularly significant in developing countries where agriculture is largely rain-fed and is the main source of employment and income for the majority of the population (Agossou et al., 2012; Delille, 2011; Enete & Onyekuru, 2011). From the home to the international scale, economic and social injustices and inequalities at all levels exacerbate these impacts, aggravating hunger and poverty in developing countries (Ramirez-Villegas & Thornton, 2015; Rawe & Deering, 2015). Cameroon is already affected by these manifestations of climate change, which are multiplying across the country's different agro-ecological zones (P. Amougou, 2016; J. Amougou, 2018; J. Amougou & Batha, 2014; J. Amougou et al., 2013; Tchindjang et al., 2017). The agricultural sector, which employs about 70 per cent of the economically active population and generates 80 per cent of the primary sector's contribution to the gross domestic product (GDP), is highly affected by climate change. Impacts on agricultural production erode the living conditions of farmers (Mamoudou, 2019; Djitie Kouatcho et al., 2019), despite their strategic actions in response.

The oil palm sector, which currently constitutes one of the strategic pillars of economic growth in Cameroon according to the nation's *Strategy Document for Growth and Employment* (Republique du Cameroun, 2009), is not spared from this reality. In the Commune of Ngwéi[1] and throughout the agricultural basins of the Littoral-Cameroon, the oil palm sector is already threatened by climate change, affecting both productivity and the living conditions of farmers. Faced with growing demand for land that fuels massive land grabbing (Sitou et al., 2014), combined with most farmers' limited access to agricultural inputs, smallholders are struggling to fit into the process of development linked to the expansion of palm groves. This situation makes the oil palm sector a sustainable development issue, especially in this area where oil palm provides 85 per cent of the local population's income.[2]

What are the forms of social injustice that characterize the oil palm sector, how does climate change affect this agricultural sector, and what alternatives exist for the development of a more climate-resilient palm sector in coastal Cameroon?

This chapter reports on our participatory research to investigate these questions, with the help of two hundred and ninety palm oil producers from twenty-nine villages in Ngwéi, where oil palm production has a long history stretching back hundreds of years.[3] Through interviews with the farmers, workshops to share knowledge about global warming and climate justice, and small-group discussions in each village, we explored the changes the farmers are experiencing, the challenges they face, and their comments about what would help to improve their livelihood options. We also documented the farmers' own informal and strategic initiatives towards protecting their livelihoods in the face of climate change. We complement the information they shared by summarizing available weather statistics and documentary research on the impacts of climate change in coastal Cameroon, and on palm oil production and climate justice in central Africa more broadly. Our concluding reflections are based on the findings, demonstrated through our research with small farmers, that land insecurity and farmers' limited access to agricultural inputs accentuate the effects of climate change for oil palm farmers in Ngwéi.

The chapter is organized as follows. Section two sets out the geographical context and methodology for our work. Survey and interview results on climate trends and impacts on small farmers are discussed in section three, along with related statistical information on rainfall and temperature. Section four explains how small farmers are adapting and changing their practices to cope with the impacts they are experiencing. Our reflections and conclusions make up section five.

Spatial Context and Methodological Approach

The Commune of Ngwéi is located in the oil palm cultivation corridor on the coastal strip of Cameroon (Map 4). It is situated at the gateway to the main outlet basins for palm oil production in Cameroon, notably 30 km from Edea and 90 km from Douala, the economic metropolis and industrial free zone of the country. This strategic position fuels strong expansion pressure within the territory. Located in the Littoral-Cameroon region, this commune covers an area of approximately 500 km² (PNDP, 2018).

Like most areas on the Cameroonian coast, the physical setting of the Commune of Ngwéi is favourable for oil palm cultivation. In terms of climatic conditions, there is fairly constant humidity and heat throughout the year. This zone has a humid, Equatorial Guinean-type climate characterized by four seasons: The short and long dry seasons extend respectively from November to January and from February to March. The short rainy season generally extends from April to June and the big one begins in July until August. The oil palm grows best in equatorial zones, which benefit from both high rainfall (at least 1800 mm per year, or 150 mm/month), and an average annual temperature of at least 26°C. Data on the monthly average trends of

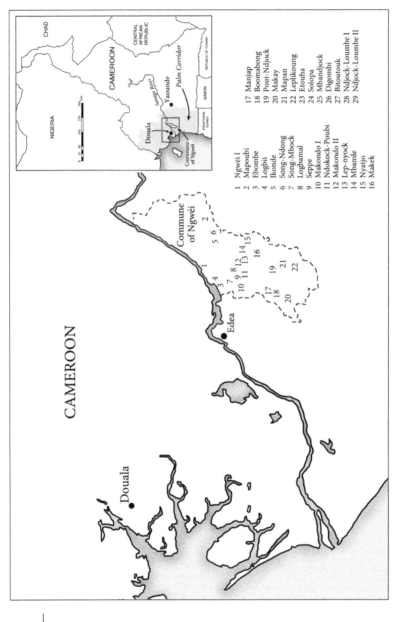

Map 4
Cameroon—
Commune
of Ngwéi

1 Ngwéi I
2 Mapoubi
3 Ebombe
4 Logbii
5 Ikonde
6 Song-Ndong
7 Song-Mbock
8 Logbamal
9 Seppe
10 Makondo I
11 Ndokock-Poubi
12 Makondo II
13 Lep-nyock
14 Mbamle
15 Nyatjo
16 Makek
17 Manjap
18 Boomabong
19 Pout-Ndjock
20 Makay
21 Mapan
22 Leplikoung
23 Etouha
24 Solopa
25 Mbandjock
26 Digombi
27 Bitoutouk
28 Ndjock-Loumbe I
29 Ndjock-Loumbe II

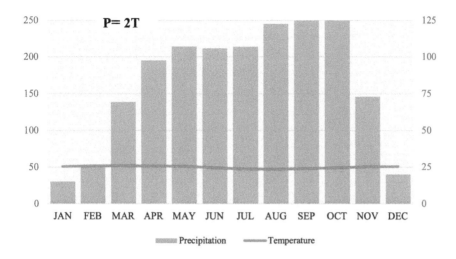

Fig. 4.1 Temperature and rainfall diagram of the Commune of Ngwéi, produced from the monthly averages over the period from 1981–2017. **Source**: Climate data from NASA (2021), Surface Meteorology and Solar Energy (SSE). Compiled by the first author.

precipitation and temperatures over the period from 1981–2017 show that the Commune of Ngwéi receives a large amount of rainfall (on average 2400 mm/year), with temperatures oscillating around 26°C (Figure 4.1).

Relatively flat land is best for oil palm cultivation (Jacquemard, 2011). The Commune of Ngwéi extends over a vast low plain with hills that decline in height farther to the south, which is dominated by river floodplains. Thus, the centre and the south of the Commune of Ngwéi, compared to the north, are more conducive to the cultivation of oil palm, which has expanded there. Map 4 shows the oil palm cultivation zone in southeastern Cameroon.

Methodology

To investigate how climate change is affecting the oil palm sector, what options palm producers have for risk reduction, and the social justice implications, we adopted a mixed methods approach blending weather and agricultural data analysis with participatory engagement with palm farmers to ask them about how climate-related impacts are affecting them and their livelihood strategies.

The background climate data used in this study was obtained from NASA, Surface Meteorology and Solar Energy (SSE) (NASA, 2021). These data cover the period from 1981–2017.[4] All the surveys carried out in Commune of Ngwéi took place from November 2019 to August 2021, as part of the first author's doctoral research.

Regarding the collection of socio-economic data, a survey questionnaire was administered to a targeted sample made up of two hundred and ninety households of oil palm producers, or ten households in each of the twenty-nine villages (Map 4). The selection of the households to be surveyed was made in 2019 as part of the implementation of a phase of the Oil Palm and Adaptive Landscape (OPAL) project in the commune,[5] a project in which we participated as investigators. The choice of producers to include was made on the basis of their age, by observation of physical conditions before confirmation of the age data, and the number of years spent in Ngwéi. Thus, producers under forty and having lived less than twenty years in the village were systematically excluded in order to focus on those with more farming experience in the area. The questionnaire focused on the characteristics of farmers and their farms, their logic and methods of access to land and agricultural inputs, their assessment of the impacts of climate change on production, and the resilience strategies they adopted. To obtain more detailed information, group interviews were organized in six villages with a total of 138 people, as well as semi-structured interviews with a sample of selected farmers who are strongly involved in oil palm growing and production.[6]

Results

The Exploitation of Oil Palm in the Commune of Ngwéi Is an Activity Strongly Marked by Inequalities of Access to Factors of Production

Depending on the size of the areas developed and how agricultural inputs are used, two main types of oil palm exploitation coexist in the Commune of Ngwéi. First is village exploitation, which is generally practiced on small areas ranging from 0.5 to 2 ha for the most part, using wild palm seed whose productivity depends essentially on the natural fertility of the soil. Its practitioners are commonly referred to as "smallholders," and are mostly Indigenous communities, plus a few farmers from neighbouring municipalities or other regions of the country. The other type of exploitation is "elitist exploitation." This is generally carried out on larger areas, ranging from 5 to more than 20 ha per farmer, using improved seeds with high-yield potential, chemical fertilizers, and regular application of phytosanitary products such as pesticides.

The oil palm activity in this commune is developing in a context of land insecurity and difficulties of access to agricultural inputs, especially for small farmers.

Massive Land Grabs Are Fuelled by Weak Land Governance

In most rural areas of Cameroon, as is the case in Ngwéi, access to and management of land is essentially governed through a customary regime. This land management system advocates the control of all land by the Indigenous population, the land being considered as the collective. It is up to each village chief to delimit plots for cultivation in proportion to the number of neighbourhoods and families. However, in certain forest areas, the effective appropriation of a plot is based on the "axe right" according to which "the land belongs to the first clearer." Such a context leads to inevitable illegal occupations of agricultural land, which most often benefits elites.[7]

Based on field interviews with village chiefs, most of the elites, because of their strong political or financial influence reputed in the village, unfairly appropriated family land and sometimes even the village land reserve. They do not hesitate to exploit for their own benefit the flaws of the customary land tenure system, but also those of local land governance, in particular the absence of rigorous management and control mechanisms. The majority of producers in our study (42 per cent), consisting mainly of farmers who practice oil palm cultivation on small areas ranging from 2 to 5 ha, and about 28 per cent practice this activity on even smaller areas, less than or equal to 1 ha (Figure 4.2).

Only 28 per cent of producers operate relatively average areas ranging from 6 to 20 ha, and less than 2 per cent have areas of palm groves greater than 20 ha, the latter category being essentially held by the elites. Indeed, the elites represent only 19 per cent of the population of producers surveyed yet hold 56 per cent of the total area of palm groves. In asserting these differences related to illegal land acquisitions and possessions, farmers do not hesitate to denounce this trend. One of them stated:

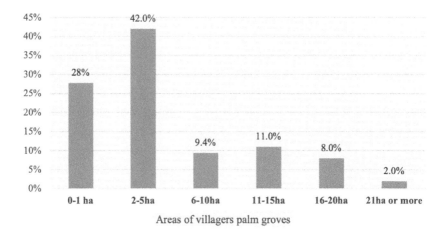

Fig. 4.2 Size of land areas farmed for oil palm in Ngwéi. **Source**: Processing of field survey data from 290 farmers in 29 villages.

We small planters in Ngwéi are finding it increasingly difficult to extend our oil palm estates on our own land. Some local elites, because (they are) very rich and influential, allow themselves everything here, and this under the gaze of the local administrative authority. We are increasingly forced to fall back on forest regrowth to create our palm groves; even if the yields per hectare are not better on these types of vegetation, we have almost no choice. (A farmer during the Focus group in Seppe village, March 2020. Translation by the authors.)

Smallholders are finding it increasingly difficult to expand their agricultural estates to improve their incomes. This is associated with their degree of limited access to agricultural inputs.

Small Producers Have Limited Access to Agricultural Inputs

High yields in palm groves and increased income for producers are largely determined by the type of oil palm seeds used and the frequency of fertilizer and phytosanitary product use in the context of the operation. However, the surveys we carried out on the methods of palm-grove exploitation in Ngwéi reveal that the practices are 83 per cent traditional in relation to the types of seeds used, the frequency of fertilizer use, and the use of phytosanitary

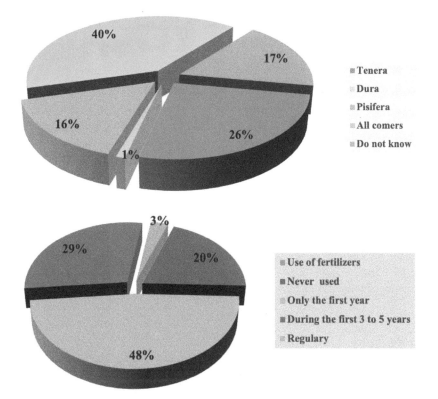

Fig. 4.3 (a) and **(b)** Type of palm seeds used and frequency chemical fertilizer use in oil palm cultivation in Ngwéi. **Source**: Processing of field survey data.

products. More than half of growers (57 per cent) use wild seed types with low yield potential (either Dura, Pisifera, or "run-of-the-mill" type[8]), and 17 per cent of growers are even unaware of the variety they are using in the field. Only 26 per cent use improved palm seed types with high yield potential, such as especially the Tenera type (Figure 4.3a). Similarly, barely 3 per cent of producers regularly use chemical fertilizers on their plantations. The rest of the producers use these fertilizers either just during the first year after the creation of the palm grove (48 per cent), or during the first three to five years (29 per cent), and 20 per cent of these farmers have never used chemical fertilizers at all (Figure 4.3b).

Two main reasons explain this: the first is the high cost of these inputs for the vast majority of small growers in view of their sometimes-derisory income. Most farmers do not have the necessary means to afford these agricultural inputs. The second reason is most farmers' ignorance about the role of agricultural inputs in terms of production. This is linked to the lack of information dissemination on the subject, in combination with the farmers' lack of access to education. Besides, the minority of peasants with means do not even know where to get farm supplies. To describe this reality, a planter explains:

> My son, almost everyone here uses wild seeds to create the palm plantations, even if the production is not always satisfactory. If the government could often help us with improved seeds that would allow us to have better production, it could increase our yields and income, because we do not have enough means for that. (A planter from Solopa village, December 2020)

Faced with these realities, most farmers in Ngwéi have to content themselves with the derisory income linked to the traditional exploitation of oil palm. The impact of climate change on production accentuates the decline in income.

Climate Change Significantly Impacts the Oil Palm Sector in the Commune of Ngwéi

Like almost all regions of Cameroon, the Commune of Ngwéi is deeply affected by the effects of climate change, and these have perceptible repercussions on the oil palm sector. We used statistics on temperature, precipitation, and their variability to trace the evolution of climate trends in Ngwéi over the thirty-six-year period from 1981 to 2017 (Figures 4.4, 4.5, and 4.6). In this analysis, we focused our gaze on the evolution of the two main climatic parameters (precipitation and temperature), because the growth and productivity of the oil palm depend essentially on them.

The evolution of rainfall on an annual scale (Figure 4.4) indicates a continuous decrease in the amount of rainfall over the study period. The annual average is 2,168 mm.

Of the thirty-six years (1981–2017) studied, seventeen years, representing 47 per cent of the series, recorded a cumulative rainfall below normal. Seven

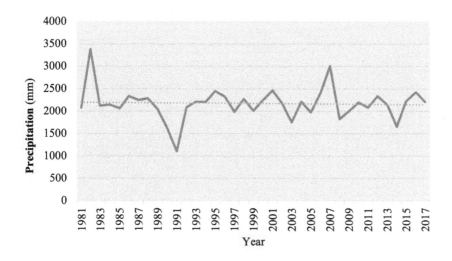

Fig. 4.4 Interannual variability and trend of precipitation between 1981 and 2017 in Ngwéi.
Source: Processing of climate data from NASA (n.d.), Surface Meteorology and Solar Energy
(SSE). Compiled by the first author.

of them, including 1990, 1991, 1997, 1999, 2003, 2005, 2008, and 2014, representing 22 per cent of the series, recorded cumulative rainfall strictly less
than 2,000 mm. The year 1991, which was the driest with 1,104 mm, showed
a deficit of 1,064 mm of rain compared to the annual average.

The analysis and interpretation of the standardized precipitation index in
Ngwéi between 1981 and 2017 makes it possible to see alternations between
surplus years and deficit years (Figure 4.5).

In fact, between 1981 and 2017, there were twenty surplus years, or 55.55
per cent of the series, with different humidity levels from one wet year to
another. Extremely humid, highly humid, and moderately humid years are
irregularly distributed and respectively represent two years (10 per cent), one
year (5 per cent), and seventeen years (85 per cent). In terms of deficit years,
there are sixteen years, or 44.44 per cent, with moisture deficits varying from
year to year. These years reflect to varying degrees the decrease in cumulative
rainfall or even drought episodes that occurred in Ngwéi between 1981 and
2017. Years with moderate, high, and extreme moisture deficits, respectively, were twelve years (75 per cent), three years (18.75 per cent), and one year
(6.25 per cent). The sub-period going from 1989 to 1992 successively recorded

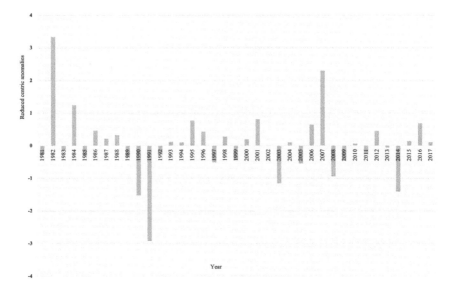

Fig. 4.5 Evolution of rainfall anomalies over the period 1981–2017 in Ngwéi. **Source:** Processing of climate data from NASA (n.d.), Surface Meteorology and Solar Energy (SSE). Compiled by the first author.

severe moisture deficits, with a peak of drought reached in 1991.[9] This climatic trend is like the recurrence of prolonged drought episodes that affected all of the coastal zone and the Cameroonian littoral between 1982 and 2010 (J. Amougou, 2018).

In addition, an inter-monthly analysis of the evolution of cumulative rainfall between 1981 and 2017 in Ngwéi shows two different trends compared to the climatological normal for rainfall (Figure 4.6).

Figure 4.6 shows that during the sub-period from 1981–2011, the Commune of Ngwéi experienced a regular increase in rainfall between seasons. The main rainy months (September and October) recorded a monthly cumulative rainfall of about 325 mm, or 25 mm more than the normal trend. On the other hand, the sub-period from 2011–2017 is marked by a considerable drop in rainfall between the seasons. These have, for example, dropped in the high season (September and October), going from 325 mm/month on average to around 230 mm/month, in particular a recorded rainfall deficit of 95 mm/month, i.e., 29 per cent lower than the 1981–2011 sub-period. This

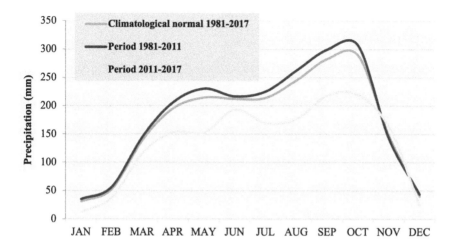

Fig. 4.6 Inter-monthly variability of precipitation compared to the climatological normal over the period 1981–2017 in Ngwéi. **Source**: Processing of climate data from NASA (n.d.), Surface Meteorology and Solar Energy (SSE). Compiled by the first author.

downward trend is in line with the findings of the majority of producers (77 per cent) in Ngwéi who perceived a general drop in rainfall for seven to ten years.

Conversely, the inter-monthly analysis of the evolution of temperatures over the same study period at Ngwéi reveals a considerable increase in temperatures compared to normal (Figure 4.7).

By separating the study period into two sub-periods, we can see that compared to the 2011 to 2017 sub-period, the hottest months in Ngwéi (December, January, February) all record a monthly temperature greater than or equal to 26°C. However, these months had never reached 26°C during the previous sub-period (1981–2011). This change in temperature probably reflects the manifestation of global warming of the climate and the establishment of a hot microclimate in Ngwéi.

The variability of these climatic conditions directly affects the production of oil palm and has a variable impact on the living conditions of producers.

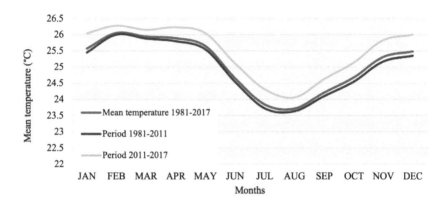

Fig. 4.7 Inter-monthly temperature variability compared to the climatological normal over the period 1981–2017 in Ngwéi. **Source**: Processing of climate data from NASA (n.d.), Surface Meteorology and Solar Energy (SSE). Compiled by the first author.

Impacts of Climatic Disturbances on Oil Palm Production Have Repercussions on the Lives of Producers in Ngwéi

At different levels of growth, disturbances linked to the variability of climatic conditions in Ngwéi (intense rains, intense and prolonged drought, recurrence of high winds, etc.) have perceptible effects on oil palm production. Nine impact indicators linked to three types of climatic disturbances are clearly perceived and identified by producers (Table 4.1).

Faced with the recurrence of episodes of intense drought during the year, farmers critically note, for example, the drying out and loss of leaves of young plants in the nursery, the lengthening of the growth period of young plants in the nursery and the delayed ripening of nut bunches. In addition to the drying up of young plants caused by the recurrence of episodes of intense drought, producers are increasingly victims of the devastation caused by locust invasions such as flies (*Diptera*), small snails (*Helix aspersa*), and stinking locusts (*Zonocerus variegatus*). These find refuge in nurseries and cause deterioration of the leaves and often the loss of many young plants in the nursery (Figure 4.8 a, b, and c).

Table 4.1 Impacts of the main climatic disturbances on oil palm production.

Climatic Disturbances	Effects on Oil Palm Production
Heavy rain/storms	- Loss of young plants due to invasions of small snails that find refuge on the leaves - Early and mass ripening of nut bunches and low yields - Seasonal upheaval in the ripening of palm nut bunches
High winds	- Turnover and destruction of young palm seedlings in nurseries or plantations - Delayed inflorescence of oilseed bunches
Intense and prolonged dryness	- Drying and loss of leaves of young plants in the nursery - Extension of the growth time of plants in the nursery - Loss of plants in the nursery due to locust attacks (flies, locusts, caterpillars) - Delayed ripening of nuts

Source: Field surveys.

In addition, the magnitude of the impacts caused by these climatic disturbances in Ngwéi is not the same from one producer to another (Figure 4.9), likely reflecting the variation in the level of vulnerability of producers.

The majority of producers (43.4 per cent) believe that their production and their activities are strongly affected by climatic disturbances, and 9.4 per cent even report a very strong effect. This trend mainly reflects the high vulnerability of smallholders in a sector where most practices (83 per cent) are traditional, which is one of the main causes of unsatisfactory yields. A handful of producers (13.2 per cent) believe that their production is somewhat affected by climate change, and just 3.8 per cent believe that they are only very slightly affected.

In addition, the producers report perceiving many social repercussions at the scale of the related to these impacts of climate change on production (Figure 4.10).

The majority of farmers surveyed (36 per cent) believe that the impacts of climatic disturbances on oil palm production contribute to the drop in their income, due to the drop in productivity that these impacts generate. A significant portion of the producers surveyed (26 per cent) believe that this climate dynamic exacerbates famine within the commune, and 21 per cent

Fig. 4.8 (a) and **(b)** Plants drying up in the nursery in a prolonged drought situation, and **(c)** degradation of the leaves of young plants in the nursery by insects. **Photo Credit**: G.D. Abassombe, February 2021.

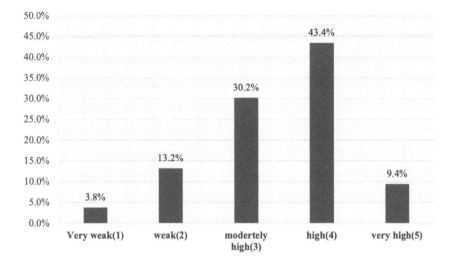

Fig. 4.9 Frequency of farmers' assessments of the level of severity of climate change on palm production in Ngwéi. **Source:** Processing and analysis of field survey data.

of producers even establish the link between these changes and the decline in local productive capacities. Indeed, like oil palm, food crops associated with palm groves are not spared the effects of climate variability. However, these crops are the main source of food for local communities and also an important source of food for neighbouring towns. A portion of the producers surveyed (12 per cent) believe that this drop in yields has a direct impact on their income and accentuates their impoverishment. These induced effects necessarily reflect the strong dependence of Ngwéi farmers on the exploitation of oil palm, especially in an area where this activity is the main source of income for local communities. Most of the peasants interviewed are finding it increasingly difficult to live essentially solely on income linked to the exploitation of palm groves.

To limit the effects of these impacts within their abilities to control their farming practices, the oil palm producers spontaneously and variably adopt a number of strategies.

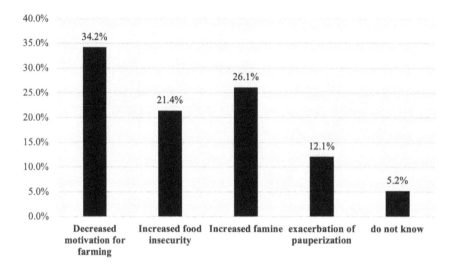

Fig. 4.10 Repercussions of the effects of climatic disturbances on the living conditions of producers/farmers in Ngwéi. **Source**: Processing and analysis of field survey data.

Adaptation Strategies Developed by Oil Palm Producers

Oil palm planters described nine types of strategic actions they have developed with the goal of minimizing the repercussions linked to climatic disturbances (Figure 4.11).

The temporary interplanting of food crops within palm groves is the first form of adaptation of oil producers to the effects of climatic disturbances in the Commune of Ngwéi because it represents 31 per cent of the sample surveyed. This high percentage surely reflects the particularly accessible nature and good mastery of this agricultural technique, no doubt because, basically, it has been endogenous cultural know-how for several decades. Currently in the Commune of Ngwéi, most planters are adopting this practice, to make their developed plots more profitable in order to compensate for the low income from the exploitation of the palm groves. A village chief involved in oil palm exploitation for more than thirty years makes revelations in this sense:

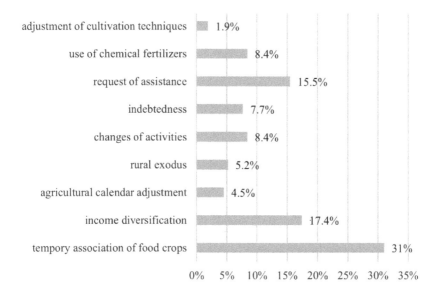

Fig. 4.11 Forms of adaptation of producers to the climate in Ngwéi. **Source**: Processing and analysis of field survey data.

Before, and surely for personal reasons, few producers planted food crops in their palm groves. Those who did this only planted plantain during the first year…. Palm oil monoculture was dominant here, especially in view of the financial security it provided. I, who am speaking to you, preferred to plant my food crop plots separately, outside the palm groves. But for more than a decade, with the decline in yields from palm groves, many planters can no longer be satisfied with the income related to their production alone. People are increasingly intercropping at least for the first three years, and several agricultural speculations are now associated with both palm groves. Above all, this allows you to have a little money to strengthen the purchase of agricultural inputs. (Boomabong village chief, January 2021)

The other form of adaptation of oil palm operators is the diversification of sources of income, adopted by 17.4 per cent of the sample of producers surveyed. The decline in yields of palm groves exacerbated by the impacts of climatic disturbances and their induced effects on farmers' incomes and living conditions feeds the need to develop other income-generating activities in parallel. Based on information collected from certain operators particularly affected by these changes, the diversification of economic activities aims to compensate for the almost derisory income from the current exploitation of palm groves. However, the diversification of income sources remains ineffective because it is unsatisfactory for most of the producers surveyed. The latter believe that the losses linked to the impacts of climate variations are enormous and the income from the diversification of economic activities is not always able to compensate for the drop in income linked to the exploitation of palm groves. A planter in distress recounts his situation:

> For nearly ten years, I have been operating a palm grove with an area of three hectares and at the same time, I extract and I also market palm wine. The income from these two activities made it easy to meet the needs of my small family. But in recent years, the observed decline in income related to the operation of my palm grove has considerably limited my ability to meet these needs. The situation even forces me to consider other sources of income to get by (a farmer from Logbii village, January 2021).

The use of foliar fertilizers and phytosanitary products is another form of adaptation by producers. It is practiced by 8.4 per cent of the sample of producers surveyed. Faced with the impacts of climatic disturbances, and to compensate for the lower yields they cause, planters with decent incomes generally resort to foliar fertilizers and insecticides (Figure 4.12 a and b).

A notable of the Solopa village clarifies these uses by brandishing these different products:

> Here are the products we use to fight against the damage caused by locust invasions in dry seasons on our nurseries. For a one litre bottle of this product, an average of 150 L of water must be used, i.e., ten sprayers. The treatment is done after one month. Apart from insecticide treatment, foliar fertilizers are also combined to restore vigour and colour to the foliage of plants that have been affected by drought or intense heat or that have been attacked by insects. It is therefore a

Fig. 4.12 Types of **(a)** foliar fertilizers and **(b)** insecticides used by farmers to fight against the attacks of locust plagues exacerbated by climate change. **Photo Credit:** G.D. Abassombe, February 2021.

treatment that is both preventive and curative. (notable person from Solopa village, January 2021)

Following these statements, one of the producers present at the discussion session took the floor with a look of astonishment and exclaimed:

I am very surprised to see that some producers have solutions to fight against these insect pests. I have never heard of these products. Obviously, I would be very surprised if such information reached us peasants, as usual, it is limited to the chieftaincy and among certain elites. Even if we don't necessarily have the means to buy it, we must at least be informed, you never know. (A beekeeper from Solopa village, January 2021)

However, and as mentioned above, these strategies, although diverse, remain insufficient, in view of numerous testimonies identified and observations made on the repercussions of the impacts of climate change on the living conditions of producer communities in the Commune of Ngwéi. More appropriate social compensation mechanisms are needed for this purpose.

Moreover, in response to the upheaval in annual rainfall patterns, some farmers opt to adjust the planting period for palm trees. This measure practiced by a minority of producers (4.5 per cent) consists of shifting the dates of the start of planting from those of the start of the rains. In a situation of declining yields and incomes exacerbated by climatic disturbances, some farmers (7.7 per cent) opt for debt, and others (15.5 per cent) desperately seek public assistance, which generally does not help them. This uncomfortable situation leads some producers to convert completely to other sectors of activity that they consider more profitable and for others to consider migrating to the city in search of better living, as is the case for 8.4 per cent and 5.2 per cent of operators surveyed.

However, and as mentioned above, these strategies, although diverse, remain ineffective, in view of numerous testimonies identified and observations made on the repercussions of the impacts of climate change on the living conditions of producer communities in the Commune of Ngwéi.

Discussion and Perspectives

The agricultural production area of Ngwéi, like other similar areas on the Cameroonian coast, is considerably affected by climatic disturbances, and these have perceptible effects on the production of oil palm. These can be summarized as essentially a general decrease in cumulative rainfall and increase in temperatures, marked by the recurrence of episodes of intense drought throughout the year. Fomekong and Ngono (2011), analyzing the effects of climate change on agricultural production and on the population in Cameroon, point to a similar trend in climate dynamics in the increasingly unstable rainfall across the agricultural basins of the different agro-ecological zones. By analyzing the impact of climate change on the agricultural sector in Côte d'Ivoire, Gbossou (2020) also presents the trend of climate change impacts. Based on analysis of climate data (temperature and precipitation between 1961 and 2014), he notes an increase in average monthly temperatures (maximum and minimum), respectively of 1.5°C and 0.5°C, and a falling precipitation trend of nearly 20 per cent since 1965.

The recurrence of these climatic fluctuations in Ngwéi is largely responsible for the decline in yields of the production of palm nut bunches and the rate of palm oil extraction decried by farmers. This aggravates the unfavourable living conditions in which the vast majority of peasants live. P. Amougou's (2016) analysis of the impact of climate change on the oil palm agricultural sector in Cameroon even evokes a disaster scenario. He notes that the peasants of the agricultural production basins of the Littoral, Southwest, and West Cameroon find themselves overwhelmed by recent events in view of the manifest decline in productivity. This trend has also been highlighted since as early as 1995 by Yao et al. (1995), who analyze the evolution of the palm oil extraction rate under conditions of climate change in the Northeast of Ivory Coast. These studies indicate that oil palm production is affected by climatic fluctuations, in particular water stress, which affects growth, yields, and consequently the quantities of palm oil extracted, implying a drop in producer income.

However, the extent of these impacts on oil palm yields varies from one producer to another, according to their level of access to arable land and especially to agricultural inputs (seeds with high yield potential, chemical and phytosanitary fertilizers, etc.). The socio-political or administrative status and purchasing power of farmers largely determine these differences. In his analysis of climate change in the rice sector in Gagnoa in Ivory Coast, Gbossou (2020) arrives at a similar result by pointing out that this agricultural sector is very affected by climate change because of several factors that lead to precarious socio-economic conditions. The analysis of CARE and Food Tank (2015) on the culture of equality for fair and sustainable agricultural systems in the context of climate change also reflects this reality. It notes and demonstrates that inequalities determine who has access to food and the resources to grow and buy it. Hunger and poverty are not accidents, they are the result of economic and social injustices, and this reality is even more true for smallholder farmers. Similarly, and always with the aim of highlighting the link between socio-economic inequalities and the differentiated effects of climate change, Development and Peace (2015), through its analysis of peasant agriculture at the heart of climate justice, highlights the same reality. This analysis reveals that the areas worked by peasant farmers rarely exceed two hectares, making peasant agriculture one of the activities most affected by climate change.

In such conditions of injustice and social and economic inequalities between farmers, the administrative authorities have a sovereign duty to reduce these inequalities according to the principle of difference (Rawls, 1971). This implies maximizing the primary goods (income and wealth, and access to land) of the weakest. This measure is not aimed principally at alleviating handicaps on an equitable basis, but instead at improving the long-term expectations of the most disadvantaged. In reference to oil palm cultivation in Ngwéi, these provisions should above all promote more rigorous and assured governance that can promote equitable management of arable land, thus stimulating the development of a more productive agricultural system by facilitating access and distribution of high-yielding palm seeds, fertilizers, and phytosanitary products. Raising awareness about the challenges of climate change, and strengthening the technical capacities of small producers for more productive and economically resilient farming practices, must be added to these measures.

Conclusion

By emphasizing the inequalities of access to factors of production among oil palm producers, this study aimed to characterize the impacts of climate change on the oil palm sector and highlight the strategic actions developed by producers in Ngwéi. The recurrence of heavy rains and episodes of intense and prolonged drought, added to high winds, affect the oil palm variably and at different stages of growth. Because of most small producers' difficulties of access to production resources, the decline in agricultural yields and their incomes caused by climatic fluctuations exacerbate poverty, especially among small producers. In this context, they have spontaneously developed several strategic actions, including the extension of the areas developed for some, inter-planting food crops and fruit trees in palm groves, and the diversification of sources of income for others. Thus, the development of a more climate-resilient sector in this agricultural production area must necessarily go through the implementation of an agricultural system that reconciles these factors: More rigorous and fair land governance, awareness-raising and training of producers, and facilitating access to and dissemination of agricultural inputs for all. Echoing the calls of local farmers, we conclude that this is the *sine qua non* condition for improving peasant resilience and advancing climate justice in this agricultural sector in Ngwéi, and by extension in Cameroon in general.

NOTES

1   In Cameroon, a "commune" is the smallest territorial unit in the administrative hierarchy—a decentralized territorial community.

2   Increased global demand for palm oil as a biofuel and substitute for petrochemicals in some uses is helping to drive the global expansion in palm oil production, especially in Southeast Asia and Latin America (Paterson & Lima, 2017; Pye, 2010; Ordway et al., 2019). In Cameroon, where the oil palm is native and many parts of the tree (not just oil-bearing nuts) have local uses, oil palm products are destined for domestic consumption (Ayompe et al., 2021). We focus here on climate justice implications from the viewpoint of local small farmers.

3   Historically, the exploitation of oil palm in the Commune of Ngwéi and in most coastal localities is a socio-cultural and colonial heritage. Oil palm has been growing there naturally for a long time (Carrere, 2013; Ndjogui et al., 2014), and was exploited for family subsistence, as the main source of dietary fats, until the colonial period. In plots intended mainly for food production, the density of naturally growing palm trees is maintained; these are spared during clearing and burning and exploited for twenty-five to thirty years. Cultivated oil palm plantations began on the Cameroonian coast in 1907 during the German protectorate and continued in 1910 with the creation in Edea of industrial plantations by the company Ferme-Suisse (Elong, 2003; Ndjogui et al., 2014).

4   Data choices were linked to the incomplete nature of the daily data series in the study area, due to the condition and quality of the measuring instruments used in the collection stations. The processing of these data consisted on the one hand in calculating the arithmetic mean over the study period (thirty-six years). This served as a reference for the assessment of the various upward or downward trends in rainfall, with a view to characterizing their evolution in Ngwéi. Dispersion parameters such as standard deviation and coefficient of variation were determined. In a complementary way, the standardized rainfall index was determined to assess the indicators of interannual rainfall variation over the study period.

5   The Oil Palm Adaptive Landscapes (OPAL) project uses natural and social sciences to create role plays that illustrate the existing realities of oil palm landscapes. Using these games, the team aims to explore alternative trajectories for oil palm with stakeholders and decision-makers in Indonesia, Cameroon, and Colombia, in order to chart the course towards a more sustainable future. Further information is on the project website: http://www.opal-project.org/.

6   The statistical series of climatic data on the study area as well as the survey data from the questionnaires were processed using Microsoft Excel 2013 software. The database of geographical coordinates of the National Institute of Cartography (INC) made possible cartographic processing using ArcGIS 10.2 software.

7   This often applies to people from the locality, who are senior executives in the public or private sector, permanently residing in the city or abroad and receiving high and regular incomes, and who have a real social, political, and economic influence.

8   This phrase is used to describe wild oil palm seeds of various types that are planted at the same time in the context of oil exploitation. These are obtained by selecting nuts from the production of an old palm grove, from which seedling nurseries are created.

9   Nicholson et al. (1988) cite a method for determining the standardized precipitation index (SPI). Better than the annual variability of precipitation alone, the SPI makes it possible to monitor the evolution of rainfall fluctuations, in particular the levels of drought severity. The SPI is one of the most recommended methods for analyzing climate variability. It is calculated from the mean of the data series and its standard deviation. Our calculations and use of the SPI are reported elsewhere.

Reference List

Agossou, D.S.M., Tossou, C.R., Vissoh, V.P., & Agbossou, K.E. (2012). Perception des perturbations climatiques, savoirs locaux et stratégies d'adaptation des producteurs agricoles béninoi. *African Crop Science Journal*, *20*(2), 565–588.

Amougou, J. (2018). Les changements climatiques au Cameroun, éléments scientifiques, incidences, adaptations et vulnérabilité. In O.C. Ruppel et E.D. Kam Yogo (Eds.), *Droit et politique de l'environnement au Cameroun-Afin de faire de l'Afrique l'arbre de vie* (pp. 687–712). Nomos Verlagsgesellschaft.

Amougou, J., Abossolo, S., & Batha, R.A.S. (2013). Dynamique du climat et impacts sur la production du maïs dans la région de l'ouest du Cameroun. *Rev. Ivoir. Sci. Technol.*, *21 & 22*, 209–234. https://revist.net/REVIST_21&22/REVIST_21&22_11.pdf

Amougou, J., & Batha, R.A.S. (2014). Dynamique spatio-temporelle des précipitations de 1960 à 2010 et essai d'élaboration d'un calendrier agricole dans la zone des hauts plateaux du Cameroun. *Rev. Ivoir. Sci. Technol.*, *24*, 153–177. https://revist.net/REVIST_23/REVIST_23_11.pdf

Amougou, P. (2016). Les changements climatiques affectent la production d'huile de palme au Cameroun. Médiaterre: Afrique Centrale. Retrieved 18 January 2022, from https://www.mediaterre.org/afrique-centrale/genpdf,20161125114714.html

Ayompe, L.M., Nkongho, R.N., Masso, C., & Egoh, B.N. (2021). Does investment in palm oil trade alleviate smallholders from poverty in Africa? Investigating profitability from a biodiversity hotspot, Cameroon. *PLoSONE*, *16*(9), e0256498. https://doi.org/10.1371/journal.pone.0256498

Baudouin, C. (2021, October 5). Alimentation, agriculture et changement climatiques. *Impacts*. https://notreaffaireatous.org/impacts-5-octobre-2021-alimentation-agriculture-et-changement-climatique/

Bélanger, G., & Bootsma, A. (2004). Impacts des changements climatiques sur l'agriculture au Québec. 65e congrès de l'ordre des agronomes du Québec.

Boko, M. (1988). Climatologie et communautés rurales du Bénin; Rythmes climatiques et rythmes de développement [Unpublished doctoral dissertation]. Université de Bourgogne Dijon.

Boko, M., Niang, I., Nyong, A., Vogel, C., Githeko, A., Medany, M., Osman-Elasha, B., Tabo, R., & Yanda, P. (2007). Africa. In M.L. Parry, O.F. Canziani, J.P. Palutikof, P.J. van der Linden, & C.E. Hanson (Eds.), *Climate change 2007: Impacts*, *adaptation and vulnerability: Contribution of working group II to the fourth assessment report of the intergovernmental panel on climate change* (pp. 433–467). Cambridge University Press.

Brown, O., & Crawford, A. (2008). *Évaluation des conséquences des changements climatiques sur la sécurité en Afrique de l'Ouest: Étude de cas nationale du Ghana et du Burkina Faso.* Institut international du développement durable. https://www. iisd.org/system/files/publications/security_implications_west_africa_fr.pdf

Capocci, H., Caudron, M., & Letocart, F. (2015, November). *Paysans résolus, réchauffement combattu* [Working paper]. Entraide et Fraternité. https://www.entraide.be/IMG/ pdf/etude_agriculture_et_changements_climatiques.pdf

CARE et Food Tank. (2015). *Cultiver l'égalité: Pour des systèmes agricoles justes et durables dans un contexte de changement climatique.* https://www.carefrance.org/ wp-content/uploads/import/reports/1/11380dd-5368-2015-11_CULTIVER-L-EGALITE_.pdf

Chanzy, A., Martin G., Colbach, N., Gosme M., Launay, M., Loyce C., Métay, A., & Novak, S. (2015). *Adaptation des cultures et des systèmes de culture au changement climatique et aux nouveaux usages.* Institut National de la Recherche Agronomique.https://nanopdf.com/download/enjs1-adaptation-des-cultures-et-des-systemes-evenements_pdf

Carrere, R. (2013). *Le palmier à huile en Afrique: Le passé, le présent et le futur.* World Rainforest Movement. https://wrm.org.uy/fr/livres-et-rapports/le-palmier-a-huile-en-afrique-le-passe-le-present-et-le-futur-2013

Delille, H. (2011). *Perceptions et stratégies d'adaptation paysannes face aux changements climatiques à Madagascar. Cas des régions Sud-ouest, Sud-est et des zones périurbaines des grandes agglomérations.* https://www.avsf.org/public/posts/704/ perceptions-et-strategies-d-adaptation-paysannes-face-aux-changements-climatiques-a-madagascar.pdf

Development and Peace (2015). *Chaud devant: Impacts des changements climatiques dans les pays du Sud et recommandations pour une action au Canada.* https://www2. devp.org/sites/www2.devp.org/files/documents/materials/rapport_chaud_devant. pdf

Djitie Kouatcho, F., Malla Alhadji, M., Yahangar, Marie., & Katchouang N. (2019). *Stratégies de résilience du secteur agro-pastoral au changement climatique dans le septentrion du Cameroun.* Retrieved 18 January 2022, from https://www.iedafrique. org/Strategies-de-resilience-du-secteur-agro-pastoral-au-changement-climatique-dans.html

Elong, J.G. (2003). Les plantations villageoises de palmier à huile de la SOCAPALM dans le bas Moungo (Cameroun): Un projet mal intégré aux préoccupations des paysans. *Les Cahiers d'Outre-Mer, 224,* 401–418. http://com.revues.org/index738.html

Enete, A.A., & Onyekuru, A.N. (2011). Challenges of agricultural adaptation to climate change: Empirical evidence from southeast Nigeria. *Tropicultura, 29,* 243–249.

Fomekong, F., & Ngono, G. (2011). *Changements climatiques, production agricole et effet sur la population au Cameroun* [Working paper]. https://uaps2011.princeton.edu/papers/110881

Gbossou, C. (2020). *Changements climatiques en Côte d'Ivoire: L'urgence de l'action.* https://www.mediaterre.org/actu,20200130113406,5.html

Jacquemard, J.-C. (2011). *Le palmier à huile.* Quae/CTA: Presses agronomiques de Gembloux. https://publications.cta.int/media/publications/downloads/1666_PDF.pdf

Mamoudou, A.-R. (2019). Pratique de l'agropastoralisme et changements climatiques: Analyse des stratégies locales de résilience dans l'Extrême-Nord du Cameroun. *Science Afrique, 1*(1). https://www.revues.scienceafrique.org/naaj/texte/mamoudou2019/

Mertz, O., Mbow, C., Reenberg, A., & Diouf, A. (2009). Farmers' perceptions of climate change and agricultural adaptation strategies in rural Sahel. *Environmental Management, 43*, 8–16.

NASA. (2021). NASA prediction of worldwide energy resources (POWER) project. Retrieved 7 October 2021, from https://power.larc.nasa.gov/

Ndjogui, T.E., Nkongho, R.N., Rafflegeau, S., Feintrenie, L., & Levang, P. (2014). *Historique du secteur palmier à huile au Cameroun.* Occasional Paper 109, Centre for International Forestry Research. https://www.cifor.org/knowledge/publication/4789/

Niang, I. (2009). Le changement climatique et ses impacts: les prévisions au niveau mondial. In Adaptation au changement climatique [Special issue]. *Liaison Énergie-Francophonie, 85*, 13–19.

Nicholson, S.E., Tucker, C.J., & Ba, M.B. (1998). Desertification, drought and surface vegetation: An example from the west African Sahel. *Bulletin of the American Meteorological Society, 79*, 815–829.

Ordway, E.M., Naylor, R.L., Nkongho, R.N., & Lambin, E.F. (2019). Oil palm expansion and deforestation in Southwest Cameroon associated with proliferation of informal mills. *Nature Communications, 10*, 114. https://doi.org/10.1038/s41467-018-07915-2

Paterson, R.R.M., & Lima, N. (2017). Climate change affecting oil palm agronomy, and oil palm cultivation increasing climate change, require emelioration. *Ecology and Evolution, 8*(1), 452–461. https://doi.org/10.1002/ece3.3610

PNDP (Programme National de Développement Participatif). (2018). *Mécanisme de contrôle citoyen de l'action publique dans la Commune de Ngwéi* [Study report].

Pye, O. (2010). The biofuel connection—transnational activism and the palm oil boom. *The Journal of Peasant Studies, 37*(4), 851–874.

Ramirez-Villegas, J., & Thornton, P.K. (2015). Climate change impacts on African crop production. *CGIAR Research Program on Climate Change, Agriculture and Food Security Working Paper No. 119*. https://cgspace.cgiar.org/bitstream/handle/10568/66560/WP119_FINAL.pdf

Rawe, T., & Deering, K. (2015). *Cultiver l'égalité: Pour des systèmes agricoles justes et durables dans un contexte de changement climatique*. CARE USA and Food Tank. https://www.carefrance.org/wp-content/uploads/import/reports/1/11380dd-5368-2015-11_CULTIVER-L-EGALITE_.pdf

Rawls, J. (1971). *A theory of justice*. Harvard University Press. https://giuseppecapograssi.files.wordpress.com/2014/08/rawls99.pdf

Republique du Cameroun. (2009). *Document de stratégie pour la croissance et l'emploi*. http://onsp.minsante.cm/fr/publication/194/document-de-stratégies-pour-la-croissance-et-lemploi-dsce

Sitou, L., Mormont, M., et Yamba, Y. (2014, May). Gouvernance et stratégies locales de sécurisation foncière: Étude de cas de la commune rurale de Tchadoua au Niger. *VertigO, 14*(1), 14 p. https://www.erudit.org/fr/revues/vertigo/2014-v14-n1-vertigo01649/1027968ar/

Tchindjang, M., Amougou, J., & Abossolo, S. (2017). Aperçu générale des changements climatiques, perception et adaptation des paysans dans les zones agro écologiques du Cameroun. In S.A. Abossolo, J.A. Amougou, M. Tchindjang (Eds.), *Pertubations climatiques et pratiques agricoles dans les zones agro écologiques du Cameroun: Changements socio-économiques et problématiques d'adaptation aux bouleversements climatiques* (pp. 23–46). Connaissances et Savoirs.

Yao, N.R., Orsot-dessi, D., Ballo, K., & Fondio, L. (1995). Déclin de la pluviosité en Côte d'Ivoire: impact éventuel sur la production du palmier à huile. *GéoProdig, portail d'information géographique*. Retrieved 19 January 2022, from http://geoprodig.cnrs.fr/items/show/61031

5

Common-Pool Resources and the Governance of Community Gardens: Experimenting with Participatory Research in São Paulo, Brazil

Kátia Carolino and Marcos Sorrentino[1]

Introduction

While the definition of *commons* can include different philosophical, political, economic, and legal conceptualizations according to each field of knowledge (Ruschel, 2018), in this chapter we associate the term with the concept of common-pool goods or natural resources. These include different natural elements, such as seas, lakes, rivers, forests and animals, among others, and they may be used in different forms, including in a communal manner (Dietz et al., 2002).

Due to their physical nature, common-pool natural resources basically present two characteristics: (i) shared use, which allows each individual to benefit from natural resources that are also used by others; and (ii) difficulty in restricting access by different users (Ostrom, 1990; Berkes, 2005). Because of these characteristics, their use can generate opportunistic behaviours in the users, especially when they extract more resources than they need and/ or use them predatorily (Dietz et al., 2002). This generally occurs when the natural resources are migratory and/or cover large areas.[2]

The governance (public, private, or communal) of these natural resources is quite complex, and for this reason different governing mechanisms have been considered and created to deal with the difficulties imposed by each context. One of the main challenges of natural resource governance is related to property regimes, since the inadequate and unsustainable use of these resources is directly linked to the manner of their access-regulation and control.

Bearing in mind that public and private property regimes are not the only possible modalities of natural resource management, this chapter gives an account of the results of a research study that involved urban farm growers from the outskirts of the city of São Paulo (Brazil; see Map 2, page 30), and presents some difficulties related to the access rights and communal governance of the land and natural resources of a communal garden. As discussed below, the questions of land use and governance are central.

This chapter is the result of a research process that made possible learning with local individuals and groups, from the perspective of the local community's technical and scientific knowledge of the area. It can be framed as community-based participatory research since it blended an academic process with the practice of observing and learning from local knowledge, focused on climate justice and common-pool resources.

In a broad sense, the study addresses the need for new research focused on rebuilding means of access and governance for common-pool natural resources, taking into consideration the social, cultural, and biological needs and peculiarities of each context. Furthermore, this communal garden study discusses how to empower the most vulnerable to better resist climate change, as it is the urban poor who face the greatest risks due to the lack of infrastructure that could allow them to adapt to climate-related impacts (Dubbeling, 2014). For example, informal dwellings located in low-lying and flood-prone areas or on steep and unstable slopes subject to landslides are those most affected by heavy rains, but which tend to be available as occupation sites for the most vulnerable populations.

The implementation of urban gardens is a strategy capable of bringing several interrelated benefits to address the effects of climate change. First, creating more open spaces reduces the impacts of high rainfall due to greater storage of excess water, more interception and infiltration in green areas, reduction of runoff and related flood risks, and better replacement of groundwater. Besides reducing flooding, the more porous soil caused by gardens favours the recharge of groundwater and reinforces groundwater flows. When associated with large plants and trees, urban gardens can reduce heat island effects by providing more shade and enhancing evapotranspiration. Producing food in and around the city, by requiring less energy to transport, refrigerate, store and package food products, can contribute to reducing urban energy consumption and greenhouse gas emissions. Agriculture also allows for the productive reuse of organic waste, reducing methane emissions from landfills and energy consumption in the production of synthetic fertilizers. Decentralized urban recycling of organic waste reduces the need for transportation, energy, and emissions to get it to landfill. The reuse of wastewater from cities in urban agriculture frees up fresh water for higher-value uses and reduces the emissions generated in water treatment (Dubbeling, 2014).

Vegetable gardens in urban areas can also play an important role in generating climate solutions for poor and vulnerable communities, by contributing to the reduction of hunger through the production of healthy and nutritious foods that improve local food security. However, for urban agriculture to be effectively practiced, it is necessary to plan and invest financial resources so that people can develop and maintain community gardens in cities, where land is often in short supply. Making this happen, by participating in the process of finding and developing solutions for socio-environmental vulnerabilities, promotes the social inclusion of people who are all too often marginalized and silenced.

In the city of São Paulo, there is significant urban poverty, and there are also many abandoned or underused public and privately owned areas that could be used by different social groups to plant community gardens. Nevertheless, measures of this kind are often impeded, for various reasons (Carolino, 2021).

Most of the time, the implantation of a community garden requires collective efforts—of the community, associations, non-governmental organizations (NGOs), and public authorities—not only with the intent of finding appropriate land for the implementation of the community garden but also in the sense of enabling adequate communal/collective governance of the area. Our study of a community garden in São Paulo explores how this process can take place.

Because we believe that property and governance regimes are crucially important to such a project's success, we take time here to outline the theoretical literature on commons governance before recounting the São Paulo community garden story in detail.

Theoretical Framework

A number of empirical studies have demonstrated that in addition to public and private property, there are other forms of managing common-pool natural resources. The literature identifies four categories: (i) open access, (ii) private property, (iii) public or state property, and (iv) common property. Feeny et al. (2001) note that, in practice, there is overlap and even conflict among these regimes. These four categories are explained in more depth in the following sections.

1) *Open Access*

An open access situation was set out in Garret Hardin's article "The Tragedy of the Commons," published in *Science* in 1968. Considered one of the most-cited scientific articles in the second half of the twentieth century, this article stimulated ample debate and a new interdisciplinary research field on common-pool resources.

Although "The Tragedy of the Commons" reinforces the arguments that open access permits greater degradation of common-pool resources than private property, the author equated the concept of communal ownership with the conditions of open access, meaning no rules to limit the use of commons (Dietz et al., 2002). Furthermore, the author ignored that in many cases, the "tragedy" only occurred after open-access conditions were created, as a consequence of the destruction of pre-existing, communal systems limiting access rights to land and marine areas (Feeny et al., 2001).

In this context, users who depended on common-pool resources for their subsistence were forcibly removed from their territories, thus provoking what McCay and Acheson (1987, as cited in Diegues & Moreira, 2001) call "The Tragedy of the Community." As a solution for avoiding this tragedy of community governance and livelihoods, the authors proposed that common-pool resources be privatized or defined as public (state) property (Feeny et al., 2001).

2) *Private Property*

Under private property regimes, only the title holders may access and use a natural resource. Therefore, this type of ownership differs from the other forms of access rights, since the rights to the natural resources are exclusive and non-transferable. In other words, on private property, the owner is at liberty to decide how to use a natural resource and who shall have access to it, albeit limited somewhat by legal norms and state control. In Brazil this may involve, for example, regulations regarding Permanent Preservation Areas and legal reserves, among others.

Exclusive use and access, seen from the logic of profit, allows owners to sell and degrade natural resources on their property. This leads to exploitation to the detriment of environmental protection (Feeny et al., 2001), and privatizes many resources that are inappropriate for private ownership (e.g., aquifers/watershed recharge areas, marine resources, corridors used by migratory species, etc.).

Although state regulation and control are relevant, they neither provide a sufficiently adequate mechanism for solving the problem of overexploitation of natural resources, nor address the problem of unjust exclusion from private areas, since the command-and-control mechanisms employed by the state are not always sufficient to monitor and control these uses. Regarding the overexploitation of natural resources, the government often lacks the will, resources, or both to adequately supervise private areas. For example, although Brazilian legislation establishes rules for the use of pesticides, lack of inspection on most Brazilian agricultural properties effectively allows the indiscriminate use of these products. As a consequence, water tables are contaminated, which in turn affects the common use of water by urban populations that depend on this water for basic sanitation. Effective inspection of these areas and/or legal compliance by the property owners would make it possible to control and avoid this environmental degradation.

The Brazilian constitution allows the state to expropriate underutilized/ unproductive private property (providing compensation to the owners in the form of government bonds) on the basis of judgements about the best "social function," and reserve it for environmental reasons or make land available to people who need land to farm. However, implementation of this provision has been inconsistent for a variety of reasons; land ownership in Brazil remains highly concentrated (Ondetti, 2021).

Within this context of ineffective government controls, it is the urban poor who arguably face the greatest risks, as most live in informal settlements located in low-lying and floodable areas, or on steep and unstable slopes, subject to landslides caused by increasingly intense rains. Thus, cities have an important role to play in mitigating and adapting to climate change and in strengthening the resilience of the most vulnerable residents. Urban agriculture can be considered an adaptation strategy capable of mobilizing several benefits in this effort.

3) *Public or State Property*

Natural resources on public or state property are protected by the state, and individuals and groups only make use of them when authorized by representative agencies of a state entity. Moreover, the state, through its legislative bodies, sets the standards that define the property regimes among sub-jurisdictions, which in turn directly manage the natural resources on behalf of the public interest.

The same standards of protection for natural resources that apply to private property also apply to public property. However, the mere existence of these standards does not necessarily guarantee the protection of natural resources, even under public ownership, since the state may not have effective control over their use. As an example, Dietz et al. (2002) explain that many state areas have been transformed into open-access areas due to the lack of inspections, associated with corruption on the part of public officials who, in turn, may receive payoffs from users wishing to exploit government-owned resources. According to Dietz et al. (2002), case studies in Africa, Latin America, Asia, and the United States indicate that policies which transform the common-pool resources of local communities into state property favour an increase in the degradation of natural resources.

In this regard, for Feeny et al. (2001), successful resource management in less developed countries is rarely associated with state ownership. According to them, the professional infrastructure responsible for the management of resources in state organizational charts is normally not well developed, and the imposition of norms is problematic. Even so, for Berkes (2005), a state ownership regime performs a key function in situations in which the resources require multiple integrated mechanisms of governance in order to be protected: for example, trans-jurisdictional hydrographic basins.

4) *Common Property*

Since the publication of Hardin's "The Tragedy of the Commons" article in 1968, common-property regimes have been associated with environmental degradation, especially when many individuals use a scarce resource communally. This generally occurs because the common property regime is confused with open access (Ostrom, 1990).

For Bromley and Cernea (1989), "this inadequate diagnosis is very serious in its consequences since it further invites inappropriate policy recommendations and misguided operational decisions." They point out:

> By confusing an open access regime (a free-for-all) with a common property regime (in which group size and behavioral rules are specified) the metaphor denies the very possibility for resource users to act together and institute checks and balances, rules and sanctions, for their own interaction within a given environment.

> The Hardin metaphor is not only socially and culturally simplistic, it is historically false. In practice, it deflects analytical attention away from the actual socioorganizational arrangements able to overcome resource degradation and make common property regimes viable (Bromley & Cernea, 1989, 6–7).

Even though Hardin concluded that only public or private governance is able to avoid the depletion of common-pool resources, empirical and theoretical studies demonstrate that evidence contrary to "The Tragedy of the Commons" exists in abundance; rather, there are alternatives to protecting natural resources—such as common property—that go beyond the public or private ownership dichotomy presented by the author.

One such study is by Elinor Ostrom, who proves in her field studies that "The Tragedy of the Commons" is mistaken. In her book *Governing the Commons*, published in 1990, she presents an alternative for natural resource protection, one that is different from those presented by the theoreticians of state or privatization since, on communal property, resources are divided in an egalitarian manner among community members (though external individuals are excluded from access) (Ostrom, 1990).

According to Ostrom, commons governance refers to the self-organization of communities which, to some extent, do not need (but do not exclude) private and state interventions. Furthermore, in cases where the governance of common-pool resources was successful, the author identified that the users built relationships of trust, cooperation, and collective action, essential for the imposition of resource use limits and maintenance responsibilities.

In this sense, communal ownership of natural resources is directly related to the concept of resilience, a concept closely related to "adaptive capacity" that in the social sciences is associated with the way people are affected by and respond to changes. According to Cinner and Barnes (2019), there are six broad social factors that create resilience. These are: 1) assets that people can draw upon, 2) flexibility to change strategies, 3) ability to organize and act collectively, 4) learning to recognize and respond to change. 5) socio-cognitive constructs that enable or constrain human behaviour, and 6) agency to determine whether to change or not.

This set of principles, created by the community and for the community, stimulates confidence and reciprocity and also encourages more cooperative conduct among the community members.

Our participatory research study based on the lived experience of a community of residents in a peri-urban neighbourhood in the East Zone of São Paulo showed us that community governance is able to reorganize spaces forgotten by the government and generate forms of collaborative work, based on personal relationships of trust and mutual aid.

Methodological Path

We gathered background information to carry out this study using primary and secondary sources as well as documentary and bibliographic research. The secondary sources included books, scientific articles, dissertations, theses, and information collected from official websites, while the primary sources comprised official documents (laws, decrees, ordinances, and policies) related to the research. We also relied on information gathered through community-based participatory research with residents of the area, interviews with community members, and participant observation.

According to Holkup, Tripp-Reimer, Salois, and Weinert (2004, 2), community-based participatory research "... provides an alternative to traditional research approaches that assume a phenomenon may be separated from its context for purposes of study." Moreover, considering that the research process should be a means of facilitating change, community-based participatory research is important because it recognizes the need to involve members of the community as active participants in every phase of the research project—crucial since after all, community gardens depend on local engagement for their maintenance into the future.

Within the scope of this research, we can highlight the roles of three specific sets of actors—community members, civil society organizations, and environmental educators in addition to other partner-collaborators—who participated in activities such as the creation of different low-cost "social technologies,"[3] including cisterns to capture rainwater, a solar dehydrator for fruits and vegetables, vertical vegetable gardens in PET bottles, earthworms in buckets, and bioconstruction techniques.

These activities allowed environmental educators and researchers to support and engage with members of the community, mainly women and young people, who together sought strategies based on the participation of all in building activities to address existing problems in their surroundings. Through discussions in community meetings and workshops, and informal conversations during garden-planting and parties, local residents shared their current and future concerns, livelihood responsibilities, and details of community dynamics. To protect everyone's privacy, names and organizations remain confidential here. This privacy is important to ensure that community members feel confident in openly sharing their thoughts and experiences.

It is also worth mentioning that during the community meetings, concepts such as critical environmental education and permaculture were included in the group's dialogues, and from their observations about their living space it was possible to build collective reflections relating social problems with regard to natural elements present in the surroundings such as trees, streams, weeds, hilltops, and types of buildings, among others.

We would like to note that, while the idea of participatory intervention is important, especially for environmental educators within the Freirean tradition (Ministério do Meio Ambiente & Ministério da Educação, 2005; Sorrentino, 2014), a valid question is whether actions carried out with the community are in fact transformative: that is, whether the actions developed have real impacts or effectively address community needs. To answer this would imply constant and consistent analysis with the community, in the context of an ongoing long-term relationship. Our relatively recent and short-lived study cannot offer conclusive information in this regard. We base this chapter on community reports and interviews, which are relevant in the sense of noting the situation of vulnerability that affects the community as a whole, and their resilience and agency in addressing those challenges over the time period documented in this account.

Results: Participatory Research with Community Gardeners in the East Zone of São Paulo

Here is the story that we were able to assemble through our documentary research, interviews, and participatory research with the community. The personal relationships built within the community began in 2002, when the Housing Company of the State of São Paulo (CDHU) created an urbanization project in the favelas of the East Zone, covering an area of approximately 980,000 m², and made part of the area available to the local residents for a community garden.

The local residents united and initiated the process of creating the community garden. However, after four years of work, the CDHU identified that the lot they had provided was private property, and the local residents were asked to vacate the area. The community, undeterred, then decided to begin a new community garden in another location.

The new area that they found had been a dumping ground for construction waste. For two years, the residents worked arduously until the area was totally recovered and revitalized, and then they began to plant. The women who worked in the garden had no sources of income; their motivation to invest their time and work in the garden was for their own and their families' subsistence. On the weekends, those who had planted the food divided the harvest among themselves. The problem was that without any money, the community could not pay for seedlings, and at the end of the month, there were unpaid bills.

The situation began to change sometime around 2012 or 2013, when an NGO that was active in the East Zone requested authorization from the CDHU to implement a social project with these residents. This social project contributed a great deal to the collective organization of the people involved with the community garden. From then on, the local residents began to receive support and training in production planning, bioconstruction, and composting.

Later, the city of São Paulo, through the program called *Programa Operação Trabalho*, or Work Operation Program (POT), awarded ten grants worth two years of full-time minimum-wage payments, so that the grant recipients could receive agriculture training and dedicate some hours of paid work to the community garden.

The NGO's work combined with the grant payments allowed the people working there to strengthen the community garden space. However, with the approaching end of the NGO project and the grants, people feared that the relationships might weaken and undermine the collective actions that were being carried out. Therefore, they recognized the need to form a collective, which they did: It was made up of four men and six women, who then began to make natural cosmetics and sell the products from the community garden at organic farmers' markets.

Later on, with the goal of highlighting the importance of female representation and leadership, nine women of various ages formed a group that, in addition to the garden produce and the cosmetics, also sold vegan food with the slogan "from garden to table" ("da horta para a mesa"). Around 60 to 70 per cent of what is planted is destined for the kitchen, which transforms the produce into food that is served in companies. The rest is sold "at the door" or distributed to partner organizations such as the Center for Reference and Assistance to Women (*Centro de Referência de Atendimento à Mulher*, CRAM).

At the time our research took place (2019), 10 per cent of the total income received was reserved for the purchase of inputs and materials, while the rest was divided among the people involved. The CDHU, which had covered costs such as electricity and water, was dissolved by the São Paulo state government in 2020, and for this reason it is unknown if the incentives will continue. Before its dissolution the CDHU made a new area available to the collective so that they could increase the community garden's production, but this new area needs environmental restoration.

Discussion

Hardin's 1968 article "The Tragedy of the Commons" points out the harms caused by open access, without, however, affirming that the problem is the absence of property rights or governance regimes, and not the shared use of common-pool resources (McKean & Ostrom, 2001). In other words, Hardin ignored that in the regimes of communal use, there are also rules and principles designed to govern life in the community, as well as avoid the overuse of natural resources.

We identified an enormous capacity for organization and reorganization within the São Paulo community we studied—a community faced with different challenges imposed by the capitalist system, which favours private ownership to the detriment of collective ownership of the land, and by the state, which offers no guarantees of settlement for the group in the territory where they live.

It is evident within this context that the possibility of a communal form of urban land use and natural resources management still depends on public and private ownership regimes, since there is no recognition of common property in Brazil, except in a few situations.[4] Moreover, we observed that community governance in the implementation of the gardens involved the establishment of rules, with the provision of rights and sanctions that were regularly readjusted, sometimes with daily agreements. These daily agreements, important for group cohesion, were made through self-management processes involving decision-making negotiations within the group in relation to any problems that arose during the development of the work. At the end of each day, the group met to discuss the strengths and weaknesses of the decisions taken, and to re-establish updated agreements to define new responsibilities. These agreements were based on shared values such as cooperation and respect for others.

We also learned that the rules of land ownership and resource use are important for the group's cohesion, and that they allow the people to work collaboratively based on the common use of space and natural resources (earth, water, seeds, fertilizer) and other things (such as tools) that are used in the common area.

Regarding group cohesion, we observed the six social factors provide resilience, according to Cinner and Barnes (2019): the group's **flexibility** (to re-construct the garden after they were displaced), **organization** (the group's cohesion), **learning** (taking advantage of work and training programs to continually plan the garden and expand their products), **agency** (e.g., of women as members of the group and the group's ability to reach out and collaborate with the NGO and the CDHU), **socio-cognitive constructs** (these community members clearly grow up in an environment of adversity with a lack of basic human needs—which makes them very active fighters for their own subsistence). The group worked to expand their **assets** such as tools and seedlings; securing the fundamental right to land to farm was the weakest link in this resilience chain.

All of their work has strengthened the community's right to permanence on the land, since in Brazil as everywhere, occupation builds usufruct rights (though as described, this is not inviolable). The community members' work contributes to the production of food for sustenance, promotes mutual aid in the commercialization of products, and even drives claims processes for other rights with the government. Most importantly, we realized from this community's experience that this strengthening is the result of a long process that involved a network of support with the shared goal of removing obstacles to the implementation of community gardens, occupying empty spaces in the city, and guaranteeing food and income for the portion of São Paulo's population that lives in a socially and environmentally vulnerable situation.

In other words, we came to understand that this group of urban farmers had begun to take an active role in society, in the sense of taking responsibility for local governance. In theory, this would be the responsibility of the state, represented by its administrative institutions which, while recognizing people's rights, in fact do nothing to contribute to the improvement of their quality of life. In this context, we believe that the initiatives of this group of farmers could be replicated in other areas, although there is a pressing need to strengthen the relationships among people and support special training and skills for communal local governance.

In reference to Armitage et al. (2007), Ostrom and Cox (2010) explain that local users have no personal stake in the success of a project in which they are not involved; they can even directly or indirectly undermine the project. When users are involved, however, they can use their local knowledge to make a governance regime more adaptive, using collaboration to promote systematic learning.

Thus, for natural resources and land to be protected, we see a need for the state to recognize and encourage new forms of governance that include the community, so that understanding and empowerment can take place, producing actions that bring local benefits while at the same time designing general guidelines with a view to protecting resources globally and/or regionally.

Final Considerations

In this chapter, we have analyzed common-pool resources using the lens of property regimes, and described the results of a participatory research study involving urban farmers from the periphery of the city of São Paulo, with the objective of presenting some challenges for communal use and governance of land and natural resources.

We have reached the conclusion that in Brazil, public and private ownership regimes—as the predominant institutional forms of regulating access to the land and the natural resources within it—do not recognize the rights or the necessities of communities in situations of social and environmental vulnerability, and this is the reason that many communities have been adopting communal forms of territorial ownership and governance.

In contrast to Hardin's view (1968) that only public or private ownership would be able to protect common-pool resources, we have observed that local communities are capable of reclaiming, organizing, and administering the territories where they live, generating sustenance and income for the community, and even utilizing the resources in a sustainable manner.

Nevertheless, this does not mean that one single correct and successful formula exists. Communal ownership, just like private and public ownership, can be either a success or a failure. What we have tried to call attention to in this chapter is the necessity of reinforcing community values, cooperation, and mutual aid in order to promote territorial governance processes through the development of collective self-governance agreements that involve greater collaboration among the members of society in public affairs and decision-making.

We also draw attention to the need for communities to produce their own food, as changing rainfall patterns can affect agricultural productivity and food availability, and shorter supply chains reduce both uncertainty and carbon emissions. More diverse local food systems will be better able to respond to eventual emergencies, helping the poor population that will be most affected by increases in food prices.

In this sense, urban agriculture is an increasingly relevant strategy to tackle climate change and reduce disaster risks for low-income urban populations.

NOTES

1   We are grateful to the Queen Elizabeth Scholars program, funded by the Social Science and Humanities Research Council and the International Development Research Centre, and the Coordination for the Improvement of Higher Education Personnel in Brazil for the doctoral scholarship.

2   For instance, water, as a common-pool natural resource, can suffer from overuse, as well as problems related to the dumping of pollutants, which consequently makes the resource unavailable and generate the need for costly treatment.

3   The term "social technology" as used in Brazil means a product, method, technique, or process designed to solve some kind of social problem while meeting the principles of simplicity, low cost, easy applicability, and proven social impact. See Bazely et al., 2015; Pozzebon et al., 2021.

4   In Brazil's Federal Constitution of 1988 (unlike the legal documents of many countries), collective land ownership is allowed, but only in the following situations: Indigenous lands, *Quilombolas*, conservation areas in the forms of Extractive Reserves (*Reserva Extrativista* [Resex]) and Sustainable Development Reserves (*Reserva de Desenvolvimento Sustentável* [RDS]), as well as land reform settlements in the form of Sustainable Development Projects (*Projeto de Desenvolvimento Sustentável* [PDS]).

Reference List

Armitage, D., Berkes, F., & Doubleday, N. (Eds.). (2007). *Adaptive co-management: Collaboration, learning and multi-level governance*. University of British Columbia Press.

Bazely, D., Perkins, P.E., Duailibi, M., & Klenk, N. (2015). Strengthening resilience by thinking of knowledge as a nutrient connecting the local person to global thinking: The case of social technology / techologia social. In S.A. Moore and R. Mitchell (Eds.), *Planetary Praxis and Pedagogy* (pp. 119–132). Brill.

Berkes, F. (2005). Sistemas sociais, sistemas ecológicos e direitos de apropriação de recursos naturais. In Vieira, P.F., Berkes, F., & Seixas, C. *Gestão Integrada e participativa de recursos naturais*. Secco-APED.

Bromley, D.W., & Cernea, M.M. (1989). *The management of common property natural resources: Some conceptual and operational fallacies* [Discussion Paper]. World Bank, WDP 57.

Carolino, K. (2021). *Agricultura em São Paulo: Uma análise sobre os programas públicos instituídos no município* [Unpublished doctoral dissertation]. Universidade de São Paulo.

Cinner, J.E., & Barnes, M.L. (2019, September 20). *Social dimensions of resilience in social-ecological systems, 1*(1). Elsevier.

Diegues, A.C., & Moreira, A.C. (2001). *Espaços e recursos naturais de uso comum*. NUPAUB, USP.

Dietz, T., Dolsak, N., Ostrom, E., & Stern, P.C. (2002). The drama of the commons. In E.E. Ostrom, T.E. Dietz, N.E. Dolšak, P.C. Stern, S.E. Stonich, & E.U. Weber (Eds.), *The drama of the commons* (pp. 1–36). National Academies Press.

Dubbeling, M. (2014, March). A agricultura urbana como estratégia de redução de riscos e desastres diante da mudança climáticas. *Revista de Agricultura Urbana, 27*.

Feeny, D., Berkes, F., Mccay, B.J., & Acheson, J.M. (2001). A tragédia dos comuns: Vinte e doisanosdepois. In A.C. Diegues & A. Moreira (Eds.), *Espaços e recursos naturais de uso comum*. NUPAUB, University of São Paulo.

Hardin, G. (1968). The tragedy of the commons. *Science, 162*(3859), 1243–1248.

Holkup, P.A., Tripp-Reimer, T., Salois, E.M., & Weinert, C. (2004). Community-based participatory research. An approach to intervention research with a Native American community. *Advances in Nursing Science (ANS), 27*(3), 162–175.

McCay, B., & Acheson, J.M. (1987). *The question of the commons: The culture and ecology of communal resources*. University of Arizona Press.

McKean, M.A., & Ostrom, E. (2001). Regimes de propriedade comum em florestas: Somente uma relíquia do passado? In A.C. Diegues, & A. Moreira (Eds.), *Espaços e recursos naturais de uso comum*. NUPAUB, University of São Paulo.

Ministério do Meio Ambiente & Ministério da Educação. 2005. *Programa Nacional de Educação Ambiental—ProNEA* (3rd ed.). MMA Editions.

Ondetti, G. (2021). Ideational bases of land reform in brazil: 1910 to the present. In L. Leisering (Ed.), *One hundred years of social protection* (pp. 343–379). Palgrave Macmillan. https://link.springer.com/chapter/10.1007/978-3-030-54959-6_10

Ostrom, E. (1990). *Governing the commons: The evolution of institutions for collective action*. Cambridge University Press.

Ostrom, E., & Cox, M. (2010). Moving beyond panaceas: A multi-tiered diagnostic approach for social-ecological analysis. *Environmental Conservation, 37*(04).

Pozzebon, M., Tello-Rozas, S., & Heck, I. (2021). Nourishing the social innovation debate with the "social technology" South American research tradition. *Voluntas: International Journal of Voluntary and Nonprofit Organizations, 32*, 663–677.

Ruschel, C.V. (2018). *Os limites do Direito Ambiental na preservação dos recursos naturais comuns: Epistemologia da sustentabilidade e estudos de caso* [Unpublished doctoral dissertation]. Universidade Federal de Santa Catarina.

Sorrentino, M. (2014). Educador ambiental popular. In L.A. Ferraro Junior (Ed.), *Encontros e caminhos: Formação de educadoras(es) ambientais e coletivos educadores* (Vol. 3, pp. 141–154). MMA/DEA. https://www.terrabrasilis.org.br/ecotecadigital/index.php/estantes/educacao-ambiental/2515-encontros-e-caminhos-vol-3

Linking Soil and Social-Ecological Resilience with the Climate Agenda: Perspectives from *Quilombola* Communities in the Atlantic Forest, Brazil

Marcondes G. Coelho Junior, Eduardo C. da Silva Neto, Emerson Ramos, Ronaldo dos Santos, Ana P.D. Turetta, Marcos Gervasio Pereira, and Eliane M.R. da Silva

Introduction

Soils play an important and diverse role for environment and humanity. The World Soil Charter states that the "soils are a key enabling resource, central to the creation of a host of goods and services integral to ecosystems and human well-being" (FAO, 2015). Soil functions provide essential ecosystem services such as provisioning services (e.g., food production), supporting services (e.g., carbon storage), regulating services (e.g., climate regulation, nutrient cycling, and flood control), and cultural services (e.g., heritage, composing the landscape aesthetic, and community identities) (Dominati et al., 2010; Adhikari & Hartemink, 2016; Jónsson & Davíðsdóttir, 2016; Rodrigues et al., 2021). Besides maintaining biodiversity and contributing to global ecosystem protection, these services are especially important for Sustainable Development Goals (SDGs) 2—Zero Hunger, 13—Climate Action, and 15—Life on Land.

Approximately double the total carbon in the atmosphere is in soil reserves (Smith et al., 2021). Thus, soils have become part of the global carbon agenda for climate change mitigation through the launch of three high-level initiatives: i) the "4 per mille initiative," signed by more than one hundred nations at the 21st Conference of the Parties (COP) in Paris in 2015; ii) the Koronivia workshops on agriculture, which included soils and soil organic carbon (SOC) for climate change mitigation and were initiated at COP23 in 2018; and iii) the RECSOIL, a United Nations Food and Agriculture Organization (FAO) program for the recarbonization of soils (Amelung et al., 2020). These all recognize the potential of soils to remove between 0.79 and 1.54 Gt C yr–1 from the atmosphere (Fuss et al., 2018).

Despite the evident value of soils for human well-being and the global climate, unsustainable human activities threaten it. In Latin America, about 50 per cent of soils are facing some type of degradation (FAO, 2015). In Brazil, soil losses are caused mainly by erosion and inadequate agricultural management, which affects soil quality (e.g., by pollution, salinization, and acidification, among others). Land use conversion from natural ecosystems to cattle pastures and expansion of agricultural crop areas has ranked Brazil fourth among the top CO_2 emitting countries (Carbon Brief, 2021). Therefore, there is no doubt that the land use model urgently needs to adapt (Ball et al., 2018).

If on the one hand this historical model of natural resource uses shows that change is urgently needed, on the other hand, sustainable livelihoods and other knowledge systems can reveal paths to more inclusive and effective conservation. The Intergovernmental Science-Policy Platform on Biodiversity and Ecosystem Services (IPBES) recognizes the contribution of Indigenous and local knowledge (ILK) to the conservation and sustainable use of biodiversity (IPBES, 2019). Knowledge about commons, ecosystems, and associated management practices has been developed and is possessed by communities that have engaged in agriculture for their livelihood and benefits over long time frames (Berkes & Folke, 1998; Folke, 2004)—and this is the case for *Quilombola* communities.

Over more than three centuries, *Quilombola* communities have been formed in Brazil by formerly enslaved Africans who migrated by force from Africa through the Atlantic slave trade and who escaped the plantation systems (Arruti, 2008). Under slavery, these people suffered labour exploitation, rights violations, torture, and prolonged punishment, which caused massive mortality rates (Gomes, 2015). To struggle against colonial exploitation, the enslaved *Quilombola* ancestors fled into the forests to create small settlements—*Quilombos*, from a Kimbundu word for "war camp"—as a strategy in their struggle for freedom (Leite, 2015; Gomes, 2015). Today, the focus of *Quilombola* struggle is no longer defense of freedom, but rather defense of land and territory. *Quilombola* communities have a unique ethnic identity and depend on the land for their physical, social, economic, and cultural reproduction. Due to the social and environmental vulnerability of most *Quilombola* territories, the *Quilombola* communities experience a critical state of living conditions, which has been aggravated during the COVID-19 pandemic (Coelho-Junior et al., 2020).

As part of their historical and cultural process, *Quilombola* communities have developed land uses grounded in traditional agricultural practices shaped by their identity processes (Gomes, 2015; Steward and Lima, 2017). In this context, this chapter discusses the social values of soils and their links to soil quality indicators (biological, physical, and chemical) in *Quilombola* communities, including decisive factors for adapting sustainable solutions and enhancing livelihood resilience while ensuring forest conservation and safeguarding cultural identity based on soil quality. We also describe a participatory research project that is ongoing in two *Quilombola* communities in the Brazilian Atlantic Forest of Rio de Janeiro State: *Quilombo do Campinho da Independência* (from now on called *Quilombo do Campinho*) and *Quilombo Santa Rita do Bracuí* (from now on called *Quilombo do Bracuí*) (see Map 2, page 30). This research is grounded in ecological economics, environmental justice, community-based management, and ethnopedology perspectives, as we aim to explore the links between soil and human well-being, approaching this from local to global levels to address the challenges of climate change in vulnerable communities.

Background

Soils' Contributions to People: Context and Novel Approach

Principle 3 of the World Soil Charter states that "soil management is sustainable if the supporting, provisioning, regulating, and cultural services provided by soil are maintained or enhanced without significantly impairing either the soil functions that enable those services or biodiversity." These soil ecosystem services are directly related to benefits that people obtain from

Table 6.1 Soils' role in delivering Nature's Contributions to People (NCP).

NCP Category	Soils' Contributions to People	Key References*
Material NCP	Food and feed	Silver et al. (2021)
	Materials and assistance	Morel et al. (2021)
	Energy	Smith et al. (2021B)
	Genetic, medicinal, and biochemical resources	Thiele-Bruhn (2021)
Non-material NCP	Learning and inspiration, physical and psychological experiences, and supporting identities	McElwee (2021)
Regulation NCP	Regulation of climate	Lal et al. (2021)
	Regulation of freshwater quantity, flow, and timing	Keesstra et al. (2021)
	Regulation of freshwater and coastal water quality	Cheng et al. (2021)
	Regulation of hazards and extreme events	Saco et al. (2021)
	Habitat creation and maintenance	Deyn and Kooistra (2021)
	Regulation of air quality	Giltrap et al. (2021)
	Regulation of organisms detrimental to humans	Samaddar et al. (2021)
	Dispersal of seeds and other propagules	Carvalheiro et al. (2021)
	Regulation of ocean acidification	Renforth and Campbell (2021)
	Formation, protection, and decontamination of soils and sediments	Sarkar et al. (2021)

* All references cited in Smith et al. (2021).

soils, as considered by the Millennium Ecosystem Assessment and further represented in the pioneering works by Dominati et al. (2010) and Adhikari and Hartemink (2016). But recently, the IPBES established a conceptual framework that attempts to contextualize "ecosystem services" by defining Nature's Contributions to People (NCP) as "all the contributions, both positive and negative, of living nature (i.e., diversity of organisms, ecosystems, and their associated ecological and evolutionary processes) to the quality of life of people" (Díaz et al., 2018).

A special issue of the journal *Philosophical Transactions B* provides an assessment of the contribution of soils to NCP. In the editorial article, "The Role of Soils in Delivering Nature's Contributions to People," Smith et al. (2021) presents the key insights from each article that make up this special issue (Table 6.1). Smith et al. (2021) also emphasize that soil management priorities should include: (i) for healthy soils in natural ecosystems, protect them from conversion and degradation; (ii) for managed soils, manage them in a way to protect and enhance soil biodiversity, health, productivity and sustainability and to prevent degradation; and (iii) for degraded soils, restore to full soil health.

Socio-Ecological Resilience Based on Soil: Implications for Ethnopedology

The concept of resilience focuses on the adaptation and change a system can undertake while remaining within critical system thresholds (Walker et al. 2006). Thus, resilience thinking proposes a systemic approach to human-environment relations that fits well with attempts to predict or model social-ecological change. Adapting this concept for social-ecological resilience (SER), we have the combination of both: i) social resilience as the ability of a social system to react to a disturbance and, afterwards, return to a state in which social functions, structures, and processes continue as before (Adger et al., 2005); and ii) ecological resilience as an ecosystem's ability to absorb or recover from disturbance and change while maintaining its functions and services (Carpenter et al., 2001). Therefore, SER can be understood as the interplay of factors involved in recovering from disturbances, re-organization, and the development of socio-ecological systems.

Applying a SER lens in soil studies, we emphasize soil as a common thread in integrating social and ecological systems. The contribution of soils (an ecological system) to human well-being (a social system) depends on land uses and management (Adhikari and Hartemink, 2016; Prado et al., 2016; Turetta et al., 2020), which are often associated with cultural values. Waroux et al. (2021) highlighted that "culture as context is thus present as a frame for land-use decisions, behaviors, and land system outcomes." In this context, traditional knowledge of soil management, inherited through generations and adapted to social-ecological changes (Krasilnikov & Tabor, 2003), frames the role of culture and land history in soil studies, bringing to light ethnopedology as an interdisciplinary field (Barrera-Bassols & Zinck, 2003). Therefore, participatory research on soils in *Quilombola* communities can reveal the cultural reasons that explain physical, chemical, and biological parameters, enabling better strategies for socio-ecological resilience to climate challenges.

Participatory Research on Quilombola Communities in the Brazilian Atlantic Forest, Rio de Janeiro State

Quilombo do Campinho

The *Quilombo do Campinho* is located in Paraty, southern Rio de Janeiro State, in a protected area (*APA do Cairuçu*) (see Map 2, page 30). The native vegetation is Atlantic Forest, a biome highly threatened by climate change (Colombo & Joly, 2010). The region's climate is of type CWa, according to the Köppen classification, with moderate temperatures and a tropical summer (Alvares et al., 2013). The *Quilombo* territory covers more than 287 ha and has a population of one hundred and fifty families, totalling approximately five hundred people.

The origin of *Quilombo do Campinho* goes back to the nineteenth century and it centres on three women—Antonica, Marcelina, and Luiza—who worked at the farmhouse of *Fazenda da Independência*, when the economic decline of the region forced the colonial farmers to abandon their lands and donate them to the enslaved people. The struggle for land continued for decades, until the *Quilombo do Campinho* became the first *Quilombola* community to receive land title in the State of Rio de Janeiro, on 21 March 1999. Their recognition as a "*Quilombo*" brought to the community the incentive for local farmers to be self-sustaining, even though many men and women work outside the community, mainly as employees in family households or in luxury resort condominiums in the region (Tavares, 2014). Currently, activities such as seedling production, agroforestry, ethnic tourism, and the community restaurant, have been developed in the community and are major income sources (Lima, 2008).

Despite their rights as a Brazilian "traditional community," *Quilombo* residents have faced challenges for many reasons: i) real estate speculation, which has increased due to the UNESCO designation of Paraty as a World Heritage Site; ii) restrictions imposed for clearing new areas for "agroforestry," since the traditional territory overlaps a protected area; iii) imminent risk of accidents and pollution related to oil and gas exploration in the Pré-Sal Pole of the Santos Basin; and iv) direct impacts of the COVID-19 pandemic on community-based tourism and the community restaurant, the main income sources of the *Quilombolas*, rendering the community even more vulnerable.

Quilombo do Bracuí

The *Quilombo do Bracuí* is also part of the Atlantic Forest and is in Angra dos Reis, southern Rio de Janeiro State (see Map 2, page 30). The community territory has an area of 616 ha that are managed by 129 families, totalling approximately 362 people (INCRA, 2015). The *Quilombo do Bracuí* is located in the middle of the *Santa Rita do Bracuí* river basin, important for regional water supply (INCRA, 2015). The climate according to the Köppen classification is type Af, rainy tropical forest climate (Alvares et al., 2013). Also, the territory of *Quilombo do Bracuí* covers the buffer zone of the Bocaina National Park, a protected area recognized as a World Heritage Site by UNESCO.

The *Quilombo do Bracuí* is located at an old farm that was used for many years as an illegal port for the African slave trade, since there was a direct path from the sea to the farm, although the slave trade was officially prohibited in 1831 (Karasch, 2000). Due to economic decline at this time, José Breves, the colonial farmer, made a will donating part of his farm to ex-slaves. Their return to this area allowed the development of a community based on the reference to enslaved ancestors' freedom in a social context known as "black proto-campesinato" (Marques, 2011).

The *Quilombo do Bracuí* has faced huge challenges to maintain itself on the territory. Threats emerged from government initiatives such as projects for the development of "hygienic tourism," the construction of the BR 101 highway, and construction of luxury condominiums (Ramos, 2018). All these "drivers" aimed to force the inhabitants to leave the *Quilombola* territory, and even induced people to sign fake documents for land titles (Ramos, 2018). The community resisted by creating a local association, *Associação dos Remanescentes de Quilombo de Santa Rita do Bracuí* (ARQUISABRA) in 1998, which was certified in 1999 by the Palmares Cultural Foundation, Brazil's federal institution supporting Black cultural, historical, economic, and social contributions. However, it was only in 2006 that the land-titling process of the *Quilombo do Bracuí* was initiated by the federal government. And almost fifteen years after the titling process began, the *Quilombo do Bracuí* still has no land title.

Currently, there are two major problems faced by the *Quilombolas:* i) real estate speculation through land invasion due to the absence of land title; and

ii) the project to install a hydroelectric plant (UHE Paca Grande I and II) on the Paca Grande River, which is part of the Bracuí River watershed. Evidence warns of the "socio-environmental disaster" arising from these hydroelectric plants, both for the *Quilombola* community and for other traditional communities (e.g., the *Guarani de Bracuhy* Indigenous territory), in addition to affecting the buffer zone of the Bocaina National Park (Alves, 2019). Another threat factor is the proximity to the Angra dos Reis Nuclear Power Plants, leaving *Quilombola* inhabitants more exposed to potential environmental disasters.

Research Design and Goals

Our participatory research on *Quilombolas*' perceptions and social values of soil and soil sampling for physical, chemical, and biological analyses, includes four steps. The main purpose of this research is to identify and evaluate the determining factors for the soils' contributions to people by linking local and scientific knowledge. Thus, we aim to address four specific objectives: i) Select a set of indicators to evaluate soils' contributions for people in *Quilombola* communities; ii) Identify the threats and opportunities related to soils' contributions to people in *Quilombola* communities; iii) Describe and organize the determining criteria for soil management practices according to local knowledge; iv) Understand and explain the perception of social values of soils in *Quilombola* communities. Our secondary goals are: i) Explore and evaluate participatory methodologies to assess the potential of soils' contributions to people for socio-ecological resilience; ii) Facilitate knowledge transfer between local and scientific knowledge holders for socio-environmental innovation. For this, we draw on interdisciplinary methods of socio-environmental research, including participant observation at community meetings; open interviews with key informants; Q-methodology, or systematic study of participants' viewpoints, on social values of soils based on local perceptions, and laboratory procedures (technical and scientific methods for soil sampling and the chemical, physical, and biological analysis of soils).

The research process started with visits and participation in community meetings and cultural events in both *Quilombola* communities (Figures 6.1–6.3). The first meetings with community leaders occurred through the residents' associations (Associação de Moradores do Quilombo Campinho da Independencia [AMOQC] and Associação dos Remanescentes de Quilombo

Figs. 6.1–6.3
Community meetings
in early stages of the
research project.

Figs. 6.4–6.6 Soil samples at Soil Genesis and Classification Laboratory, UFRRJ.

de Santa Rita do Bracuí [ARQUISABRA]) and also through collaborative work by the Observatory of Sustainable and Healthy Territories of Bocaina (OTSS), an institution formed from the partnership between Fundação Oswaldo Cruz (Fiocruz), a Rio de Janeiro scientific institution for research and development in public health and biological sciences, and Fórum de Comunidades Tradicionais de Angra dos Reis, Paraty e Ubatuba (FCT), a local traditional communities organization. At these meetings, the project was designed, considering the specific demands of these local communities regarding soil quality and the potential of community engagement as an opportunity for participatory research with local impacts (especially, for the physical, chemical, and biological characterization of the soils, to guide them in improving management practices).

The research project was approved by the Research Ethics Committee at the Federal Rural University of Rio de Janeiro (UFRRJ), and the communities signed informed consent forms, indicating their awareness of the study, and gave their permission to use images and sounds from their territories. All interested participants were informed about the objectives and steps of this study at the beginning of this process. In each community, a local researcher was selected to join the fieldwork and to be a community spokesperson. An OTSS technical officer was also selected to assist fieldwork and data analysis. Finally, an assistant professor from UFRRJ and several undergraduate students were invited to collaborate on soil sampling, laboratory analysis, and data analysis. This collaborative work enabled an experience of sharing throughout the whole research process, enhancing the scope of participatory research in socio-environmental studies. However, due to the COVID-19 pandemic, which caused unsafe conditions in Brazil, this teamwork had to be suspended temporarily to comply with UFRRJ's biosecurity guidelines.

Initial soil samples were sent to the Soil Genesis and Classification Laboratory at UFRRJ, where analysis began (Figures 6.4–6.6).

Discussion and Conclusions

Local Soil Knowledge in Traditional Territories

As soil is a vital entity (Ball et al., 2018) that integrates water security, agricultural production, energy, climate, and biodiversity (McBratney et al., 2014), all impacts on soil have indirect effects on other systems, such as health and human well-being (Prado et al., 2016). To study soils of traditional or specially protected areas, such as *Quilombola* territories, we must consider that traditional ecological knowledge is transmitted through generations, sharing experiences, and is adapted to the socio-ecological changes that occur in time and space (Krasilnikov & Tabor, 2003). The relationship between these communities and the soil derives most strongly from subsistence agriculture.

Local soil knowledge can be defined as "the knowledge of soil properties and management by people living in a particular environment for some period of time" (Winklerprins, 1999). This knowledge implies a lot of trial and error, but also includes scientific processes (Barrera-Bassols & Zinck, 2003). It has also been described as "both skill and knowledge" and "the heritage from practical daily life, with its functional demands." This characterizes a mixture between knowledge and practice, in general causing a difficulty in distinguishing the threshold between them (Sillitoe, 1998).

Local soil knowledge in traditional communities can provide major contributions to science. For instance, one key contribution is the lessons it can provide for understanding land use over different time scales, supporting strategies for sustainable agriculture. Traditional soil and crop management practices are based on local knowledge, obtained through experimentation by generations of people working on the land in a specific environment. Therefore, these practices reveal how to maintain the use of resources and the environment in a sustainable way. Recognizing this, there is surely no reason to ignore this knowledge/practice as a technology for advancing soil conservation.

Overview of Findings

The *Quilombola* communities in this study divide their territories into family areas (each family has a limited area for land use). Thus, different land uses integrating permanent crops, temporary crops, and agroforests can be highlighted (Figures 6.7–6.9). The agroforests in *Quilombola* communities demonstrate traditional soil management practices and produce food while promoting Atlantic Forest conservation and delivering ecosystem services (Tubenchlak et al., 2021). For example, in *Quilombo do Campinho*, Tavares et al. (2018) found that the agroforestry systems maintained high levels of total organic carbon, as well as providing the same conditions for soil aggregation as the forest. Thus, the authors concluded that the formation of biodiverse

Figs. 6.7–6.9 Different land uses and agroforestry systems in *Quilombola* communities.

agroecosystems by *Quilombolas* contributed to maintaining soil quality. These results correspond with literature that assembles evidence regarding benefits of agroforestry for global climate, food security, water supply, and forest conservation with direct impacts on land use sustainability (Verchot et al., 2007; Schroth et al., 2011; Miccolis et al., 2019).

During our fieldwork, we observed the intrinsic link between landscape conservation and sustainable soil management practices. Also, our dialogues with *Quilombola* farmers revealed the role of culture in soil management: "This crop area here belonged to my grandfather, it passed to my father, and I am training my grandchildren to take care of it as well." The oral transfer of cultural practices over generations is a characteristic of *Quilombola* peoples (Alves, 2019). Waroux et al. (2021) also present different cases to highlight how aspects of culture influence land systems in myriad ways.

We also conducted training on soil sampling for socio-environmental studies. It was possible to combine scientific and traditional knowledge

Figs. 6.10–6.13 Participatory soil sampling for knowledge transfer in *Quilombola* communities.

during this experience, strengthening participatory research (Figures 6.10–6.13). The experience with ethnopedology made it possible to understand soil beyond its environmental characteristics. Soil, or "land," has a relational value that makes the *Quilombolas* feel part of the soil system, managing a live system—soil, that gives life and the community power. A site in the *Quilombo do Bracuí* that represents these social values of soils is called *Aiê Eleteloju* (@ eleteloju.aie on Instagram), which means "Fertile Land" in the Afro Yoruba language. This space is divided into areas with crops (cassava, corn, beans), agroforestry systems, and conventional and medicinal vegetable gardens. In addition, it includes the *Terreiro de Candomblé*—a ceremonial meeting place in the Afro-Brazilian religion. According to Ramos (2018), the goal of *Aiê Eleteloju* is to be a space for dialogue and sharing of traditional knowledge, as well as for training on agroecological practices, social learning, and religious and cultural celebrations.

Quilombolas' Struggle for Land Tenure and Environmental and Climate Justice in Brazil's Atlantic Forest

Injustice in land access in Brazil is a consequence of the colonization process that generated a high concentration of land in few hands (Robles, 2018). Brazil has one of the highest rates of non-productive large estates in the world (Paulino, 2014) while the country has a huge number of people waiting for the opportunity to have and work their own land (Reydon et al., 2015). Also, it is important to highlight that the current structure of land ownership in Brazil acquired its form in the 1960s through the implementation of the Green Revolution and the modernization of large estates for agriculture and livestock production (Sauer and Leite, 2012). Agrarian reform for a more equitable distribution of rural land is the basis for a process of social justice and democratization in the country (Leite et al., 2004).

A critical point on inequality in access to land in Brazil is its Land Law itself (Law No. 601/1850), signed by Emperor Dom Pedro II in September 1850. The first restriction imposed by this law is in Article 1, which determines that only land purchases grant access to land, thus rendering it impossible for poor, Black, and *Quilombola* people to acquire land due to their socio-economic conditions. As the law was established under the slavery regime, its intention was to make it impossible for Black people to access land, in an attempt to hinder the slavery abolition movement, which only succeeded in

1888 (Amorim & Tárrega, 2019). However, the transition from slavery to free labour was characterized by numerous social and economic changes that directly interfered with former slaves' interaction with land (Smith, 1990).

Only one hundred years after the abolition of slavery did the Brazilian government recognize *Quilombolas*' right to continue living on their territories, by means of Article 68 in the Federal Constitution of 1988. It establishes that "the descendants of *Quilombola* communities who are occupying their lands are recognized as having definitive land title, and the State must provide their respective titles" (Brasil, 1988). Beyond the right to land tenure, the Federal Constitution also legitimized the cultural rights of the Afro-descendant *Quilombola* communities and other traditional peoples, in Articles 215 and 216 (Brasil, 1988). Despite such institutional advances in the Federal Constitution, the implementation of Article 68 for access to land required an additional definition of *"Quilombo,"* since the Federal Constitution did not specify this (Thorkildsen & Kaarhus, 2017). This legal "gap" became an arena for political disputes over the guarantee of *Quilombola* rights to their territories. After many years' delay, the Brazilian government published Federal Decree No. 4,887/2003, which regulates the process of identification, recognition, delimitation, demarcation, and titling of *Quilombola* lands (Brasil, 2003).

Recent history shows that recognition by legislation alone does not guarantee social equity for *Quilombolas*. The attacks suffered by *Quilombola* communities are directly related to their defense of permanence in their territories, historically denied by the land tenure system in Brazil and consolidated through the denial of land access and the absence of social reparations to Black people for more than three hundred years of slavery (Terra de Direitos & CONAQ, 2018). Also, *Quilombola* communities in Brazil's Atlantic Forest are facing environmental regulatory barriers that prohibit their cultural practices of soil management due to environmental racism[1] and institutional racism.[2] Restrictions on cultural practices have generated notifications of environmental infractions for *Quilombolas*, putting them at risk of being arrested just for developing their traditional practices. This is despite much evidence on the role of *Quilombola* communities in Atlantic Forest conservation (Diegues et al., 2000; Diegues & Viana, 2004; Pereira & Diegues, 2010; Penna-Firme & Brondízio, 2007; Adams et al., 2013; Thorkildsen, 2014; Thorkildsen & Kaarhus, 2017).

According to Almeida (1989), the territories used by the *Quilombolas* are "lands of common use," since the use of land and natural resources is not carried out individually, but collectively by the community, which creates specific management rules commonly agreed upon by the families living on the land, and different from state legislation based on private property. Soil studies from this perspective ("lands of common use") provide evidence of a range of contributions soils make to people and ecosystems, as well as ways of understanding the nexus of soil quality, management practices, and *Quilombola* rights. These rights also include the right to contribute to the climate agenda. Participatory ethnopedagogy with *Quilombolas* creates an opportunity to shift research back towards the basis of sustainability as evidenced in traditional territories—the healthy soil.

NOTES

1   Environmental racism refers to any environmental policy, practice, or directive that differentially affects or disadvantages (whether intended or unintended) individuals, groups, or communities based on race or colour (Bullard, 1999).

2   Institutional racism is manifested through mechanisms, explicit or not, that hinder the presence of Black people in governmental spaces, as well as the formulation of effective public policies to combat racial inequalities (Giacomini & Terra, 2014).

Reference List

Adams, C., Munari, L.C., Van Vliet, N., Murrieta, R.S.S., Piperata, B.A., Futemma, C., Pedroso Jr., N.N., Taqueda, C.S., Crevelaro, M.A., & Spressola-Prado, V.L. (2013). Diversifying incomes and losing landscape complexity in Quilombola shifting cultivation communities of the Atlantic rainforest (Brazil). *Human Ecology, 41*(1), 119–137.

Adger, W.N., Hughes, T.P., Folke, C., Carpenter, S.R., & Rockström, J. (2005). Social-ecological resilience to coastal disasters. *Science, 309*(5737), 1036–1039.

Adhikari, K., & Hartemink, A.E. (2016). Linking soils to ecosystem services—A global review. *Geoderma, 262*, 101–111.

Almeida, A.W.B. (1989). Terras de preta, terras de santo, terras de Índio: Uso comum e conflito. In E.M.R. de Castro & J. Hébette (Eds.), *Na trilha dos grandes projetos: modernização e conflito na Amazônia* (pp. 163–196). NAEA (Núcleo de Altos Estudos Amazônicos)/UFPA (Universidade Federal do Pará).

Alvares, C.A., Stape, J.L., Sentelhas, P.C., Gonçalves, J.D.M., & Sparovek, G. (2013). Köppen's climate classification map for Brazil. *Meteorologische Zeitschrift, 22*(6), 711–728.

Alves, E.C.S. (2019). *Tornar-se uma escola Quilombola: caminhos e descaminhos de uma experiência no Quilombo Santa Rita do Bracuí, Angra dos Reis-RJ* [Unpublished doctoral dissertation]. Pontifícia Universidade Católica do Rio de Janeiro.

Amelung, W., Bossio, D., de Vries, W., Kögel-Knabner, I., Lehmann, J., Amundson, R., Bol, R., Collins, C., Lal, R., Leifeld, J., Minasny, B., Pan, G., Paustian, K., Rumpel, C., Sanderman, J., van Groenigen, J.W., Mooney, S., van Wesemael, B., ... & Chabbi, A. (2020). Towards a global-scale soil climate mitigation strategy. *Nature Communications, 11*(1), 1–10.

Amorim, L.P., & Tárrega, M.C.V.B. (2019). O acesso à terra: a Lei de terras "1850" como obstáculo ao direito territorial Quilombola. *Emblemas, 16*(1), 10–23.

Arruti, J.M. (2008). Quilombos. *Jangwa Pana, 8*(1), 102–121.

Ball, B.C., Hargreaves, P.R., & Watson, C.A. (2018). A framework of connections between soil and people can help improve sustainability of the food system and soil functions. *Ambio, 47*, 269–283.

Barrera-Bassols, N., & Zinck, J.A. (2003). Ethnopedology in a worldwide view on the soil knowledge of local people. *Geoderma, 111*, 171–195.

Berkes, F., & Folke, C. (1998). Linking social-ecological systems for resilience and sustainability. In F. Berkes, C. Folke, & J. Colding (Eds.), *Linking social-ecological systems: Management practices and social mechanisms for building resilience* (pp. 1–25). Cambridge University Press.

Brasil. Constituição da República Federativa do Brasil de 1988. *Senado Federal.*

Brasil. Decreto No. 4.887, de 20 de novembro de 2003. Regulamenta o procedimento para identificação, reconhecimento, delimitação, demarcação e titulação das terras ocupadas por remanescentes das comunidades dos Quilombos de que trata o art. 68 do Ato das Disposições Constitucionais Transitórias. *Diário Oficial da União.*

Bullard, R.D. (1999). Dismantling environmental racism in the USA. *Local Environment, 4*(1), 5–19.

Carbon Brief. (2021). *Analysis: Which countries are historically responsible for climate change?* Retrieved December 10, 2021, from https://www.carbonbrief.org/analysis-which-countries-are-historically-responsible-for-climate-change

Carpenter, S., Walker, B., Anderies, J.M., & Abel, N. (2001). From metaphor to measurement: resilience of what to what? *Ecosystems, 4*(8), 765–781.

Coelho-Junior, M.G., Iwama, A.Y., González, T.S., Silva-Neto, E.C., Araos, F., Carolino, K., Campolina, D., Nogueira, A.S., Nascimento, V., Santos, R., Perkins, P.E., Fearnside, P.M., & Ferrante, L. (2020). Brazil's policies threaten Quilombola communities and their lands amid the COVID-19 pandemic. *Ecosystems and People, 16*(1), 384–386. https://doi.org/10.1080/26395916.2020.1845804

Colombo, A.F., & Joly, C.A. (2010). Brazilian Atlantic Forest lato sensu: The most ancient Brazilian forest, and a biodiversity hotspot, is highly threatened by climate change. *Brazilian Journal of Biology, 70*, 697–708.

Díaz, S., Pascual, U., Stenseke, M., Martín-López, B., Watson, R.T., Molnár, Z., Hill, R., Chan, K.M.A., Baste, I.A., Brauman, K.A., Polasky, S., Church, A., Lonsdale, M., Larigauderie, A., Leadley, P.W., Van Oudenhoven, A.P.E., Van Der Plaat, F., Schröter, M., Lavorel, S., ... & Shirayama, Y. (2018). Assessing nature's contributions to people. *Science, 359*(6373), 270–272.

Diegues, A.C., Arruda, R.S.V., da Silva, V.C.F., Figols, F.A.B., & Andrade, D. (2000). *Os saberes tradicionais e a biodiversidade no Brasil.* NUPAUB (Núcleo de Pesquisas Sobre Populações Humanas e Áreas Úmidas Brasileiras)/USP (Universidade de São Paulo).

Diegues, A.C.S.A., & Viana, V.M. (2004). *Comunidades tradicionais e manejo dos recursos naturais da Mata Atlântica: Coletânea de textos apresentados no Seminário Alternativas de Manejo Sustentável de Recursos Naturais do Vale do Ribeira.* Editora HUCITEC.

Dominati, E., Patterson, M., & Mackay, A. (2010). A framework for classifying and quantifying the natural capital and ecosystem services of soils. *Ecological Economics, 69*(9), 1858–1868.

FAO (Food and Agriculture Organization of the United States). (2015). *Revised world soil charter.*

Folke, C. (2004). Traditional knowledge in social-ecological systems. *Ecology and Society, 9*(3), 7.

Fuss, S., Lamb, W.F., Callaghan, M.W., Hilaire, J., Creutzig, F., Amann, T., Beringer, T., Garcia, W.O., Hartmann, J., Khanna, T., Luderer, G., Nemet, G.F., Rogelj, J., Smith, P., Vicente, J.L.V., Wilcox, J., Dominguez, M.M.Z., & Minx, J.C. (2018). Negative emissions—part 2: Costs, potentials and side effects. *Environmental Research Letters, 13*(063002).

Giacomini, S.M., & Terra, P. (2014). Participação e movimento negro: Os desafios do "racismo institucional." In J.S.L. Lopes & B. Heredia (Eds.), *Movimentos sociais e esfera pública: O mundo da participação* (pp. 189–210). Colégio Brasileiro de Altos Estudos. https://www.ipea.gov.br/participacao/images/pdfs/2014%20-%20 movimentos%20sociais%20-%20seminario%20participacao.pdf

Gomes, F.S. (2015). *Mocambos e Quilombos—Uma história do campesinato negro no Brasil.* Claro Engima.

INCRA (Instituto Nacional de Colonização e Reforma Agrária). (2015). *Relatório técnico de identificação e delimitação do Quilombo Santa Rita do Bracuí.*

IPBES (Intergovernmental Platform on Biodiversity and Ecosystem Services). (2019). *Report of the plenary of the intergovernmental science-policy platform on biodiversity and ecosystem services on the work of its seventh session.* https://ipbes. net/sites/default/files/ipbes_7_10_add.1_en_1.pdf

Jónsson, J.O.G., & Davíðsdóttir, B. (2016). Classification and valuation of soil ecosystem services. *Agricultural Systems, 145*, 24–38.

Karasch, M.A. (2000). *Vida dos escravos no Rio de Janeiro (1808–1850).* Companhia das Letras.

Krasilnikov, P.V., & Tabor, J.A. (2003). Perspectives on utilitarian ethnopedology. *Geoderma, 111*(3–4), 197–215.

Leite, I.B. (2015). The Brazilian Quilombo: "race," community and land in space and time. *The Journal of Peasant Studies, 42*(6), 1225–1240.

Leite, S., Heredia, B., Medeiros, L., Palmeira, M., & Cintrao, R. (2004). *Impactos dos assentamentos: Um estudo sobre o meio rural brasileiro.* Instituto Interamericano de Cooperação para a Agricultura-Núcleo de Estudos Agrários do Desenvolvimento Rural.

Lima, L.R. (2008). *Quilombos e políticas de reconhecimento: O caso do Campinho da Independência* [Unpublished master's thesis]. Universidade de São Paulo.

Marques, C.M. (2011). *À margem da economia: Cachaça e protocampesinato negro no litoral sul fluminense (1800–1888)* [Unpublished master's thesis]. Universidade Federal Fluminense.

McBratney, A., Field, D.J., & Koch, A. (2014). The dimensions of soil security. *Geoderma, 213*, 203–213.

Miccolis, A., Peneireiro, F.M., & Marques, H.R. (2019). Restoration through agroforestry in Brazil: Options for reconciling livelihoods with conservation. In M. van Noordwijk (Ed.), *Sustainable development through trees on farms: Agroforestry in its fifth decade* (pp. 209–231). World Agroforestry (ICRAF) Southeast Asia Regional Program.

Paulino, E.T. (2014). The agricultural, environmental and socio-political repercussions of Brazil's land governance system. *Land Use Policy, 36*, 134–144.

Penna-Firme, R., & Brondízio, E. (2007). The risks of commodifying poverty: Rural communities, Quilombola identity, and nature conservation in Brazil. *Habitus— Revista do Instituto Goiano de Pré-História e Antropologia, 5*(2), 355–373.

Pereira, B.E., & Diegues, A.C. (2010). Conhecimento de populações tradicionais como possibilidade de conservação da natureza: Uma reflexão sobre a perspectiva da etnoconservação. *Desenvolvimento e Meio Ambiente, 22*, 37–50.

Prado, R.B., Fidalgo, E.C.C., Monteiro, J.M.G., Schuler, A.E., Vezzani, F.M., Garcia, J.R., de Oliveira, A.P., Viana, J.H.M., Pedreira, B.C.C.G., Mendes, I.C., Reatto, A., Parron, L.M., Clemente, E.P., Donagemma, G.K., Turetta, A.P.D., & Simões, M. (2016). Current overview and potential applications of the soil ecosystem services approach in Brazil. *Pesquisa Agropecuária Brasileira, 51*(9), 1021–1038.

Ramos, E.L. (2018). *Aiê Eletuloju: A porta para o sagrado na Comunidade Quilombola de Santa Rita do Bracuí* [Unpublished undergraduate thesis]. Universidade Federal Rural do Rio de Janeiro.

Reydon, B.P., Fernandes, V.B., & Telles, T.S. (2015). Land tenure in Brazil: The question of regulation and governance. *Land Use Policy, 42*, 509–516.

Robles, W. (2018). Revisiting agrarian reform in Brazil, 1985–2016. *Journal of Developing Societies, 34*(1), 1–34.

Rodrigues, A.F., Latawiec, A.E., Reid, B.J., Solórzano, A., Schuler, A.E., Lacerda, C., Fidalgo, E.C.C., Scarano, F.R., Tubenchlak, F., Pena, I., Vicente-Vicente, J.L., Korys, K.A., Cooper, M., Fernandes, N.F., Prado, R.B., Maioli, V., Dib, V., & Teixeira, W.G. (2021). Systematic review of soil ecosystem services in tropical regions. *Philosophical Transactions Royal Society B, 8*, 201584.

Sauer, S., & Leite, S.P. (2012). Agrarian structure, foreign investment in land, and land prices in Brazil. *The Journal of Peasant Studies, 39*(4), 873–898.

Schroth, G., da Mota, M.D.S.S., Hills, T., Soto-Pinto, L., Wijayanto, I., Arief, C.W., & Zepeda, Y. (2011). Linking carbon, biodiversity and livelihoods near forest margins: the role of agroforestry. In B.M. Kumar & P.K.R. Nair (Eds.), *Carbon sequestration potential of agroforestry systems* (pp. 179–200). Springer.

Sillitoe, P. (1998). Knowing the land: Soil and land resource evaluation and Indigenous knowledge. *Soil Use and Management, 14*(4), 188–193.

Smith, P., Keesstra, S.D., Silver, W.L., & Adhya, T.K. (2021). The role of soils in delivering Nature's Contributions to People. *Philosophical Transactions Royal Society B, 376*, 20200169.

Smith, R. (1990). A transição no Brasil: A absolutização da propriedade fundiária. In R. Smith, *Propriedade da terra & transição: Estudo da formação da propriedade privada da terra e transição para o capitalismo no Brasil* (pp. 237–338). Brasiliense.

Steward, A.M., & Lima, D.M. (2017). "We also preserve": Quilombola defense of traditional plant management practices against preservationist bias in Mumbuca, Minas Gerais, Brazil. *Journal of Ethnobiology, 37*(1), 141–165.

Tavares, P.D. (2014). *Qualidade do solo em sistemas agroflorestais na Mata Atlântica* [Unpublished master's thesis]. Universidade Federal Rural do Rio de Janeiro.

Tavares, P.D., Silva, C.F., Pereira, M.G., Freo, V.A., Bieluczyk, W., & Silva, E.M.R. (2018). Soil quality under agroforestry systems and traditional agriculture in the Atlantic Forest biome. *Revista Caatinga, 31*(4), 954–962.

Terra de Direitos & Coordenação Nacional de Articulação das Comunidades Negras Rurais Quilombolas—CONAQ. (2018). Racismo e violência contra Quilombos no Brasil. *Terra de Direitos.*

Thorkildsen, K. (2014). Social-ecological changes in a Quilombola community in the Atlantic forest of Southeastern Brazil. *Human Ecology, 42*(6), 913–927.

Thorkildsen, K., & Kaarhus, R. (2017). The contested nature of Afro-descendant Quilombo land claims in Brazil. *The Journal of Peasant Studies, 46*(4), 792–810.

Tubenchlak, F., Badari, C.G., Strauch, G.F., Moraes, L.F.D. (2021). Changing the agriculture paradigm in the Brazilian Atlantic Forest: The importance of agroforestry. In M.C.M. Marques & C.E.V. Grelle (Eds.), *The Atlantic forest: History, biodiversity, threats and opportunities of the mega-diverse forest* (pp. 369–388). Springer.

Turetta, A.P.D., Pedrosa, B., Eufemia, L., Bonatti, M., & Sieber, S. (2020). Agriculture and supporting ecosystem services: A trade-off analysis using open data in Brazil. *Sustainable Agriculture Research*, *9*(1), 1–9.

Verchot, L.V., Van Noordwijk, M., Kandji, S., Tomich, T., Ong, C., Albrecht, A., Mackensen, J., Bantilan, C., Anupama, K.V., & Palm, C. (2007). Climate change: linking adaptation and mitigation through agroforestry. *Mitigation and Adaptation Strategies for Global Change*, *12*(5), 901–918.

Walker, B., Gunderson, L., Kinzig, A., Folke, C., Carpenter, S., & Schultz, L. (2006). A handful of heuristics and some propositions for understanding resilience in social-ecological systems. *Ecology and Society*, *11*(1), 13.

Waroux, Y.I.P., Garrett, R.D., Chapman, M., Friis, C., Hoelle, J., Hodel, L., Hopping, K., & Zaehringer, J.G. (2021). The role of culture in land system science. *Journal of Land Use Science*, *16*(4), 450–466.

Winklerprins, A.M.G.A. (1999). Insights and applications local soil knowledge: A tool for sustainable land management. *Society & Natural Resources*, *12*(2), 151–161.

Commons Governance and Climate Resilience: Intergovernmental Relationships in the Guapiruvu Community, Brazil

Aico Nogueira

What human beings seek to learn from nature is how to use it to wholly dominate both it and human beings. Nothing else counts. Ruthless toward itself, the Enlightenment has eradicated the last remnant of its own self-awareness. Only thought which does violence to itself is hard enough to shatter myths. (Horkheimer & Adorno, 2002, p. 2)

Introduction

The effects of climate change, and its differentiated impacts on distinct social groups, are increasingly documented across the world (Gardiner, 2011; Shepard & Corbin-Mark, 2009; Porter et al., 2020). In this context, developing countries, which already suffer from serious problems of inequitable income distribution, low levels of education, hunger and malnutrition, poor access to healthcare, and lack of infrastructure, are also the ones that suffer most from climate change, which deepens social inequalities and further exposes the gap between rich and poor. The concept of climate justice expresses an environmental justice response to climate change, contemplating the complex interconnections between environmental and social justice issues, and above all emphasizing the umbilical relationship between global warming and an economic system guided exclusively by growth. Studies on environmental justice movements emphasize how communities affected by climate change organize actions that mitigate its negative effects on people's lives, highlighting initiatives often developed in the interstices of society, in response to the negative effects of the current economic development model.

Examples include cooperation projects linking producers and consumers, fair trade arrangements, community gardens, alternative currencies, free open-source software, and many others, which proliferate in different parts of the world. They are based, above all, on a culture of cooperation, mutual support, shared responsibility, and cultural diversity, as well as social, economic, and environmental justice (Miller, 2010, p. 1). Among these initiatives, with particular reference to the rural areas covering most of the globe where nearly half the world's population lives, two things stand out: 1) the important role of associations and cooperativism in successful rural development projects (Develtere, 1998; Frantz, 2012; Pelegrini, Shiki, & Shiki, 2015); and 2) agro-ecology and alternative agricultural systems (Rosset, 2011; Rosset & Martínez-Torres, 2012; Wezel et al., 2009) as keys to asserting identities, safeguarding livelihoods, and defending disputed territories (Fernandes, 2008; Van der Ploeg, 2009).

Such initiatives have been particularly challenging for groups living in and around territorial areas that are protected by states for conservation or other reasons (Protected Areas or PAs), as they are usually subject to restrictive environmental laws that often have negative impacts on local people's traditional lifestyles (Andrade & Rhodes, 2012; Lane, 2001; Pretty & Smith, 2004; Wilshusen et al., 2002). In effect, PAs often reserve land-based ecological services for those living farther away from the territory, at the expense of the interests of those living closer (see Temper et al., 2020). In order to deal with these challenges, several studies have demonstrated the importance of strengthening community institutions, as a way of empowering local actors in decision-making processes to guarantee community autonomy, self-management, and access to common resources through effective inter-institutional dialogue with the official institutions that operate in and around protected areas. However, the real transformative potential of participation and empowerment of local groups has also been critically examined by researchers, who emphasize how this can decontextualize and over-simplify local social structures (Eversole, 2003; Henry, 2004; Loker, 2000; Sesan, 2014). These groups sometimes express their agency by subverting the proposed objectives of an official or outside-determined project, showcasing their ability to mobilize their identity relationships effectively around specific issues (Gilmour et al., 2013; Sampson et al., 1997; Durham et al., 1997; Newman & Dale, 2005; Nogueira, 2018). Through this process strategies are created, and advantage taken of political opportunities, in support of their own demands for development, which are not always in line with officially defined objectives. Not widely discussed in the literature is the way some groups develop the ability to incorporate sustainability narratives in order to strengthen their dialogue with other levels of governance, eventually becoming an instrument of compliance and reproduction of the dominant agrifood or other regime. This in effect subverts or subsumes their locally grounded traditional governance, and cultural and risk-reduction strategies based in collectivism, mutual aid, and sustainable agricultural practices.

To explore these complex issues, I have conducted research in Vale do Ribeira in the State of São Paulo, Brazil (see Map 2, page 30). This area has been under various forms of environmental protection since the 1950s, as it comprises the main contiguous areas of Atlantic Rainforest remaining in the country.

My case study focuses on the Guapiruvu community in the Municipality of Sete Barras, where over the last thirty years environmental challenges and the implementation of two large conservation areas neighbouring the community (all in the context of an ongoing struggle for land ownership) have guided social processes of development. The Guapiruvu community, which includes people who have lived there for hundreds of years, has built local social organizations capable of establishing effective dialogue among themselves and with other levels of governance operating in the area; it is recognized as an effective community working towards its own development (Bernini, 2009; Grigoletto, 2018; Valentin, 2006).

In this chapter, I explore this story and how participatory research in the community has allowed me to include the viewpoints of many different community members in a relatively complex process of rural transition. The use of "environmental" discourse by some community leaders has allowed initially contentious relations with state agencies to be gradually converted into a more cooperative relationship. These community leaders' claims to be transitioning from conventional agriculture to agro-ecology, to take advantage of a niche urban market for agroecologically grown produce, has come to represent the peak of this process—a strategy that has become the foundation of some local organizations' actions.

In contrast, most community members are committed to continuing the reproduction of the dominant agrifood regime, by producing for local markets where low prices are more important than an "organic" designation or agroecological production processes. Even for these farmers, the community is part of a new process of legitimation and consolidation of short, sustainable circuits of production, commercialization, and consumption. These circuits are extremely important in times of climate change for promoting agro-sustainable production, and for consolidating new patterns of responsible consumption, especially in and around large urban areas, such as the city of São Paulo (Bava, 2012; Feenstra, 2002).

This transition, with all its complexities, relies on the community's strong social capital and the underlying commitment of its leaders to environmentally sustainable processes, with resulting benefits for the community in terms of pollution control, income generation, education, health, and infrastructure. The political strategies constructed by the community rely on their ability to communicate and organize (Levidow et al., 2014; Smith & Raven, 2012). This shows the importance of local groups' internal structures in formulating public policies and dealing with higher levels of government, as well as the agency of local people and leaders in response to a lack of support and leadership from other public authorities.

My research draws on recent theoretical perspectives on sustainability transitions, agro-ecology, food security, multilevel governance/inter-institutional dialogue, and the participation of local communities in the management of common natural resources, especially in and near protected areas. My field work was carried out using qualitative methods, with primary and secondary data collected between September 2019 and April 2020 from three sources: published government documents and academic works, informal and semi-structured interviews, and field observations in Guapiruvu.

During visits and interviews with Guapiruvu residents and through documentary research about the community, I attempted to observe the internal organization and relations of the community with local society, society as a whole, and formal and informal institutions structuring cooperation and conflicts.

I conducted semi-structured interviews with representatives of federal, and state-government organizations present in the community, namely the National Institute of Colonization and Agrarian Reform (INCRA), the Fundação Florestal/Forest Foundation (FF) and the Intervales State Park (PEI). These interviews were organized around the following themes: 1) inter-institutional dialogue at the local level and with other levels of government, 2) the main obstacles faced by the stakeholders in these discussions/processes, 3) the main obstacles faced by the stakeholders in implementing policies, 4) the main advances and challenges in the process of converting traditional agriculture to sustainable agro-ecological systems in the territory.

Protected Areas (PAs), Local Communities, and Agro-Ecology

Many areas sensitive to biodiversity loss and in need of conservation are also areas of high social vulnerability. They are generally characterized by elevated levels of poverty, repressive and unstable anti-democratic regimes, and problems linked to the struggle for land tenure (Brechin et al., 2002; Myers, 1988; Myers et al., 2000; Brüggemann et al., 1997). Often, ecologically sustainable human-land inter-relationship systems, sometimes evolved over millennia by Indigenous peoples, are under pressure from "outside" populations, extraction, and political considerations. Furthermore, these areas are frequently arenas of conflict (Ostrom, 2005), with disputes between groups representing such diverse interests as tourism, mineral and oil exploration companies, guerrilla groups, and drug cartels (Brechin et al., 2002). Such factors make these spaces a complex mixture of social, economic, and political disputes, which present further challenges for the management of environmental conservation programs. The question of who has access and rights in such spaces is therefore central.

There is a vast literature dealing with the often-contentious relationship between the management of PAs and the populations living in and around them. Researchers focus mainly on the impacts caused by conservation programs and policies on the traditional ways of life of local people, particularly regarding changes to their access to natural resources (Andrade & Rhodes, 2012; Bennett et al., 2017; Bernini 2009; Brüggemann et al. 1997; Chape et al., 2008; García-Frapolli et al., 2009; Pretty & Smith, 2004). The frequent prohibition of communities' access to important natural resources, and even the removal of some of these groups from their lands, has in many cases harmed rather than helped these communities, which sometimes brings the conservation programs into question (Anthony, 2007; Hamilton et al., 2000; Jim & Xu, 2002; Lane 2001).

In developing countries where there is unequal land distribution and ownership, the rural population's restricted access to resources such as water, land, energy, and environmental services builds pressure on these resources, driving social conflicts. The wealth these resources generate is often appropriated by a limited number of actors, further widening social inequities. Resource inaccessibility leads to environmental degradation in areas where local populations do have access, and to increasing inequality, constituting a persistent source of instability, and demonstrating the strong relationship between equity and sustainability (Guzmán Casado et al., 2000), which depends "critically on the institutional settings that structure interactions among agents" (Baland et al., 2018, p. 8).

Thus, the importance of local institutions, the participation of local actors in the management and conservation of biodiversity, and the transition to sustainable societies are increasingly recognized in the literature (Hagedorn, 2015; Ostrom, 1990, 2005; Pretty & Smith, 2004). These analyses show the difficulties faced when local communities are not co-participants in conservation processes (Andrade & Rhodes, 2012; Anthony, 2007; Grainger, 2003; Pretty & Smith, 2004). In these studies, especially those by Elinor Ostrom (1990, 2001, 2009a, 2009b, 2010) and her adherents, existing social dynamics, and processes that either allow or hamper the construction of appropriate institutional arrangements, designed to manage shared natural resources, have been identified in many places (Leroy, 2016; Perkins et al., 2017; Santana & Fontes Filho, 2010). However, the degree of participation of local populations in governance, as a way of ensuring better compliance with conservation policies (Wilshusen et al., 2002), and the factors that most influence communities' agreement with these actions, have been attributed generally to local specificities, especially the communities' capacity to engage in inter-institutional dialogue. The capacity of local actors to engage in discussions across levels of government is crucial for conflict resolution, especially due to the lack of legitimacy that external regulations may have, as they are often contrary to the customary practices of traditional communities (Brechin et al., 2002). This ability to enter dialogue is at the basis of resolving conflicts in governance, which are seen as "processes of interaction and decision-making among the actors involved in a collective problem that led to the creation, reinforcement, or reproduction of social norms and institutions" (Hufty, 2011, p. 405). Although authors also recognize difficulties, mainly due to the multiple power relations that may exist in these communities, and the great heterogeneity of the groups involved in terms of class, ethnicity, and religious and political orientation, they point to the importance of incorporating governance diversity in conservation initiatives (Brechin et al., 2002, Ostrom 1990, Ostrom et al., 1994).

The Guapiruvu Community and Environmental Issues in Vale do Ribeira

Guapiruvu is in the Vale do Ribeira, a remote area strongly marked by the presence of conservation units and restrictive environmental laws. The pioneer settler families of the community, the Alves, Teixeira, and Pereira families, have struggled for recognition of their ownership rights on land they have occupied for more than one hundred years. The area is located in the buffer zone of a large state park, the Alto Ribeiro State Park.

In 1996, Guapiruvu was recognized by the non-governmental organization (NGO) Vitae Civilis[1] for its leadership related to disadvantaged groups in the area, especially dispossessed families, and the community was selected to lead an Agenda 21[2] pilot project to create local solutions for global socio-environmental problems.[3] One of the first initiatives was the creation of the Solidarity Economy and Sustainable Development Association of Guapiruvu, known as AGUA, in 1997.

The launch of Agenda 21, in 1998, also led to a closer relationship between the Guapiruvu community and public authorities, addressing provision of basic services that are theoretically guaranteed by law, such as income-generation projects and activities related to eco-tourism and environmental preservation. Following its creation in 1997, AGUA started a series of programs such as eco-tourism activities, production and commercialization of medicinal plants, courses on agroforestry, support for the creation of the municipal secretariat for rural development, the mapping of tourist trails in the PEI, the creation of guided activities, and fundraising from various sources for activities aimed at environmental sustainability. In 2000, AGUA started supporting the creation of a rural settlement in the area, where the community's colonial history could be recognized (Grigoletto, 2018). AGUA was thus responsible for bridging the gap with other institutions outside the community, such as local public authorities and the agencies of the federal and state government, which allowed for the formulation of public policy demands and support for sustainable development in the area. Also in 2000, AGUA, with the support of the Forestry Institute and Vitae Civilis, presented a proposal to INCRA to create a sustainable development project (PDS) in the area, using alternative forms of rural settlement developed by INCRA in the Amazon region to mitigate land conflicts (Paula & Silva, 2008).

Proximity to the park largely determines the community's relationship with the environment and its forms of local social organization, profoundly impacting the traditional practices of the local groups, as they are prevented from making their livelihood from the protected forest and land. Access to traditional resources has been limited by checkpoints and inspections carried out by the police inside and outside the park, seeking to prevent poaching of prohibited species and animal-hunting, especially the illegal extraction of *juçara* (heart of palm, *Euterpe edulis*) for family consumption and mainly for sale. Given the importance of the *juçara* tree, whose fruits are essential for the diet of birds and mammals in the Brazilian Atlantic Rainforest ecosystem, and due to the fact that after the extraction of the heart of palm the tree is totally discarded and does not regenerate, its removal became an environmental crime in Brazil. Some local residents who had depended on heart-of-palm extraction became targets of repression and even arrests (Bernini, 2009), while also deprived of one of their main means of subsistence.

In Guapiruvu, the interaction between local institutions and federal and state bodies happens through the various official agencies representing the community. At the state level, the main regulatory body for the conservation units is the FF of the State of São Paulo. It also manages the PEI, and its remit, as stated in its management plan, is that it "establishes specific rules regulating the occupation and use of land in its buffer zone and suggests ways to integrate the unit into the Continuum of Paranapiacaba[4]; promoting the socioeconomic integration of the surrounding communities and valuing their traditional knowledge as principles of governance" (Furlan et al., 2008). Federal actions in the community are carried out by INCRA, the agency responsible for the division of plots, selection and settlement of families, land credit, construction of houses, opening of roads, electricity, and technical assistance in the rural settlements. Once settled, the families in the community cannot sell, lease, rent, lend, or give the plots to private individuals.

Community's Socio-Productive Structure, Agro-Ecology, and Interinstitutional Dialogue

As indicated by the classic work of Ostrom (1998), the cooperation mechanisms and the internal structuring of the community are key to the communication channels built by the subjects with other institutional levels. Hence, to understand how the process of internal community organization and dialogue with other institutions take place in Guapiruvu, it is important to analyse the community's leadership.

An analysis of the narratives collected from the community, as an essential source of shared mental representations (Hoff & Walsh, 2018), revealed subtle aspects of the existing social classification system shaping local organizations and determining leaders. Within this structure, six basic criteria are used by community inhabitants to mentally categorize each other within the community and to allocate everyone to a cognitive model that works not only to order, rank, and map each person in the broader group, but also to guide their likely reactions to specific situations. These criteria are: 1) whether people are born in the district (insiders or outsiders), 2) their socio-economic level (class), 3) their educational level, 4) the size of their property, 5) whether the agrarian reform allows them to be "settled" or not, 6) whether they use conventional agrarian practices or support a move towards sustainable development.

Two groups of leaders stand out in the community. On the one hand, there are those who are considered outsiders, meaning they were not born in the district and have no links to the pioneer families in the area, but instead acquired lands more recently and are linked, above all, to large banana producers in the region. They tend to have a higher economic, educational, and cultural level, and support social inclusion and agro-ecological transition. On the other hand, there are other leaders who are natives of Guapiruvu, generally have lower socio-economic, social, and educational levels, were mainly settled through the agrarian reform, and are thus part of the largest portion of the community's population. They tend to advocate for increasing investment in traditional agriculture and strongly criticize the high costs of organic production, lack of government support for farming activities, and absence of nearby markets.

The community's local institutions end up expressing not only the interests of these specific groups, but also the socio-educational and economic divisions of the neighbourhood and different views of development. AGUA is the locus of action and expression of ideas led by the local "elite," and COOPERAGUA is the space controlled by the poorest, oriented toward the consolidation and reproduction of conventional forms of development.

Although the first group is a minority and is composed of "outsiders" in the community, they are responsible for much of the local social organization and agro-ecological production. In addition, they are the main agents of interaction with higher-level government structures, and the main agents of the community's resilience, ecological transition, and environmental justice.

AGUA became responsible for the commercialization of the neighbourhood's organic production and contributed greatly to setting up a system of selling the family agricultural organic products of the town of Sete Barras and integrating it with the growing alternative agri-food systems in large urban centres. COOPERAGUA, on the other hand, is responsible for marketing the community's traditional agricultural production. With COOPERAGUA as a model, and with the support of the municipal council for rural development of Sete Barras, in 2011 the Family Agriculture Cooperative of Sete Barras (COOPAFASB) was created. Its objective is to promote the solidarity economy, inspired by the principles of self-management, cooperation, economic viability, equal relations, and sustainability (Singer, 2002, 2008), by seeking market opportunities and supplying products to institutional and conventional markets.

Conclusion

The literature on transitions from current models of conventional rural development and agriculture to more sustainable rural development emphasizes the vital role of the state in facilitating this process.

However, this study shows that in the presence of elements such as local capacity for inter-institutional dialogue, social capital, and community agency (regardless of the community's socio-economic and cultural divisions), people can overcome the obstacles brought about by the absence of official support while creating alternatives for the production and marketing of agro-sustainable products.

The experience of agro-ecological transition initiated in Guapiruvu surpassed the limits of the community, influencing sustainable agriculture practices in the broader municipality and contributing to the strengthening of an agrifood system that transcends Sete Barras, extending to the niches of consumer markets in large urban centres in the state of São Paulo.

In Guapiruvu, strong community social capital and agency, combined with an efficient appropriation of sustainability discourse, acts to reduce conflict, and facilitate inter-institutional dialogue. However, the community's socio-economic and cultural divisions make local institutions a reflection of these internal separations, whose actions result in a double movement. On the one hand, the community subscribes to conventional patterns of production and commercialization through growth and strong insertion in the markets; on the other hand, it also expresses resistance to the deepening of market forces, as stated by Polanyi (1980).

Examples such as the Guapiruvu community show us the creative power of local groups to promote environmental justice and social inclusion, amidst the uncertainties and adversities arising from climate change and an absence of government support for sustainable development initiatives. One way to overcome these problems may lie not in the easiest and most immediate option, conventional agriculture, but in a process of changes to sustainable production, marketing, and consumption practices based on rural/urban partnerships—social solidarity.

The interdisciplinary and participatory research approach, through collective self-reflection, cooperation, and participation, associated with ethnographic research, semi-structured interviews, document analysis, and focus groups, was fundamental in obtaining this understanding of the complex situation in Guapiruvu. This approach allowed for inclusion of local social processes, which are crucially important in commons theory. Local people don't often have an opportunity to reflect or comment on their own social processes, such as the complex networks of local social classification and their effects on the management of local social organizations, which centrally determine the community's ongoing socio-economic-ecological transition.

NOTES

1 For an overview of Vitae Civilis' work in Guapiruvu, see the video: https://www. youtube.com/watch?v=q3n53Hg3X-k.

2 See: https://acervo.socioambiental.org/sites/default/files/documents/22D00056.pdf.

3 See: https://sustainabledevelopment.un.org/content/documents/Agenda21.pdf.

4 The Ecological Continuum of Paranapiacaba is an Atlantic Forest corridor of more than 120,000 hectares, formed by the Intervales Park, the Carlos Botelho State Park, the Alto Ribeira Tourist State Park (PETAR), the Xitué Ecological Station, the Serra do Mar Environmental Protection Area, and the Atlantic Forest Biosphere Reserve.

Reference List

Andrade, G., & Rhodes, J. (2012). Protected areas and local communities: An inevitable partnership toward successful conservation strategies? *Ecology and Society, 17*(4) , art. 14.

Anthony, B. (2007). The dual nature of parks: Attitudes of neighbouring communities towards Kruger National Park, South Africa. *Environmental Conservation, 34*(3), 236–245.

Baland, J.-M., Bardhan, P., & Bowles, S. (2018). *Inequality, cooperation, and environmental sustainability.* Princeton University Press.

Bava, S.C. (2012). Circuitos curtos de produção e consumo. In *Um campeão visto de perto—Uma análise do modelo de desenvolvimento Brasileiro.* Heinrich Böll Stiftung.

Bennett, N.J., Roth, R., Klain, S.C., Chan, K., Christie, P., Clark, D.A., Cullman, G., Curran, D., Durbin, T.J., Epstein, G., Greenberg, A., Nelson, M.P., Sandlos, J., Stedman, R., Teel, T.L., Thomas, R., Verissimo, D., & Wyborn, C. (2017). Conservation social science: Understanding and integrating human dimensions to improve conservation. *Biological Conservation, 205*, 93–108.

Bernini, C.I. (2009). *De posseiro a assentado: A reinvenção da comunidade do Guapiruvu na construção contraditória do assentamento agroambiental Alves, Teixeira e Pereira, Sete Barras-SP* [Unpublished master's thesis]. University of São Paulo.

Brechin, S.R., Wilshusen, P.R., Fortwangler, C.L., & West, P.C. (2002). Beyond the square wheel: Toward a more comprehensive understanding of biodiversity conservation as social and political process. *Society & Natural Resources, 15*(1), 41–64.

Brüggemann, J., Ghimire, K.B., & Pimbert, M.P. 1997. *Social change and conservation: Environmental politics and impacts of national parks and protected areas.* Earthscan.

Chape, S., Spalding, M., Taylor, M., Putney, A., Ishwaran, N., Thorsell, J., Blasco, D., Vernhes, J.R., Bridgewater, P., Harrison, J., & McManus, E. (2008). History, definitions, value and global perspective. In S. Chape, M. Spalding, & M. Jenkins

(Eds.), *The world's protected areas—Status, values and prospect in the 21st century* (pp. 1–35). University of California Press.

Develtere, P. (1998). *Économie sociale et développement: Les coopératives, mutuelles et associations dans les pays en développement*. De Boeck Université.

Durham, C.C., Knight, D., & Locke, E.A. (1997). Effects of leader role, team-set goal difficulty, efficacy, and tactics on team effectiveness. *Organizational Behavior and Human Decision Processes, 72*(2), 203–231.

Eversole, R. (2003). Managing the pitfalls of participatory development: Some insight from Australia. *World Development, 31*(5), 781–795.

Feenstra, G. (2002). Creating space for sustainable food systems: Lessons from the field. *Agriculture and Human Values, 19*(2), 99–106.

Fernandes, B.M. (2008). Questão agrária: conflitualidade e desenvolvimento territorial. In A. M. Buainain (Ed.), *Luta pela terra, reforma agraria e gestão de conflitos no Brasil* (pp. 173–224). Editora Unicamp.

Frantz, W. (2012). *Associativismo, cooperativismo e economia solidária*. Editora Unijuí.

Furlan, S.A., Leite, S.A., Marinho, M.d.A., & Leonel, C. (2008). *Plano de manejo: Parque Estadual Intervales*. Fundação Florestal.

García-Frapolli, E., Ramos-Fernández, G., Galicia, E., & Serrano, A. (2009). The complex reality of biodiversity conservation through Natural Protected Area policy: Three cases from the Yucatan Peninsula, Mexico. *Land Use Policy, 26*(3), 715–722.

Gardiner, S.M. (2011). Climate justice. In J.S. Dryzek, R.B. Norgaard, & D. Schlosberg (Eds.), *The Oxford handbook of climate change and society* (pp. 309–322). Oxford University Press.

Gilmour, P.W., Dwyer, P.D., & Day, R.W. (2013). Enhancing the agency of fishers: A conceptual model of self-management in Australian abalone fisheries. *Marine Policy, 37*, 165–175.

Grainger, J. (2003). "People are living in the park." Linking biodiversity conservation to community development in the Middle East region: a case study from the Saint Katherine Protectorate, Southern Sinai. *Journal of Arid Environments, 54*(1), 29–38.

Grigoletto, F. (2018). *O bairro Guapiruvu como lugar-organização: Uma abordagem institucional do organizar*. Fundação Getúlio Vargas—Escola de Administração de Empresas de São Paulo.

Guzmán Casado, G., González de Molina, M., & Sevilla Guzmán, E. (2000). *Introducción a la agroecología como desarrollo rural sostenible*. Ediciones Mundi-Prensa.

Hagedorn, K. (2015). Can the concept of integrative and segregative institutions contribute to the framing of institutions of sustainability? *Sustainability, 7*(1), 584–611.

Hamilton, A., Cunningham, A., Byaruguba, D., & Kayanja, F. (2000). Conservation in a region of political instability: Bwindi Impenetrable Forest, Uganda. *Conservation Biology, 14*(6), 1722–1725.

Henry, L. (2004). Morality, citizenship and participatory development in an Indigenous development association: The case of GPSDO and the Sebat Bet Gurage of Ethiopia. In S. Hickey & G. Mohan (Eds.), *Participation: From tyranny to transformation? Exploring new approaches to participation in development* (pp. 140–156). Zed Books.

Hoff, K., & Walsh, J. (2018). The whys of social exclusion: Insights from behavioral economics. *The World Bank Research Observer, 33*(1), 1–33.

Horkheimer, M., & Adorno, T.W. (2002). *Dialectic of enlightenment.* Stanford University Press.

Hufty, M. (2011). Investigating policy processes: The governance analytical framework (GAF). In U. Wiesmann & H. Hurni (Eds.), *Research for sustainable development: Foundations, experiences, and perspectives,* (pp. 403–424). Geographica Bernensia.

Jim, C.Y., & Xu, S.S.W. (2002). Stifled stakeholders and subdued participation: Interpreting local responses toward Shimentai Nature Reserve in South China. *Environmental Management, 30*(3), 327–341.

Lane, M.B. (2001). Affirming new directions in planning theory: Comanagement of protected areas. *Society & Natural Resources, 14*(8), 657–671.

Leroy, J.P. (2016). Mercado ou bens comuns: O papel dos povos indígenas, comunidades tradicionais e setores do campesinato diante da crise ambiental. FASE.

Levidow, L., Pimbert, M., & Vanloqueren, G. (2014). Agroecological research: Conforming-or transforming the dominant agro-food regime? *Agroecology and Sustainable Food Systems, 38*(10), 1127–1155.

Loker, W.M. (2000). Sowing discord, planting doubts: Rhetoric and reality in an environment and development project in Honduras. *Human Organization, 59*(3), 300–310.

Miller, E. (2010). Solidarity economy: Key concepts and issues. In E. Kawano, T. Masterson, & J. Teller-Ellsberg (Eds.), *Solidarity economy I: Building alternatives for people and planet* (pp. 25–41). Center for Popular Economics.

Myers, N. 1988. Threatened biotas: "Hot spots" in tropical forests. *Environmentalist, 8*(3), 187–208.

Myers, N., Mittermeier, R.A., Mittermeier, C.G., Da Fonseca, G.A.B., & Kent, J. (2000). Biodiversity hotspots for conservation priorities. *Nature, 403*(6772), 853.

Newman, L., & Dale, A. (2005). The role of agency in sustainable local community development. *Local Environment, 10*(5), 477–486.

Nogueira, A.S. (2018). Institutionalization of rural social movements in the Lula government and the decline of land reform in Brazil: Co-option, political identity and agency. *Análise Social, 53*(227), 362–387.

Ostrom, E. (1990). *Governing the commons: The evolution of institutions for collective action.* Cambridge University Press.

Ostrom, E., Gardner, R. & Walker, J. (1994). *Rules, games, and common-pool resources.* University of Michigan Press.

Ostrom, E. (1998). A behavioral approach to the rational choice theory of collective action: Presidential address, American Political Science Association, 1997. *American Political Science Review, 92*(1), 1–22.

Ostrom, E. (2001). Commons, institutional diversity of. In S.A. Levin (Ed.), *Encyclopedia of biodiversity* (Vol. 1, pp. 777–791). Academic Press.

Ostrom, E. (2005). *Understanding institutional diversity.* Princeton University Press.

Ostrom, E. (2009a). A general framework for analyzing sustainability of social-ecological systems. *Science, 325*(5939), 419–422.

Ostrom, E. (2009b). A polycentric approach for coping with climate change. *The World Bank Policy Research Paper no 5095.* https://core.ac.uk/download/pdf/6305219.pdf

Ostrom, E. (2010). Beyond markets and states: polycentric governance of complex economic systems. *American Economic Review, 100*(3), 641–672.

Paula, E.A.d. & Silva, S.S.d. (2008). Floresta, para que te quero? Da territorialização camponesa a nova territorialidade do capital. *Revista Nera, 12*(11), 86–97.

Pelegrini, D.F., Shiki, S.d.F.N., & Shiki, S. (2015). Uma abordagem teórica sobre cooperativismo e associativismo no Brasil. *Extensio: Revista Eletrônica de Extensão, 12*(19), 70–85.

Perkins, P.E., Cesar, M., dos Santos, N. B., Bohn, S., & Luna, I. (2017). *Brazil's traditional and new commons.* [Unpublished manuscript]

Polanyi, K. (1980). *A grande transformação: As origens de nossa época.* Editora Campus.

Porter, L., Rickards, L., Verlie, B., Bosomworth, K., Moloney, S., Lay, B., Latham, B., Anguelovski, I., & Pellow, D. (2020). Climate justice in a climate changed world. *Planning Theory & Practice, 21*(2), 293–321.

Pretty, J., & Smith, D. (2004). Social capital in biodiversity conservation and management. *Conservation Biology, 18*(3), 631–638.

Rosset, P. (2011). Food sovereignty and alternative paradigms to confront land grabbing and the food and climate crises. *Development, 54*(1), 21–30.

Rosset, P.M. & Martínez-Torres, M.E. (2012). Rural social movements and agroecology: Context, theory, and process. *Ecology and Society, 17*(3).

Sampson, R.J., Raudenbush, S.W., & Earls, F. (1997). Neighborhoods and violent crime: A multilevel study of collective efficacy. *Science, 277*, 918–924.

Santana, V. & Fontes Filho, J. (2010). *Elementos de Gestão Local: A perspectiva de Elinor Ostrom aplicada ao Parque Estadual da Ilha do Cardoso* [Paper presentation]. Encontro de administração pública, Vitória, ES.

Sesan, T. (2014). Peeling back the layers on participatory development: Evidence from a community-based women's group in Western Kenya. *Community Development Journal, 49*(4), 603–617.

Shepard, P.M. & Corbin-Mark, C. (2009). Climate justice. *Environmental Justice, 2*(4), 163–166.

Singer, P. (2002). *A recente ressurreição da economia solidária no Brasil*. Civilização Brasileira.

Singer, P. (2008). Economia solidária. *Estudos avançados, 22*(62), 289–314.

Smith, A. & Raven, R. (2012). What is protective space? Reconsidering niches in transitions to sustainability. *Research Policy, 41*(6), 1025–1036.

Temper, L., Avila, S., Del Bene, D., Gobby, J., Kosoy, N., Le Billon, P., Martinez-Alier, J., Perkins, P., Roy, B., Scheidel, A., & Walter, M. (2020). Movements shaping climate futures: A systematic mapping of protests against fossil fuel and low-carbon energy projects. *Environmental Research Letters, 15*(12), 123004. https://doi.org/10.1088/1748-9326/abc197

Valentin, A. (2006). *Uma civilização do arroz: Agricultura, comércio e subsistência no Vale do Ribeira (1800–1880)* [Unpublished doctoral dissertation]. Universidade São Paulo.

Van der Ploeg, J.D. (2009). *The new peasantries: Struggles for autonomy and sustainability in an era of empire and globalization*. Earthscan.

Wezel, A., Bellon, S., Doré, T., Francis, C., Vallod, D., & David, C. (2009). Agroecology as a science, a movement and a practice: A review. *Agronomy for Sustainable Development, 29*(4), 503–515.

Wilshusen, P.R., Brechin, S.R., Fortwangler, C.L., & West, P.C. (2002). Reinventing a square wheel: Critique of a resurgent "protection paradigm" in international biodiversity conservation. *Society & Natural Resources, 15*(1), 17–40.

PART III

Water and Fisheries Commons

Mining and Water Insecurity in Brazil: Geo-Participatory Dam Mapping (MapGD) and Community Empowerment

Daniela Campolina and Lussandra Martins Gianasi[1]

Introduction: Environmental and Climate (In)Justice, Mining, and Water (In)Security

Brazil (see Map 2, page 30) can be considered a water power, given that it possesses 12 per cent of the total available fresh water on the planet, 90 per cent of its rivers are perennial, and 90 per cent of its territory receives regular rainfall. Brazil houses several aquifers, including the Guarani aquifer,[2] as well as large extensions of important planetary wetlands, including the extensive ecosystems of the Pantanal and Amazon. Brazil also houses a major portion of the biggest watershed in the world—the watershed of the Amazon River (Rebouças et al., 2002). Yet despite this apparent abundance, water is not evenly distributed in all states and cities. Moreover, as a country of huge size, Brazil encompasses regions with great water wealth while other regions experience water scarcity. In addition, due to poor management and usage of water, supply problems are increasing in urban centres. In Brazil's major cities, with high demand for water and patterns of land occupation and use that disregard impacts on the watersheds and waterways, there is increasing scarcity of fresh, potable water. Another situation of concern in Brazil with regard to water is the major mining disasters provoked by ruptures of mine tailings dams that have occurred in recent years, killing hundreds of people and contaminating entire watersheds.

In this chapter, we show how water-related environmental injustices in Brazil are worsening, due in part to climate change. These climate injustices are predictable, resulting as they do from a combination of overt government policy, the inefficiency of the Brazilian government in its implementation of management and inspection systems, corporate impunity, and private-sector interventions to sway public opinion. We focus on situations in Brazil in which mining activity impacts water quantity (destroying areas of water storage and replenishment) and water quality (through mine tailings dam disasters). Mining impacts the water security of thousands of people. Many of these people do not have the slightest notion of the risks and violations associated with mining in terms of their right to water access. Many regions affected by mining are far removed from the actual site of extraction, meaning that mining is invisible as a component of people's daily economic reality. The mining companies themselves are promoters of environmental injustices, especially in times of climate change-related rainfall events which worsen disasters like breaches of mine tailings dams.

Fighting such planned climate injustice requires naming and exposing it, combatting corporate obfuscation and government failures through public education, organizing politically, and building international solidarity. We describe some movements and methods that are part of this struggle, based in our own experience as participatory researchers and educators in Minas Gerais, Brazil.

The following section of this chapter overviews recent mine-related water disasters in Brazil and their roots in regulatory and enforcement failures. Section three shows how this mismanagement is organized and planned, using disinformation as a concerted strategy. In section four we describe ways of countering disinformation and strengthening local awareness of climate injustices and risks, such as geo-participatory mapping. The chapter's conclusion situates popular education, organizing, and global solidarity as the political context for mining and water-related climate justice.

Mine Disasters: Climate Injustice Produced by Regulatory and Enforcement Failures

In November 2015, the Samarco mining company's Fundão tailings dam ruptured. Samarco is located in Mariana in the state of Minas Gerais and is jointly owned by multinationals Vale S.A. and BHP Billiton. The Fundão dam contained a volume of about 60 million m³ of toxic mud tailings. The impact of the spill, however, went far beyond the mine site, with a flow path extending for more than 600 km along the Doce River system until it reached the Atlantic Ocean. On arrival at the coastline, it travelled several kilometers out into the ocean and affected areas along 80 km of the Brazilian coast. Throughout this journey, the spill of toxic waste affected thirty-nine municipalities in the states of Minas Gerais and Espírito Santo (ES).

Just over three years later, in January 2019, the dam at the Córrego do Feijão mine collapsed in the city of Brumadinho, also in Minas Gerais state. This mine was owned solely by Vale S.A., with a tailings dam holding a volume of 12 million m³ of mine waste. This spill extended along 300 km of the Paraopeba River, a tributary of the Sao Francisco River, one of the longest rivers in Brazil (Zonta & Trocante, 2016; Pinheiro et al., 2019).

In addition to being among the largest in the world in terms of tailings volume, the Fundão dam in Mariana was also the most extensive in the world. The Brumadinho disaster had the second largest number of fatalities, and the highest number of workplace fatalities of the twenty-first century (Zonta & Trocante, 2016; Wanderley et al., 2016), in addition to being the largest "workplace accident" in Brazilian history (Espindola & Guimarães, 2019). These were not the first tailings dam collapses to occur in Brazil and, it seems, will likely not be the last (Zonta & Trocante, 2016; Pinheiro et al., 2019; Campolina, 2021; Campolina, Gianasi, et al., 2021).

A total of 291 people lost their lives in these two recent disasters (19 in the Samarco-Vale-BHP disaster and 272 in the Vale S.A. disaster). Two important river basins were destroyed, resulting in a variety of impacts on ecosystems, public health, and economic activities throughout the various municipalities and in the region as a whole. Shortly after the collapse of the dams, one of the problems immediately identified was the quantity and quality of water supply to several urban areas. This was due to contamination, especially in the Paraopeba River. Even three years later, at the beginning of 2022, environmental agencies were still recommending that the water not be used for drinking, animal watering, fishing, leisure activities, or gardening. Most of the affected population depended on the river and its resources for survival, meaning that their food supply and economic security was severely impacted. This population, already vulnerable in socio-economic terms, has experienced various situations that violate their basic rights since the dam collapsed, especially their right of access to water.

Many of these communities were unaware that there were mine tailings dams located upstream from their cities. They had no idea of their risks in the event of a possible rupture. Even more alarming is the fact that, upstream from these same populations, there are dozens more tailings dams, some of them also operating at critical safety levels, categorized as "high risk" by government monitoring authorities.

We can therefore see that millions of people in Brazil are experiencing situations of environmental injustice. Many others risk being victims of future tailings dam ruptures, especially taking into consideration climate change scenarios with predictions of increasing extreme weather events in regions where the dams are concentrated. According to Milanez and Fonseca (2011, pp. 93–94), the concept of "climate justice" emerges as an integral part of the paradigm of "environmental justice." Given that existing social inequalities define a social group's degree of exposure to environmental risks, it becomes clear that the impacts of climate change affect particular social groups with differing forms and degrees of intensity.

Acselrad, Mello, and Bezerra (2009, p.9) conceptualize environmental injustice as the "phenomenon of disproportionate imposition of environmental risks on populations less endowed with financial, political and informational resources." The authors, when observing the mechanisms that lead to the production of environmental injustice, start from the assumption that environmental inequality manifests itself in two ways: unequal access to environmental resources and unequal environmental protection.

Unequal access to environmental resources can occur in the spheres of both *production* and *consumption*. While the sphere of *consumption* refers to access to natural resources that are already transformed into manufactured goods, the sphere of *production* relates to different ways of appropriating nature for creation of the basis for sustaining life itself (Acselrad et al., 2009, p. 73). In relation to *production*, then, what we see is the continuous destruction of non-capitalist ways of appropriating nature, such as artisanal fishing, family farming, or a "commons" of shared resources. Diverse territories are affected by the environmental impacts arising from large enterprises implanted in frontier areas where capitalism is expanding. Monocultures, dams and mining enclaves create major destabilizing effects on activities carried out on traditionally occupied lands, destroying the resource base that sustains such forms of life (Acselrad et al., 2009).

This is also reflected in the unequal environmental protection that emerges as environmental policies are implemented—or the omission of such policies as a result of neglect and/or action by market forces. All of this generates disproportionate environmental risks, intentional or unintentional, for the most vulnerable. The vulnerable are characterized as lacking financial and political resources: they are among the poorest of the poor, least covered by public policies, residents of devalued areas and of marginalized ethnicities. This unequal exposure to environmental risks and impacts does not result "from any natural condition, geographic determination or historical causality, but from social and political processes that unequally distribute environmental protection" (Acselrad et al., 2009, p. 73).

With respect to environmental protection, it is worth highlighting the relatively recent achievements in Brazilian legislation, both in relation to water management and dam safety. These may be threatened, however, by a proposal for a New Brazilian Mining Code, discussion of which began in December 2021.

Brazil's National Environmental Policy (PNMA, Law L6938 9) was adopted in 1981 and established legal instruments to monitor environmental impacts. Enterprises that generate substantial environmental impacts must prepare and submit Environmental Impact Studies and Environmental Impact Reports (EIA/RIMA), as a requirement in the licensing process. The EIA/RIMA encompasses both environmental and socio-economic impacts. Its main objective is to orient inspection bodies and affected communities with regard to the type of project to be carried out and feed into their decision-making on the feasibility of issuing an environmental license.

With regard to water management, the country has recently moved from a model of centralizing legislation—which gave priority for water use to the energy and industrial sector—to a proposal to build democratic management of water. In 1997, a National Water Resources Policy (PNRH, or Política Nacional de Recursos Hídricos) was instituted in Brazil, known as the "Water Law." According to the Water Law, "the management of water resources must be decentralized and involve the participation of public authorities, users and communities."[3] It must be carried out by collegial bodies, designated as Hydrographic Basin Committees (CBH), or watershed committees. These are spaces for discussion and decision-making on the uses of water, in addition to planning actions to maintain the quality and quantity of this resource. According to the PNRH, water is a public good which has economic value, and its management must include multiple uses, but, in case of scarcity, priority must be given to human supply and animal watering. The committees' management territory is determined by the hydrographic basins which, according to the law, must be "basic units for planning the use, conservation and recuperation of natural resources" (Brasil, 1997).

Despite introduction of the Water Law in 1997, the water resources management system has not yet been fully implemented in Brazil. Many of the management instruments such as Water Resources Plans, which are meant to contain a large compendium of information about each watershed such as water demand and predicted flow capacity, as well as plans for water usage, do not mention mine tailings dams. There is no mention of the probable impact of these dams over large areas of the watersheds. Moreover, there is almost no coordination of the policies involving water security (Campolina, 2021).

With respect to dam safety, a National Policy on Dam Security (Política Nacional de Segurança de Barragens—PNSB, Law L 12334) was established in 2010, although it did not begin to be implemented effectively until 2020. One of its requirements was for mining companies to develop Mining Dam Safety Plans (Plano de Segurança de Barragem—PSBM) that contained, among other pieces of information, flood maps based on all available information, including estimates of worst-case scenarios from a dam rupture. The requirements of the PSBM included technical information regarding the construction of the dam, probable causes of breaches, means of monitoring and controlling possible failures, and steps to be followed in the event of emergencies and/or rupture of the dam. This body of information was to guide the elaboration of a Contingency Plan (PLANCON) by municipal bodies responsible for civil defense of the affected cities as defined by the flood maps study. Moreover, actions were also to be defined based on the geographical delineation of possible areas affected in the event of a breach. This was to include impact on water security throughout the river basin during a spill event (Campolina, 2021; Campolina, Iwama, & Gianasi, in press).

The legal requirement for flood studies and flood maps is also found in the 2017 legislation of the former National Department of Mineral Production,

DNPM, which in 2018 became the National Mining Agency. Article 2 defined a flood study as a study to adequately characterize the potential impacts from flooding originating in the rupture or functional failure of a mine tailings dam. The flood study had to be carried out by a qualified professional using best available methodology as defined by the mining corporation and the professional. The flood map produced by the flood study had to establish the geographic limits of the areas potentially affected in the event of a rupture and delineate possible scenarios, including worst case scenarios. The objective of these studies and maps was to facilitate efficient notification and evacuation of people in the affected areas.[4]

In Brazil, thus, national—and state—policies exist, with their respective management instruments. These policies are designed to calculate and provide warnings about potential risks affecting water security among the different populations along a hydrographic basin. Looking back at the last two major mine tailings disasters, however, what is remarkable is the ineffectiveness of these instruments. No estimate or document prepared in advance came anywhere close to capturing the dimensions of the actual impacts provoked by the breaches at Mariana and Brumadinho. Nor was there any anticipation of how these impacts would be further intensified by the crisis they created in terms of water supply. A study of the documentation on which Hydrographic Basin Committees make local decisions regarding water management reveals a more serious issue. The existence of mine tailings dams in the territories they manage is not even mentioned, much less the safety risks presented by these dams and the possibility of impacts affecting extensive regions.

As for the impacts of mining on water security, in addition to mining disasters, it is necessary to highlight the cases of regions in Brazil where mining destroys ecosystems that are essential for climate maintenance. The Amazon Forest in the north of the country is one such case. Another is the Iron Quadrangle in the central region of Minas Gerais state, an area rich in iron deposits extending over 7000 km^2. Here mining is destroying the aquifers that store water in the midst of the iron deposits (Matschullat et al., 2000; Varejão et al., 2011; Teixeira et al., 2017), which are vital both for human use and for entire ecological systems.

The New Mining Code that is currently under discussion in Brazil contains proposals contrary to the principles of environmental justice. It sets up automatic approval processes for technical and environmental impacts of mining and proposes to establish mining as "an activity of public utility, of national interest and essential to human life." If the New Mining Code is approved as it stands, it will call into question existing conservation areas and demarcation of Indigenous lands. In addition, the New Mining Code provides for flexibility in environmental rules. In practice, this change may allow for exemptions from environmental licensing and automatic approval of processes that have been stopped for more than a year at the National Mining Agency (ANM) (Bispo, 2021). If this proposed law advances, more than 90 thousand mining concession processes could be authorized without due investigation. Data from the Amazônia Minada project reveals that 2,478 current requests to mine in Brazil are on Indigenous lands, and at least 254 of these requests are for artisanal mines (Potter, 2021). It should be noted that many of the regions with mineral deposits are heavily forested areas, essential for climate maintenance, and their destruction would tend to intensify extreme weather events, making the water security situation in the country even more delicate. More mining projects are problematic, both in regions prone to drought where new mines would increase the water demand, and in regions with excessive and intense rainfall which could increase the possibility of tailings dam disasters. In other words, what is at stake is not just a proposed law and approval of its text, but the impacts that will be felt across the country should this New Mining Code be approved. It seems the country is heading towards a battle in which popular pressure could be a means to force the debate to include the effects of environmental and climate injustices.

In a study carried out on earlier tailings dam failures between 1910 and 2010, Azam and Li (2010) identified two main causes of failures. The first was adverse weather conditions (which increased from being contributing factors in 25 per cent of the dam failures in the period before 2000, to 40 per cent after 2000); the second was mismanagement of dams (which grew from 10 per cent before 2000 to 30 per cent after 2000). "Adverse weather conditions" were described mainly as unusual rainfall, attributed to recent climate changes (Azam & Li, 2010). As for the "poor management of dams," the authors took into consideration inadequate choice of procedures in dam construction, inadequate maintenance of drainage structures, and ineffective or non-existent inspections. Bulletin 121 of the International Commission on Large Dams (ICOLD, 2001) also indicates, among the main causes of tailings dam failure, factors related to inadequate management of structures.

The likelihood of disastrous tailing dam collapses becomes increasingly imminent, not only in view of the complications regarding climate change that will tend to worsen in the coming years, but also when we consider the increasing number of tailings dams being constructed throughout the world (Davies et al., 2002; Zonta & Trocate, 2016). The collapses that have occurred and the possibility of new disasters show the importance of diverse measures, carried out by different bodies (companies, government, civil society), with the aim of avoiding dam failures, and of warning systems so that new tailings disasters can be averted.

However, such proactive measures are actively fought by mining companies which benefit from lax and poorly-enforced environmental rules. One of their strategies is to spread disinformation to influence public opinion.

Organized Disinformation and Climate Risk

Many researchers have presented studies and evidence that support the argument that dam failure disasters are not isolated events, but processes, cycles of actions and omissions that are foretold long before the moment of the dam collapse and endure for many years after, as illustrated in Figure 8.1 (Zhouri, 2017; 2018; Zonta & Trocate, 2016; Carmo et al., 2017; Marshall, 2019; Campolina, Rodrigues, & Silva, 2021; Campolina, 2021).

Sometimes companies refuse to provide information essential for water management in territories where the presence of mining complexes means the definitive destruction of aquifers and/or possibilities of tailings dam failures. Furthermore, they may carry out processes of "organized disinformation" (Campolina, 2021). Following mine-related disasters, mining companies sometimes develop marketing campaigns involving schools in order to promote "organized disinformation" about mining (Campolina et al., 2020).

Acselrad, Mello, and Bezerra (2009, p. 81) define "organized disinformation" as taking place when "those responsible for the production of risks avoid making public the dangers they create." This makes it difficult to "perceive the causal relationships between corporate actions and environmental impacts and risks for affected populations."

In this context, schools have been the focus of organized disinformation processes that range from mining companies designing teacher training courses and curriculum activities on mining, to activities with students including art and writing contests and field trips / mine visits. The mining companies may further enhance their image by financing school equipment and infrastructure (Campolina, Gianasi, et al., 2021; Campolina, 2021;

Fig. 8.1 Tailings dam collapses as a process: actions and omissions before, during and after the collapse.

Campolina, Rodrigues & Silva, 2021; Campolina, Gianasi, et al., 2021). Many of the actions undertaken by mining companies with schools are carried out through partnerships between mining companies and local educational management bodies—the Municipal Education Departments. These activities in schools are featured in the annual Sustainability Reports published by the mining companies, complete with numbers and indicators. Mining initiatives focussed on the school system are presented positively to company shareholders and the international market (Campolina, Gianasi, et al., 2021; Campolina, Rodrigues & Silva,, 2021).

The mining company narrative propagated in the schools has a pronounced bias towards linking mining projects to local development and job creation. Although mining is surrounded by controversies and negative impacts, the mining company rhetoric is uniformly positive and serves to legitimize mining activity (Campolina, 2021; Campolina, Gianasi, et al., 2021, Campolina, Rodrigues, & Silva, 2021). The mining companies tout the existence of "magic" technological solutions to solve any problematic consequences caused by mining.

Coelho (2012; 2014), for example, in his dissertation on mining dependency in the region of the Iron Quadrangle Aquifer in Minas Gerais, has developed a concept that he calls Discourse on Development through Mining (DDM). He shows how DDM has been propagated in territories where there is likelihood of a mining project. The discourse presents a highly positive vision of community and territorial development based on the employment and economic gains to be generated through implementation of the mining project. This vision of mining's contribution to socio-economic development serves as a powerful argument for community consent.

According to Coelho (2012; 2014), among the arguments that support DDM are supposedly high rates of job creation and local development, increased tax collection by cities, belief that science and technology can mitigate or even eliminate all negative impacts of mining activity, dissemination of an image of social responsibility on the part of the mining company, and belief in the hypothetical sustainability of mining as a lasting activity in the region. He counters DDM with several arguments, among them the questionable number of jobs compared to other economic activities, such as tourism, and the various negative socio-environmental and even economic impacts that the mining project will generate in the region. This is in addition to the overload of public infrastructure and services; the inability of science and technology to mitigate damage that is irreversible, such as the definitive destruction of aquifers; and the limited duration of the activity, as the resources by definition are not renewable.

Coelho developed the DDM concept based on his research on mining activities in Brazil, but this same discourse is to be found in other countries where mining companies are active. Promotion of this discourse is often carried out in activities involving schools (Campolina, Gianasi, et al., 2021;

Campolina, Rodrigues, & Silva, 2021). One example from Canada is a national organization called *Mining Matters*, a registered charity that claims as its mission "educating young people to develop knowledge and awareness of Earth sciences, the minerals industry and their roles in society" (Mining Matters, n.d.). Mining Matters' main financial backers are mining companies themselves, but financial support also comes from Canadian government departments responsible for matters pertaining to Indigenous communities. Rich mineral deposits are located on Indigenous lands and there are serious conflicts over extractive sector projects. After Vale made major investments in nickel mines in Canada in 2006, it very quickly took its place in the top donor circle for Mining Matters.

PDAC (the Prospectors & Development Association of Canada) is an important mouthpiece for the global mining industry and a long-standing partner of Mining Matters. PDAC holds an annual international convention in Toronto, considered one of the biggest mining industry gatherings in the world. Mining company executives from Vale and BHP (both responsible for major tailings spills in Brazil) are among the attendees (Campolina, Gianasi, et al., 2021; Campolina, 2021). PDAC and Mining Matters jointly organize special events for teachers and students during these conventions. In 2010, just four years after Vale's purchase of important nickel mines in Canada, Vale was being lauded for its support in a Mining Matters newsletter. "Vale dreams big. The company, headquartered in Brazil and currently the second largest mining company in the world, aims to be the largest. ... At PDAC Mining Matters, we're excited that Vale is helping us to dream big, too. We are extremely grateful for the company's generous commitment of $75,000 over the next three years" (Mining Matters, 2010).

These actions promoted by mining companies in the field of education are a *programmed business modus operandi*, using access to young people through the schools as a way to gain community support for mining (Campolina & Gianasi, 2020). Figure 8.2 maps a sequence of actions that have taken place in cities downstream from tailings dams in Brazil. The mining company discourse, DDM, has been disseminated in these cities with actions undertaken by mining companies through "partnerships" with municipal education departments. This results in a "culture of silence"[5] around mining. On the one hand, actions by mining companies, including in schools, propagate a uniformly positive narrative about the benefits of mining. On the other hand, communities, teachers, students, and even universities demonstrate

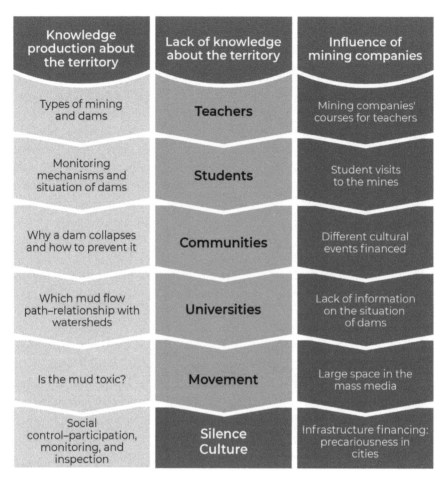

Knowledge production about the territory	Lack of knowledge about the territory	Influence of mining companies
Types of mining and dams	**Teachers**	Mining companies' courses for teachers
Monitoring mechanisms and situation of dams	**Students**	Student visits to the mines
Why a dam collapses and how to prevent it	**Communities**	Different cultural events financed
Which mud flow path–relationship with watersheds	**Universities**	Lack of information on the situation of dams
Is the mud toxic?	**Movement**	Large space in the mass media
Social control–participation, monitoring, and inspection	**Silence Culture**	Infrastructure financing: precariousness in cities

Fig. 8.2 Sequence of actions that generate the "culture of silence." **Source**: Campolina, Gianasi, and Perkins, 2020.

a lack of knowledge about the controversies surrounding mining. Critical perspectives on mining are rarely addressed as topics in initial or continuing teacher-training courses for science and environmental studies teachers. These kinds of disinformation about mining are one of the factors in the processes that lead to tailings dam collapses.

Fighting Disinformation with Geo-Participatory Mapping—(Re)Learning the Territoriality of the Disasters to Gain Critical Perspective

Education processes that provide credible information, training (for community members, young students, and teachers), and actions that enhance social participation and citizenship are imperative to help learners build on their lived experiences about mining's impacts. These processes can provide ways for the population in general, as well as in schools, to become aware of the risks to which mining subjects them and to prepare themselves to face possible future disasters related to tailings dam failures—also equipping them to take actions that reduce the risks and harms.

The starting point is to address people's limited knowledge on the location of dams, their watershed impacts, and the local-territorial impact of possible dam ruptures. We have adapted the widely-used methodology of geo-participatory mapping of hydrographic basins and applied it to geo-participatory mapping of tailings dams in a methodology we call MapGD: Geo-participatory Mapping of Dams (Mapeamento Geoparticipativo de Barragens or MapGB in Portuguese) (Campolina et al., 2013; Gianasi & Campolina, 2016; Campolina & Gianasi, 2019; Campolina, 2019). This adaptation serves as a diagnostic strategy for territories in which there are dams along watersheds. MapGD has among its objectives increased understanding of environmental injustices and construction of popular knowledge in order to contribute towards empowerment of teachers, communities, and activist movements facing water insecurity and tailings dam disasters.

As noted above, policies and instruments exist that *should have been* able to estimate damages caused by dam breaches or even the impact of mining on water quantity and quality. The collapses, however, revealed not only the ineffectiveness of these instruments, but also the vulnerability of diverse communities located below the dams.

Starting from the principle of dam failures as processes, and considering schools as an important focal point for mining companies in the propagation of the Discourse on Development through Mining and "organized disinformation," we see schools as spaces for training and construction of important information for empowerment. In this context, MapGD goes beyond just identifying dams located upstream from communities and schools. It is necessary to understand the territoriality of disasters, as well as the risk of disasters. This territoriality is related to the fact that the rupture of a dam—or even the possibility—can change the entire dynamics of a territory, causing effects from the health of the residents (including mental health) to the community's mode of production, each within its particular economic and cultural dynamics (Campolina, 2021; Campolina, Iwama, & Gianasi, in press).

The concept of *territoriality of mining disasters* emphasizes the importance of recognizing the path of the spill and how it coincides with the water path along hydrographic basins (Campolina, 2021). As a basic education teacher in a public school located below multiple mines and the flow path of about thirty tailings dams, the first author has experienced—and continues to experience—how organized disinformation campaigns hinder the processes of mobilization, information, and training in territories with a strong mining influence.

Identifying the location of the dam and the flow path in the event of a dam failure is part of the MapGD methodology. We developed and applied this method during continuing education courses for teachers at the Federal University of Minas Gerais in 2018 and 2020, within the scope of the first author's doctoral research. The thesis was entitled *Mining and Socio-Scientific Controversies with Strong Local Impact in Continuing Education Programs for Teachers*. It proposed a conceptual approach to socio-scientific controversies in the field of science education, considering territoriality, with mining as a central theme. During the research, two courses were developed in teacher-training programs dealing with the theme of socio-scientific controversies in mining. Methodologically, part of each course was constituted as data collection for the broader research. MapGD was not centrally involved with the research objectives and general questions, but it was configured as one of the steps in the courses taught (Campolina, 2021).

MapGD is an adaptation of the methodology we used and developed in "Geo-participatory Watershed Mapping: 3P—Problems, Potentialities and Possibilities"—a series of extension projects at the Federal University of Minas Gerais (UFMG) between 2011 and 2017[6] (Campolina et al., 2013; Gianasi & Campolina, 2016). 3P Geo-participatory Mapping consists of a geo-environmental analysis of the surroundings of a school or a community, using the territory of hydrographic micro-basins as a methodological/theoretical reference point. The spatial technique is consistent with popular mapping, drawing on community knowledge and local lore in addition to geographic objects and areas that can be observed in satellite images through free software such as Google Earth. Free satellite images of the region are used, highlighting the hydrographic network. Field work and walking trips allow students, teachers, and communities to also include in the map what they consider the *problems* and the socio-environmental *potential* of the territory under study.

Based on analysis and discussion of the problems and potentialities identified, the group lists *possibilities* for intervention or action in the region, taking into consideration the proposals of the Hydrographic Basin Master Plan (linked to the National Water Resources Policy—PNRH), but within the scope of micro-basins. The logic underlying 3P Geo-participatory Mapping is that students, schools, and/or communities will not only increase their knowledge and capacity to discuss problems and potentials of the region through mapping, they can also produce localized information that can assist with participatory water management. Submission of maps to the Hydrographic Basin Committees and to local, municipal, and state water managers is one of the goals of the methodology.

Both MapGD methodology and 3P Geo-participatory Mapping use the territorial contour of the watershed to locate the path of a spill, because geographically that is where the water mixed with mine tailings and other components will travel. The tailings dams in Brazil are usually located on hilltops, close to the mines. They are built to take advantage of the valleys as part of the design of the tailings reservoirs. Tailings reservoirs are usually located above water sources and small rivers, with a foundation and supposed waterproofing. The valley reservoirs are then blocked by dams, which function like dikes. In the event of a rupture, the flow follows the natural path of the rivers.

The main difference between 3P and MapGD is that MapGD is aimed at geo-participatory mapping of the tailings dams themselves. At the end of the mapping activity, those who produced the map are confronted with a geographic reality and the question of whether or not they feel threatened by the presence of tailings dams in their territory. As the mapmakers (community members, students, and teachers) construct the map, tracking the actual dams and their location in that territory, observing the level of safety and risk of the dams being mapped, delineating the possible flow paths in the event of a dam collapse, following it along the lines in the map that make up the streams and rivers, they are also deepening their knowledge on the subject and increasing their ability to discuss it with their peers.

In addition, MapGD analyzes the safety situation of dams by consulting information about them in the Integrated System of Mining Dams—the

Public SIGBM, using its Portuguese acronym—which is part of the National Information System on Dams that was made available as of January 2020.[7] But this data, although it marks an improvement in the availability of information on dams in Brazil, has data by city and not by hydrographic basin, therefore making it difficult for the population to identify whether or not there are dams above the towns and cities where they live. The Public SIGBM does not include hydrographic networks, making it impossible to verify the flow path of a spill through the system and the communities affected by this flow path. The fact that dam location information data is not available also makes it difficult to understand that when there are dam complexes, where several dams are located close to each other, the location of one dam with a rupture in process can also compromise the safety of dams that are located downstream. When the tailings flow from the dams located above reach the downstream dams, the structure of the downstream dams could well be compromised, leading, therefore, to the rupture of multiple dams in the process.

There is other important data regarding water security that is also not available in the SIGBM, namely the location of water collection points for water supply systems that would be affected in the event of ruptures. This topic is also addressed in MapGD where a methodology is included for mapping water catchment points that supply community drinking water and determining whether they would be affected in the event of a dam rupture.

Conclusion: Building Information-Training-Action Networks to Confront Disasters, Organized Disinformation, and Environmental and Climate Injustice

Mining disasters and environmental and climate injustices are topics that are still fairly unfamiliar to the general public, despite the fact that many communities have already suffered the impacts arising from these phenomena. Even being affected, many communities find it difficult to recognize the impact or defend their right to compensation for the damages suffered. Among the various effects, the violation of access to water is what permeates and creates the potential for damages in interconnected fields: environment, health, food security, economy, culture, and ways of life.

In opposition to the cycle of actions and omissions perpetrated by mining companies and governments in Brazil, which has contributed to the processes of tailings dam disasters and threatened the population's water security, it has become imperative to develop participatory methodologies that help communities fight for their rights. In such a context, we believe that schools have a latent potential for information production. Teachers can be agents of collective knowledge production and facilitate the transfer and exchange of critical knowledge about the territory in which they work, in the face of the territoriality of disasters.

Faced with the negligence of mining companies and governments, schools, together with communities, activist movements, universities, and non-governmental organizations, must unite in the construction of grounded information and action networks. This becomes even more urgent in the face of climate change and increasingly critical situations of water insecurity.

Given the importance of creating information-training-action networks, we see the development of the MapGD methodology as an instrument that can contribute to community empowerment. We understand that MapGD is just the starting point for producing quality materials and up-to-date geospatial data, both for education and for community struggles. We see in this methodology the possibility, through the production of collective and popular knowledge, to contribute to the construction of more participatory water management. New knowledge is created in the midst of the process of delineating the territory, as the mapmakers construct the map, identify the dams and their position in that territory, observe the level of safety and risk of the mapped dams, and trace the possible flow path of a spill along the lines that make up the streams and rivers. This new knowledge opens new discussions and new perspectives for collective action. We believe that the lack of information about the territoriality of disasters, the location of dams and the flow path, in the event of ruptures, intensifies the vulnerability of the affected population. Critical knowledge and popular pressure are possible ways to minimize or perhaps eliminate environmental injustice, climate injustice, and mining disasters.

In order to create a database that can be accessible to other teachers, researchers, and the general population, we have organized a research group called Education, Mining and Territory—EduMiTe—and two observatories: Mining Dams Observatory (OBM) and Education and Mining Observatory (OEM). With these proposals and actions, we aim to structure a network of collaborators who can contribute to the construction of materials, methodologies, and practices in the classroom that empower people and enable the teacher to work with these themes in depth, with actual data and high quality materials.

This work opens up possibilities for interaction and dialogue about what has been happening in Minas Gerais, ranging from mining dam disasters to dialogues around water security and water management. We see in these initiatives a vitally important space to discuss the connections between mining and water security, affecting the quantity and quality of water for countless populations in different parts of Brazil and the world.

We extend an open invitation to others to strengthen this movement through the creation of international networks for the co-creation of materials, exchange of successful experiences, and knowledge of phenomena such as those we describe here in Minas Gerais.

NOTES

1   This chapter was translated from the Portuguese by Judith Marshall, to whom we express our great thanks.

2   The Guarani aquifer is one of the largest in the world, covering 1.2 million km², with an estimated volume of 370,000 km²; 70 per cent of the Guarani aquifer is located in Brazil (Ribeiro, 2009).

3   Water users include all those who directly use the surface or groundwater of a watershed. The user can be a natural or legal person, private or public, who without needing a license for water use, captures water directly from cisterns, dams, streams, rivers, lakes, or releases effluents (sewer, industrial, agricultural or domestic) directly into water bodies.

4   See: Portaria DNPM n 70389 de 17 de maio de 2017—SEGURANÇA DE BARRAGENS —Português (Brasil) (www.gov.br).

5   Paulo Freire (2017) defines the culture of silence as arising from an education that prioritizes the oppressor, the dominant groups that have been perpetuated in Brazil from colonization to military dictatorships. Freire says that the history of Brazil has brief snippets of democracy, which makes a culture of participation and mobilization difficult, as there is a tendency towards a culture of silencing in the face of injustices and violations of rights.

6   These projects included: "Monitoring of Watersheds"; "Fapemig—Training of Teachers, Production, and Dissemination of Knowledge on Urban Micro-Basins of the Velhas River Basin as an Instrument for Participatory Environmental Management"; "Geo-Participatory Mapping and Instruments for Participatory Environmental Management"; "PROEXT/MEC Program: Environment, Education, Health, and Citizenship for the Urban Watersheds and Basins of the Rio das Velhas"; "Proext Mec 2014—Geo-Participatory Mapping and Monitoring of Hydrographic Microbasins"; "Extension and Research: Geotechnologies in Water Management Education: 3Ps Geoparticipatory Mapping"; and "Mapping and Visualization of Didactic Practices and

Challenges for Teachers in the Rio Doce Basin Affected by the Collapse of Samarco's Fundão Dam."

7   Although the National Information System on Dams has been in the making since the establishment of the National Policy on Dam Security (PNSB) in 2010, only years later were norms established and only in 2020 did this system begin to make dam information available publicly, such as the locations of dams by municipality.

Reference List

Acselrad, H., Mello, C.C.A., & Bezerra, G.N. (2009). *O que é justiça ambiental.* Garamond.

Azam, S., & Li, Q. (2010). Tailings dam failures: A review of the last one hundred years. *Geotechnical News*, 50–53. Retrieved 20 October 2021, from https://pdfs.semanticscholar.org/e57e/bdac0a801b412cefd42017c2dded29cafd41.pdf

Bispo, F. (2021, December 2). Novo código de mineração propõe aprovação automática de milhares de processos parados na ANM. *Infoamazonia.* https://infoamazonia.org/2021/12/02/novo-codigo-de-mineracao-propoe-aprovacao-automatica-de-milhares-de-processos-parados-na-anm/

Brasil. (1997). Ministério do Meio Ambiente dos Recursos Hídricos e da Amazônia Legal. Lei n. 9.433: Política Nacional de Recursos Hídricos. Brasília: Secretaria de Recursos Hídricos.

Campolina, D. (2019, March). Educação e formação de professores: A urgência do tema barragens no ensino. *Revista Manuelzão*, (84), 23. Retrieved 29 October 2021, from https://manuelzao.ufmg.br/biblioteca/revista-manuelzao-84.

Campolina, D. (2021). *Mineração e controvérsias sociocientíficas de forte impacto local na formação continuada de professores* [Unpublished doctoral dissertation]. Universidade Federal de Minas Gerais.

Campolina, D., & Gianasi, L. (2019). Mapeamento Geoparticipativo 3P: informação, formação e empoderamento. In S. Claudino, X.M. Souto, M.A.R. Domenech, J. Bazzoli, R. Lenilde, C.L. Gengnagel, L. Mendes, A.T.B. Silva (Eds.). *Geografia, educação e cidadania.* Centro de Estudos Geográficos da Universidade de Lisboa.

Campolina, D., & Gianasi, L. (2020). Processo de Rompimento de Barragens [Video]. Controvérsias sociocientíficas e território: Barragens de rejeitos. Universidade Federal de Minas Gerais. https://www.edumite.net/videos-1

Campolina, D., Gianasi, L., Marshall, J., Perkins, P.E., & Oliveira, B.J. (2021, October). Mineração, desastres, formação crítica: Casos no Brasil e no Canadá. *Revista Universidade Federal de Minas Gerais, 27*(3). https://periodicos.ufmg.br/index.php/revistadaufmg/article/view/21466/28886

Campolina, D., Gianasi, L., & Perkins, P.E. (2020, June 1–2). *Mining, territory, and education* [Paper presentation]. Environmental Studies Association of Canada—Virtual Annual Conference.

Campolina, D., Gianasi, L., & Pinheiro, T.M.M. (2013). *Gestão das águas no Brasil: Vamos participar? Mapeamento geo-participativo, participação social e gestão das águas na bacia hidrográfica do ribeirão Onça.* Universidade Federal de Minas Gerais/ Instituto Guaicuy. https://manuelzao.ufmg.br/wp-content/uploads/2018/08/gestao-das-aguas-no-brasil.pdf

Campolina, D., Iwama, A.Y., & Gianasi, L. (in press). A territorialidade dos desastres: O extrativismo e o caso da lama e do óleo invisíveis. In A.Y. Iwama, V.A. Muñoz, & F. Barbi (Eds.), *Riscos ao Sul: Diversidade de riscos de desastres no Brasil.* La Red de Estudios Sociales en Prevención de Desastres en América Latina (LA RED).

Campolina, D., Rodrigues, C., & Silva, F.A.R. (2021). Controvérsias sociocientíficas e mineração: formação cidadã crítica no enfrentamento aos processos de desastres. In M. Baumgarten; J. Guivant (Eds.), *Caminhos da ciência e tecnologia no Brasil: políticas públicas, pesquisas e redes* (pp. 127–152). Editora UFRGS. https://lume. ufrgs.br/handle/10183/225849

Carmo, F.F, Kamino, L.H.Y., Junior, R.T., Campos, I.C.D., Carmo, F.F., Silvino, G., Castro, K.J.D.S.X.D., Mauro, M.L., Rodrigues, N.U.A., Miranda, M.P.D.S., Pinto, C.E.F. (2017). Funded tailings dam failures: The environment tragedy of the largest technological disaster of Brazilian mining in global context. *Perspectives in Ecology and Conservation, 15*(3), p. 145–151. https://www.sciencedirect.com/science/article/ pii/S1679007316301566

Coelho, T.P. (2012). *Mineração e dependência no quadrilátero ferrífero-aquífero: O discurso do desenvolvimento minerador e o Projeto Apolo* [Unpublished master's thesis]. Universidade do Estado do Rio de Janeiro.

Coelho, T.P. (2014). *Projeto Grande Carajás: trinta anos de desenvolvimento frustrado.* Ibase.

Davies, M., Martin, T., & Lighthall, P. (2002). *Mine tailings dams: when things go wrong.* AGRA Earth & Environmental Limited. Retrieved 30 October 2021, from https:// damsafety.org/content/mine-tailings-dams-when-things-go-wrong

Espindola, H.S., & Guimarães, D.J.M. (2019, January/April). História ambiental dos desastres: Uma agenda necessária. *Revista Tempo e Argumento, 11*(26), 560—573. Retrieved 15 November 2021, from https://revistas.udesc.br/index.php/tempo/ article/view/2175180311262019560

Freire, P. (2017). *Pedagogy of the oppressed.* Paz e Terra.

Gianasi, L.M., & Campolina, D. (2016). *Geotecnologias na educação para gestão das águas: Mapeamento geoparticipativo 3P.* Fino Traço. Retrieved 30 October 2021, from https://www.researchgate.net/publication/340935990_Geotecnologias_na_ educacao_para_gestao_das_aguas_mapeamento_geoparticipativo_3P

ICOLD—International Commission on Large Dams. (2001). *Tailings dams: Risk of dangerous occurrences: Lessons learnt from practical experiences.*

Marshall, J. (2019, August 8). Tailings dam collapses in the Americas: Lessons learned? *Policynote* (Canadian Centre for Policy Alternatives). https://www.policynote.ca/ tailings-dam-collapses-in-the-americas-lessons-learned/

Matschullat, J., Borba, R.P., Deschamps, E., Figueiredo, B.R., Gabrio, T., & Schwenk, M. (2000). Human and environmental contamination in the Iron Quadrangle, Brazil. *Applied Geochemistry, 15*(2), 181–190.

Milanez, B., & Fonseca, I.F. (2010, July). Justiça Climática e eventos climáticos extremos: O caso de enchentes no Brasil. *IPEA (Instituto de Pesquisa Econômica Aplicada Secretaria de Assuntos Estratégicos da Presidência da República)—Boletim Regional, Urbano e Ambiental, 4*, 93–101.

Mining Matters. (n.d.). Miningmatters.ca

Mining Matters. (2010). Teck and Vale: Investing in our future. https://miningmatters. ca/docs/default-source/mining-matters---newsletters/industry---mm---newsletter---2010.pdf?sfvrsn=4893c202_8

Pinheiro, T.M.M., Polignano, M.V., Goulart, E.M.A., & Procópio, J.C. (2019). *Mar de lama da Samarco na bacia do rio Doce: em busca de respostas*. Inst. Guaycui. Retrieved 2 November 2021, from https://manuelzao.ufmg.br/biblioteca/o-livro-mar-de-lama-ja-esta- Available-em-formato-digital/

Potter, H. (2021, March 2). Isolated Indigenous lands targeted by half of all mining claims. *Rainforest Journalism Fund.* https://rainforestjournalismfund.org/projects/ amazonia-minada-mined-amazon

Ribeiro, W.C. (Ed.). (2009). *Governança da água no Brasil: Uma visão interdisciplinar.* Annablume, Fapesp, CNPq.

Rebouças, A.C., Braga, B., & Tundisi, J.G. (2002). Águas doces do Brasil: Capital ecológico, uso e conservação. (2nd ed.). Escrituras, Editora e Distribuidora de Livros Ltda.

Teixeira, M.C., Santos, A.C., Fernandes, C.S., & Ng, J.C. (2017). Arsenic contamination assessment in Brazil—past, present and future concerns: a historical and critical review. *Science of the Total Environment, 730*(138217). https://doi.org/10.1016/j. scitotenv.2020.138217

Varejão, E.V.V., Bellato, C.R., Fontes, M.P.F., & Mello, J.W.V. (2011). Arsenic and trace metals in river water and sediments from the southeast portion of the iron Quadrangle, Brazil. *Environmental Monitoring and Assessment, 172*, 631–642.

Wanderley, L.J., Mansur, M.S., Milanez, B., & Pinto, R.G. (2016). Desastre da Samarco/ Vale/BHP no Vale do Rio Doce: Aspectos econômicos, políticos e socioambientais. *Cienc. Cult.* [online], *68*(3), 30–35. http://dx.doi.org/10.21800/2317–66602016000300011

Zhouri, A. (Ed.). (2017, May–August). Dossier: Mining, violence and resistance. *Vibrant: Virtual Brazilian Anthropology/Associação Brasileira de Antropologia, 14*(2). http:// vibrant.org.br/downloads/v14n2/Vibrant14n2.pdf

Zhouri, A. (2018). Produção de conhecimento num "campo minado." In A. Zhouri (Ed.)., *Mineração, violências e resistências: Um campo aberto à produção de conhecimento no Brasil.* Editorial iGuana—ABA, 8–27.

Zonta, M., & Trocate, C. (Eds.). (2016). *Antes fosse mais leve a carga: Reflexões sobre o desastre da Samarco/Vale/BHP Billiton.* Editorial iGuana. https://www.ufjf.br/ poemas/files/2016/11/Livro-Completo-com-capa.pdf

Investigating Citizen Participation in Plans for Lamu Port, Kenya

Solomon Njenga

Introduction: Lamu Port Development and Climate Justice

The government of Kenya is developing a new deep-sea port at Lamu, on the Indian Ocean coast of Kenya at Manda Bay (Map 5). The port, 240 km north of Mombasa, is to be the terminus of the 891-km-long Kenya Crude Oil Pipeline (now set to open in 2023), bringing oil from Lokichar in northwest Kenya.[1] Extensions to South Sudan and Ethiopia are also planned. The Lamu Port—South Sudan—Ethiopia—Transport Corridor project (LAPSSET) is also to include road, railway, and fibre-optic links as well as an airport, oil refinery, and Special Economic Zone for manufacturing and industry near the port in Lamu. The first berth of Lamu Port was completed in October 2019 and the second and third berths officially opened in May 2021, to export oil and other cargo using the Kenyan road system. As of September 2021, only five container ships had docked there (Maritime Executive, 2021). Plans include thirty-two berths for large ships. The port is being built by China Communications Construction Company (Bachmann & Kilaka, 2021) with funding from the Kenyan government.

The ongoing dredging of long 18-foot-deep berths along the Manda Bay coastline, an oil terminal for loading and offloading of tankers, and a 1.5 km

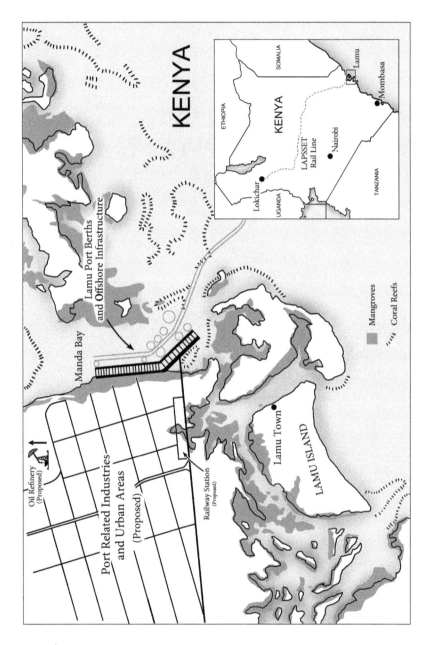

Map 5
Kenya—
Lamu

long causeway connecting the harbour to the town along the shore of the Indian Ocean, are causing a significant shift to marine ecosystems, especially mangroves (Map 5). Manda Bay mangroves have been intact for centuries but, due to the construction of Lamu Port, massive destruction of mangrove forests has been witnessed in the area (Taljaard et al., 2021). Mangroves play myriad roles including carbon sequestration, fish and turtle breeding habitat, coral reef support, and storm surge protection, and their destruction has multiple impacts on the community that depends on the environment for a living (Aalders et al., 2021). Communities in Lamu, backed up by international non-governmental organizations (NGOs) and United Nations (UN) organizations, have documented the declining volume and quality of fish in Lamu region along with impacts on the traditional and socio-economic existence of the local population, who depend on fishing and tourism for their livelihoods. Lamu Island is recognized for native fishing methods, and the port threatens this heritage. Artisanal fishing is the main source of livelihood for over 70 per cent of Lamu's population (Uku et al., 2021). Oil spills from transhipment of fuel and accidents are also of grave concern (Praxides, 2021). Due to all these impacts, there is clearly a need for dialogue by all parties involved (Mkutu et al., 2021).

A proposed coal-fired power plant at Lamu, which would have been the country's first, was stopped in late 2020 when three Chinese state enterprise investors pulled out after a community-driven protest movement convinced a court to suspend the project's permit (Yi, 2021; Obura, 2019). The local community argued that the coal plant project would be detrimental to the marine ecosystem and the environment at Lamu, with long-standing consequences for residents, particularly those living closest. Local residents also delivered a protest letter to the judiciary stating that Kenya (Figure 9.1) had emerged as a front-runner in clean and renewable energy in Africa, and those advances ought not to be lost through coal plant approvals in Lamu.

The lawsuit by the local community, arguing that port dredging has violated the cultural, fishing, and health rights of more than 4,600 people, resulted in a 2018 court order for US$170 million, which is to be used for more powerful fishing boats, landing sites, training, cooling services for fish catch, and a loan program (Njunge, 2019). Tourism, mainly from Europe, has also been impacted by the port development; just offshore Lamu Island is a fourteenth-century Swahili settlement and UNESCO World Heritage site.

Al-Shabab attacks and kidnappings in the area are a threat to local residents, tourists, and infrastructure (Mwangi, 2022).

This chapter discusses the extent to which citizen participation was included in Lamu Port planning processes, based on a study that involved household surveys, questionnaires, interviews with key informants, and focus-group discussions. A variety of coastal stakeholders including government officials, coastal businessmen, civil society organizations, and local residents were involved in the study. I carried out this research in collaboration with Climate Justice International (CJI), an NGO which has been working in Lamu for years and has a number of climate justice related projects in East Africa.

CJI's vision is "to liberate vulnerable and Indigenous coastal communities from environmental bondage through ecology conservation and stewardship" (CJI, 2022). I was based at CJI for a period of two years (2019–2021) as a research fellow, with a special focus on Lamu Port infrastructure and its implications for the community in pursuit of climate justice. I participated in community advocacy, awareness, training, and facilitations that correspond to the vision and mission of CJI; facilitated five community engagement meetings and dialogue initiatives on the impact of Lamu Port; attended three court sessions at Lamu Law Court for hearings on the ongoing cases by the community arising from the Lamu Port development and the proposed Lamu coal plant; joined as a participant observer with Lamu community members in public protests airing their grievances to the world in pursuit of climate justice; participated in a CJI field visit to the Lamu Port construction site, as well as visits to mangrove sites, coral reefs, and fishing areas; participated in reviewing Kenya's national laws and policies that relate to climate justice including REDD+; prepared a modest literature review for CJI on climate justice in Kenya, Africa, and globally; and conducted stakeholder interviews, facilitated focus-group discussions, community briefings, and report writing (Figure 9.2). This connection with CJI greatly facilitated my research, contacts, background understanding of the local situation, and mobilization of my research results.

As discussed in this chapter, I was able to document, along several parameters, that the involvement of the local community in decision-making about Lamu Port development has been very low, and to date the community is not satisfied with information-sharing regarding Lamu Port plans, transparency, or accountability. Moreover, local residents are not satisfied with

Fig 9.1 A 2014 community-led lawsuit against destructive dredging at Lamu Port resulted in a successful court verdict in 2018, though it is under appeal by the government. Disputes continue over the allocation of US$170 million decreed by the court in compensation for 4,700 fisherfolk (Lesutis, 2022, p. 2445).

the kind of education and sensitization they have received about Lamu Port plans, construction, or decision-making.

My findings were meant to document and highlight the importance of more effective citizen participation in sustainable planning of such large-scale development projects with significant livelihood implications for local people and ecosystems. Theories and much global research over many years underscore the importance of citizen participation for sustainable political, social, and environmental planning, as recognized in Kenya's constitution. This study adds to that literature, providing important contextual details on a number of climate justice-related impacts of the controversial Lamu Port and LAPSSET project.

The next section of this chapter overviews a subset of the literature on public participation that provides the theoretical framework for this research, focusing on the Kenyan context in comparison to other African and international port developments. Section three outlines the methodology for this study, and its results are presented and discussed in section four. The final section discusses a number of climate justice implications.

Literature Review and Theoretical Framework: Participation and Climate Justice in Kenya

Public participation is an important aspect of citizen engagement in environmental governance and planning (Gera, 2016). Globally, there is a growing need to engage citizens in decision making (Callahan, 2007),[2] and a number of examples from around the world indicate this is very salient in relation to port development. In Canada, the participatory public governance and community-based management has contributed greatly in the management of ocean resources (Kearney et al., 2007). Lack of citizen participation in an Alaska port city region in the USA has contributed greatly to the existing unhealthy relationship between the government and the community in the region (Jordan et al., 2013). This is the same in the coastal region of Lekki Port in Nigeria (Jimoh, 2015) where violence has been documented between the coastal community and the government for failure to include them in coastal decision-making platforms (Lawanson & Agunbiade, 2018). In Madagascar, the government is empowering coastal communities to effectively manage port and coastal resources through citizen inclusivity and transparency (Danielsen et al., 2009). In South Africa, the government, with the help of the private sector and civil society, has developed a policy on sustainable ports (Glavovic, 2000).

Citizen participation is enshrined in the Constitution of Kenya (Kenya Law, 2013). In sustainable port planning, citizen participation is a requisite, partly because ports can have a significant effect on coastlines, landscape, marine ecosystems, and community livelihoods. Siami-Namini (2018) states that access to information in public participation for decision-making is crucial in sustainable port planning and management; participation involves "enabling people to realize their rights to participate in, and access information relating to, the decision-making processes that affect their lives," and specifically, to influence details of policy legislation, and to monitor their implementation (Siami-Namini, 2018, 5). Gusinsky et al. (2015) similarly point out that transparency and accountability through citizen participation in the public sector are crucial in sustainable planning. This requires addressing all project-related issues, and being answerable for the project's plan, actions, and justifications (Gusinsky et al., 2015).

In Kenya, coastal resources are not to be exploited without proper public participation (Yeri, 2018). The Kenya Vision 2030 (Nyangena, 2012) also requires public participation for a sustainable offshore "blue economy" relying on coastal resources (Benkenstein, 2018; Novaglio et al., 2021).

This is particularly important at Lamu since, besides the ecosystem impacts noted above, concerns regarding the viability of Lamu Port include emerging challenges such as sea-level rise, oil pollution, invasive species, and storms; uncertainties about the port's impacts and their effects on the public; and other implications (Wanderi, 2019; Mohamed, Abdel-Salam, & Bakr, 2021). Years after the landmark court judgment on Lamu Port (Wanderi, 2019), the $170 million settlement has not yet been disbursed, and violations and failure to involve the community in decision-making continue, while construction goes on unabated (Owino, 2020). This has led to ongoing contestation between the government and the Lamu community.

Thus, the level of citizen participation in Lamu Port planning is nested in layers of climate justice struggles. Establishing its inadequacy from the community's perspective can be seen as a step towards building better processes, in line with Kenyan law, and acknowledging the importance of the Lamu community's engagement in climate action.

Methodology: Participatory Research on Participatory Governance

To investigate community members' views and document their input on how they were consulted about the port, I used a multi-methods approach (Mohamed, Jafari, & Hammad, 2021). The study included household surveys, questionnaires, interviews with key informants, participant observation, and focus-group discussions. This allowed me to build the story of community members' engagement, piece by piece, and to combine information from various sources. I had many opportunities to engage with local residents as a participant observer due to my work with CJI.

Fig. 9.2 Marine conservation meeting in Lamu, 2020.

I used purposive sampling to select one key informant from each of ten identified professional categories: the Kenya Forest Service (KFS), Kenya Marine and Fishery Research Institute (KMFRI), Beach Management Units, Fisher folks, County Government of Lamu Officials, LAPSSET/ Lamu Port Management Unit and allied workers, Selected Civil Society Organization representatives, Kenya National Human Rights Commission (KNHRC)- Lamu branch, Lamu Business Community (LBC) and National Police Service (NPS) & Kenya Defence Force officials. I thus picked ten individuals to act as key informants. Using interview schedules to guide the discussion, I also employed focus-group discussions (FGDs) for each of the ten professional categories of potential respondents. These FGDs brought together careful- ly selected groups of about six people from each group, for a ninety-minute guided discussion to gather information and provide data on specific ques- tions related to this research.

I carried out a household survey between November 2020 and February 2021 in Lamu County, Kenya (Hoyle, 2001), which has a population of 143,920 (Republic of Kenya, 2019, p. 7). Adopting the formulae of Kothari (2013) to determine the sample size in relation to the population of Lamu County,[3] the survey had a sample size of 195. I relied on semi-structured questionnaires for the household surveys. The questionnaire was composed of three parts. The first part was the consent form; the second part include demographic questions; and the third part contained four sets of participation-related questions: a) on the decision to establish a port; b) on information access and sharing; c) on transparency and accountability; d) on education, sensitization, and sustainable planning. The selection of these factors was based on my review of literature on public participation, including Bartoletti and Faccioli (2016); Horgan and Dimitrijević (2019); and Johann (2012). The questionnaire used a Likert scale—a psychometric scale where peoples' responses can range from 1 (strongly disagree) to 5 (strongly agree)—which allowed me to categorize and present the overall responses (Pimentel, 2010). Two research assistants and village elders helped me to distribute and explain the questionnaires. The respondents filled out the questionnaires themselves. The data collection using questionnaires took place over a period of twenty days. Results and research participants' views and comments are summarized and discussed in the following section.

Results and Discussion: Building the Account of Citizens' Views on Participation in Governance[4]

A total of 195 questionnaires were distributed, of which 163 were completed adequately for the purpose of data analysis, giving an acceptable response rate of 83 per cent.

The results from the questionnaire's demographic questions about respondents are summarized in Table 9.1. A slight majority (51.5 per cent) of the respondents were male while 48.5 per cent were female. The majority of the respondents (84 per cent) had a primary, secondary, or diploma certificate while 16 per cent were university graduates or post-graduates. Most (90.8 per cent) of the respondents had lived in Lamu for over fifteen years, with the rest (9.2 per cent) having lived there for less than ten years.

Finally, in terms of occupation, 80 per cent of the respondents were either fish traders, fishermen, teachers, government representatives, hotel

Table 9.1 Respondents' Social-Demographic Characteristics.

Gender	Frequency	Percentage
Females	79	48.5
Males	84	51.5
Total	**163**	**100.0**
Education level	**Frequency**	**Percentage**
Primary	29	17.8
Secondary	66	40.5
Diploma/Certificate/Artisan	42	25.7
Bachelors	21	12.8
Postgraduate	5	3.2
Total	**163**	**100.0**
Length of Stay in Lamu	**Frequency**	**Percentage**
Less than 1 year	3	2.0
1–5 years	5	3.1
5–10 years	7	4.1
10–15 years	52	32.2
15 and above years	96	58.6
Total	**163**	**100.0**
Occupation of Respondents	**Frequency**	**Percentage**
Fish Trader	14	8.5
Teacher	35	21.6
Government Employee	17	10.1
Hotel Employee	35	21.7
Fisherfolk	18	11.1
Farmer	11	7.0
Others	33	20.0
Total	**163**	**100.0**

employees, farmers, or government/county employees, while the rest (20 per cent) worked in other civil society occupations, NGOs, etc. This shows that respondents were distributed across government, community, and various stakeholder groups.

Table 9.2 Decision to establish Lamu Port.

Statements on Involvement of the Public in Decision-Making	Mean
There was no team of locals selected to participate in the initial discussion of establishing Lamu Port infrastructure at Manda Bay	1.6503
The Lamu County government was involved in the decisions-making process	1.5828
The vulnerable people were involved and were guided on how to adapt	1.5583
The experts from the community are involved in technical teams	1.5583
The port management carried out public opinion research before the port began	1.5521
The opinions and views of the community takes centre stage in decisions made	1.5460
The Lamu Port management organizes focus-group discussions with locals	1.5215
The port management has formed round table discussion with locals	1.5031
The members of the local community influence decisions made	1.4785
The public was involved in mitigating impacts of the port activities*	1.3681
Overall Mean	**1.5319**

* This question relates to whether the community was involved in deciding and setting up ways to mitigate port impacts such as those due to dredging, oil spills, and other effects of the ongoing construction.

Since an important purpose of this study was to explore the level of citizen participation regarding the decision to establish Lamu Port infrastructure at Manda Bay, the survey respondents were asked to indicate whether they agreed or disagreed, and to what extent, with a range of statements about participation. In general they reported a very low degree of consultation, along many parameters. In Tables 9.2, 9.3, 9.4, and 9.5, the statements are listed in order from those which averaged most to least agreement. A mean score of 1 indicates "strongly disagree" while 2 indicates "disagree."

On the question of whether the public was consulted about the decision to build Lamu Port at all, the results are shown in Table 9.2.

Table 9.2 shows that the most-supported statement was that there was no team of locals selected to participate in the initial discussion of establishing Lamu Port infrastructure at Manda Bay. This response was tending towards disagreement with the statement, implying that, even if there were representatives picked to represent the community, respondents felt that there was no fair representation of the local citizens. The statement that received the lowest rating based on mean was that the public was involved in mitigating impacts

Table 9.3 Information access and sharing.

Statements on Information Sharing and Access	Mean
The port management uses newspapers to post critical information	2.2945
The Lamu Port management has a website where it shares information	2.1534
The information shared is usually made at the right time	1.6503
The port management shares notification on your mobile phones on the happenings at the Lamu Port	1.6196
The public relations officer of the Lamu Port management gives frequent timely updates on happenings at the port	1.5828
The port activities are posted frequently on notice boards within Lamu County	1.5583
The management of Lamu Port collaborate with Lamu County government to pass on crucial information to members of the public frequently	1.5521
The Chief's Baraza's* are used frequently by the port management to pass on information about happenings at the port of Lamu	1.4969
The management uses the radio media frequently in Lamu to discuss and pass on information about issues around the port construction	1.4785
The information passed on by the management is adequate for the community	1.3681
Overall Mean	**1.67545**

* A Baraza is a traditional, semi-formal, public open-air meeting convened by a chief to address local issues and communicate government agendas and policies to the grassroots.

of the port activities. Their response tended towards strong disagreement meaning they weren't involved. The overall mean score on the statements also tended towards strong disagreement (the overall mean score was 1.5319), revealing that the involvement of the local community in decisions being made about the Lamu Port has been very low.[5]

Table 9.3 reports the results regarding whether local community members had access to information about port planning. The respondents were asked to rate a number of statements about this, again listed in the table from those which averaged most to least agreement.

The most supported statement was that the port management uses newspapers to post critical information to the public. The response was leaning towards disagreement with the statement implying that the use of newspapers to relay information about port activities was not very effective. The statement that was supported the least by the respondents was that the information passed on by the Lamu Port management is adequate for the community. The

Table 9.4 Transparency and accountability.

Statements on Transparency and Accountability	Mean
There is periodic reporting of what is ongoing at the Lamu Port	1.6503
The management of Lamu Port are upright individual and law-abiding members of the society	1.6503
Lamu Port activities is not done under strict secrets and information blackout to the local community	1.5828
The representatives of the local community are treated as equal partners in decision-making regarding port activities	1.5521
The management of Lamu Port usually abides by all court judgments	1.5460
The compensation plan for the local community affected by port activities is fair and adequate	1.5460
The management of the port does not get involved in any form of bribes to community representatives to buy their cooperation	1.5031
The management of the port does not conceal any crucial information to the public	1.4969
The management of the port treats the community with respect and dignity	1.4785
The management of the port are accountable for their actions and activities to the community	1.3283
Overall Mean	**1.53741**

overall mean score on the statements tended towards strong disagreement (1.67545), implying that the respondents were not satisfied with information sharing at the Lamu Port.

Responses to statements about citizens' level of participation regarding transparency and accountability in the planning of Lamu Port infrastructure are given in Table 9.4.

The statement with the highest agreement (though still very low) was that there is periodic reporting of what is going on at the Lamu Port. The respondents' tendency towards disagreement with the statement may reflect community members' views that the community only had the opportunity to learn about port developments through their court petitions; there were limited reviews and petition mechanisms put in place by the port developers. The statement with which there was least agreement was that the management of the port is accountable to the community for their actions and activities. The overall mean score on the statements also tended towards strong

Table 9.5 Education and sensitization.

Statements on Education and Sensitization	Mean
The management of the port sensitizes the community on the benefits the port will have to the local community	2.2515
The port management has been organizing forums to educate the public on possible impacts of the port to the livelihood of the local community	2.1718
The local community received printed materials for learning at home	1.5828
The public was educated on how to reduce the impacts of port construction	1.5583
Education received was very useful in understanding the risks	1.5399
Port management organizes seminars to teach the public how to adapt to changes brought by port construction	1.5215
The management accommodated the local community's needs during the seminars	1.4785
The public has been educated on alternative means of survival in case their traditional economic activities are affected	1.4663
The educators were rich in knowledge and experience	1.4601
The education was received at the right time when the community needed it	1.3129
Overall Mean	**1.63436**

disagreement (mean of 1.53741), meaning that respondents feel the port project developers have not conducted themselves with the transparency and accountability expected in law and by the local community.

Table 9.5 lists the statements on community education and sensitization about sustainable planning for the Lamu Port.

As shown in the table, the most-supported statement was that the management of the port sensitizes the community on the benefits the port will have to the local community when it is complete. The response was tending towards disagreement with the statement, implying that port developers' statements were attempting to convince local residents of the port's benefits. The statement that was supported the least by the respondents, tending towards strong disagreement, was that education and sensitization was received at the right time when the community needed it. The response implies that the education and sensitization received was not adequate as it came at the wrong time, after the port construction had started. The overall mean score on the statements tended towards strong disagreement (mean of 1.63436), implying that the respondents were not satisfied with the kind of education and sensitization they received about Lamu Port.

The respondents also commented on the extent to which the port builders and local police were fully aware that the community was not in support of port construction.

Fatma (not her real name) was one of the local opinion leaders who confided that "we had to hold public vigils at night because the implementers of the Lamu Port decided to bring construction materials at night in fear of the community's backlash! We demonstrated many times on the street and government county offices demanding disclose of information about what was taking place inside the Lamu Port. Unfortunately, many a time we were dispelled by the police. They instilled fear in us, to the extent that many members got scared to engage them. The port is still ongoing and yet, no full disclosure of what is happening to date."

Since about 20% of respondents were fisherfolk or fish-sellers, I also include here some of the results from those sectors that are important for the community's livelihood, as well as participants' comments on the security situation in the community in relation to the port development and its impact on land tenure issues. These are important climate justice impacts of the port construction on which local residents had plenty to share: views and knowledge from their lived experience which, given the lack of consultation with Lamu community members, can only come to light through participatory research.

Impact of Lamu Port on Fish and Fishing Activities

The study investigated the impact of Lamu Port construction on fish and fishing activities in the community. Respondents were asked to rank their responses using the Likert scale. The data collected and associated analysis is given in Table 9.6, which shows respondents' assessment of the impact of port construction on fish and fishing activities in Lamu County. The table presents the mean response before and after port construction began, the difference in mean and percentage change in mean. The most impacted fish and fishing activity based on difference in mean response was that after port construction began, sea grass (which is a major food supply for fish) became no longer predominant around the spot where port construction is ongoing ($\mu x = -2.99$).

Table 9.6 Impact of Lamu Port on fish and fishing activities.

	Before	After	Difference in Mean	Percentage Δ in Mean
	Mean Response	Mean Response		
Weight of fish caught	4.07	1.62	-2.45	-60.19
Affordability of fish	4.20	1.51	-2.69	-64.04
Average number of fish species	4.26	1.56	-2.7	-63.38
Low prices in the market	4.40	1.50	-2.9	-65.90
Coral reef growth	4.37	1.58	-2.79	-63.84
Number of fisherfolk	4.26	1.48	-2.78	-65.25
Demand of fish	4.28	1.55	-2.73	-63.78
Shortage of fish	4.35	1.50	-2.85	-65.51
Predominance of sea grass	4.36	1.37	-2.99	-68.57
Deep sea fishing	4.42	1.55	-2.87	-64.93
Overall Mean Score	**4.29**	**1.53**	**-2.767**	**-64.39**

The impact of Lamu Port construction on fish and fishing activities is evidenced further by information collected in FGD with key informants from fisherfolk, who were of the opinion that the construction of the port has had major negative impacts on fishing and fishing activities. One forty-six-year-old fisherman stated:

> Before the port activities began, I used to get 50 kg in one catch but since the exercise started I cannot get even 20 kg. The fishmongers don't want to understand when you raise the price because they will tell you they have the option of buying Chinese fish.

The fisherfolk claim they have been unable to meet demand for fish, due to the disruption of fish-breeding sites by a sand-harvesting exercise in nearby Tiwi Beach. The sand is for the construction of an oil terminal at the port of Mombasa, another project being constructed by the China Communications Construction Company. However, an officer from KMFRI was more optimistic when discussing the impact of Lamu Port construction on fish and fishing activities. The officer stated,

> Even though most fishermen are complaining about the construction of Lamu port on their fishing activities, the truth of the matter is that the government has already provided funds to support the fishing activities of the affected fishermen…. The government has approved funds to cater for the purchase of boats and engines capable of fishing in the deep sea, the development of new fish landing sites, training, cooling services and a loan scheme. [Perhaps he was referring to the court settlement of the community's lawsuit, discussed above.]

Information from a discussion with a key informant on beach management revealed a fight over beachfront land affecting fishing activities. The key informant stated that the construction of the beach has denied fisherfolk spaces for fishing, as most beachfront lands have been taken by the port construction. Private developers have also been grabbing beach lands, hence denying the fisherfolk spaces for fishing. The key informant stated,

> Of late, we as a beach management group have had many cases of fishermen complaining of beach lands being fenced off by unknown private developers from Nairobi. The port itself has also taken a big portion of our fishing grounds. This has forced many fishermen to go deep into the ocean waters to have their catch. Deep in the waters, they face stiff competition from companies with well-developed facilities for fishing. They also get exposed to the problem of crossing into waters of neighbouring countries.

Impact of Lamu Port on Security and Terrorism Incidences

The study explored the impact of Lamu Port construction on security and terrorism incidence in the community. As above, respondents were required to rank their responses for before and after port construction, using the Likert scale. Table 9.7 shows community members' assessment of the relationship between port construction and security and terrorism incidences in Lamu

Table 9.7 Impact of Lamu Port on security and terrorism incidences.

	Before	After	Difference in Mean	Percentage Δ in Mean
	Mean Response	Mean Response		
Few cases of general insecurity	3.95	1.55	-2.4	-60.75
Few police posts	4.2	1.50	-2.7	-64.28
Few terrorist attacks	4.35	1.36	-2.98	-68.57
No tourist abduction by terrorist	4.53	1.65	-2.88	-63.57
Few police patrols in the areas	4.45	1.55	-2.9	-65.16
Army activities in area was very rare	4.19	1.56	-2.63	-62.76
Few cases of security curfew	4.22	1.52	-2.7	-63.98
Arsonist attacks not a problem	4.3	1.56	-2.74	-63.72
Few sanctions by the government	4.33	1.58	-2.75	-63.51
No fear or uncertainty	4.3	1.48	-2.82	-65.58
Overall Mean Score	**4.283**	**1.532**	**-2.751**	**-64.23**

community. The table presents the mean response before and after port construction began, the difference in mean, and percentage change in mean. The most impacted security factor based on difference in mean response was that before port construction there were fewer terrorist attacks compared to after port construction, with difference in mean of = -2.98). The least impacted activity based on difference in mean was for the indicator "few cases of general insecurity," with the insecurity not changing much before and after the start of port construction.

The impact of Lamu Port construction on security and terrorism activities was a source of great concern, as expressed during the FGD with key informants and the community. One member from the business community stated:

Before the port construction began, this place was very safe; we could walk freely doing our businesses without any fear from anybody. The villagers could go fishing into the ocean without fearing that they would be attacked by the "bad people" [*implying terrorists*] ... I remember I could wake up in the early morning to go take fish from the fishermen ready for the market.

A key informant from the Kenya police stated in the discussion that

The attacks by Al-Shabaab extremists in Kenya have threatened to wipe out the gains made from tourism and [this] has threatened to destabilize livelihood earning activities in some parts of the country such as Lamu County. However, the government is working around the clock to flush out the militia's group hiding in [the] Boni Forest[6] and very soon, they will be concurred.

Even with the positive opinion of the key informant from the Kenyan police service, a key informant from the Kenyan National Commission on Human Rights painted a gloomy picture of the situation on the ground. The activist stated:

Al-Shabaab militias are roaming freely like they are in their home under the noses of our security apparatus. This port construction has attracted the interest of these animals who come here and ambush residents whenever they like and then disappear in the Boni Forest in the full knowledge of our police. In fact, since the port construction begun, there have been dozens of attacks on the locals who are now living in fear in their own land left to them by their ancestors.

Impact of Lamu Port on Land Conflicts and Land Use

The study also investigated the impact of Lamu Port construction on land conflicts and land use in the Lamu community through a number of questions about land issues. Table 9.8 shows the impact of port construction on land conflicts and land use in Lamu community, as assessed by community members. The table presents the mean response for before and after port construction begun, as well as the difference in mean and percentage change in mean. The most impacted land issue based on difference in mean response was that before port construction, there were few tourist hotels being constructed around the port construction area but since the construction began,

Table 9.8 Impact of Lamu Port on land conflicts and land.

	Before	After	Difference in Mean	Percentage Δ in Mean
	Mean Response	Mean Response		
Few people complaining of losing land	4.33	1.55	-2.78	-64.20
Few displaced people in the area	4.34	1.50	-2.84	-65.43
Few incidences of land conflicts	3.94	1.37	-2.57	-65.22
Few land cases in court	4.23	1.65	-2.58	-60.99
The price of land affordable to many	4.37	1.55	-2.82	-64.53
Low demand of land for development	4.42	1.56	-2.86	-64.70
Few tourist hotels being constructed	4.48	1.52	-2.96	-66.07
Few cases of eviction with no compensation	4.50	1.56	-2.94	-65.33
Most lands were under cultivation	3.88	1.51	-2.37	-61.08
No encroachment in gazette water towers	4.12	1.56	-2.56	-62.13
Overall Mean Score	**4.26**	**1.53**	**-2.72**	**-64.02**

most lands had been taken for development of tourist hotels and villas with difference in mean of -2.96. The least impacted activity based on difference in mean was lands under cultivation, meaning most lands are still under cultivation, even they though they are slowly changing to development for housing construction and other construction, with difference in mean of -2.37.

The impact of Lamu Port construction on land conflicts and use appears even more serious in the light of comments from the FGD with key informants. An informant from the business community stated:

Before the government bought this project to us, we did not have much problem with our land ownership … even though there were small issues here and there about land, we did not have the issue of land being grabbed by big men in government from Nairobi. Our fathers could till the land for small scale agriculture without much problem.

However, since the port construction begun, there have been issues of land being grabbed by unknown people from other parts of the country. Since construction of the port began, developers have been streaming in to buy land or dispossess the local owners, working with unscrupulous land officers in Lamu. A key informant from KMFRI stated:

> There has been stiff competition for prime land surrounding present fisheries which are leading to a loss of beach access routes by fishers.... The construction of tourist hotels along the coast, operations of private individuals outside the community, and large scale marine exploitation projects have also been prohibiting locals from accessing their resources.

Even though the key informant from KMFRI did not mention names, it was clear that Lamu and LAPSSET were partly to blame for loss of land for fishing activities, considering the magnitude of the port project and the obvious effects it will have on Kenya's marine and fisheries. Another key informant, a Lamu landowner residing in the upscale enclave of Shela, reported on condition of anonymity that an acquaintance takes "phone orders" for land he has expropriated, and for which he has manufactured title deeds. The landowner added:

> You are welcome here in Lamu, but tell the truth about Lamu when you are at home [in Nairobi]. Lamu is our paradise and those government thieves in Nairobi want to steal it from us. I am an Indigenous Lamu man who has never left here. Foreigners who have been placed here by the government now own my land and most of us are squatters on the land—the same land our fathers and their fathers and [grand]fathers worked for food.

Conclusion: Climate Justice Implications of Lamu Port Development

This investigation of citizen participation in the planning of Lamu Port in Lamu County, Kenya showed that Lamu citizens and the local community perceive the lack of citizen engagement opportunities in Lamu as a serious breach of their constitutional and environmental rights, and they have called upon the Lamu Port implementers to involve them and stop working in isolation. They placed the onus for citizen participation on central government and county officials, and expressed their desire to be engaged through applicable local avenues, entities, and frameworks. They emphasized the need to engage all stakeholders in Lamu, facilitate information-sharing and access regarding Lamu Port, spearhead transparency and accountability, and encourage continuing education by every possible medium to enhance citizen participation for sustainable planning of the ongoing Lamu Port infrastructure.

This study was preliminary and has its limitations. The survey only included residents of Lamu County within a radius of 20 km from the port of Lamu. It would have been ideal to conduct interviews and FGD with residents and key informants from a wider area—perhaps the whole coastline of Kenya. Because of the methodology, which involved a large number of self-administered questionnaires, requiring more educated participants, I made certain adjustments that may have skewed the results in favour of more-educated and more-articulate participants.[7] The survey questions might have been expanded to include more indicators that would have allowed additional insights from the study. Nonetheless, the results of this study offer unprecedented insights into citizen participation in sustainable infrastructure planning in Kenya, and climate justice impacts of the construction of Lamu Port. The investigation relied on self-reported practices and views, which are central to its results; the participatory methodology allowed for triangulation to validate and enhance the conclusions.

This study provides rich detail on several matters that are central to climate justice and associated struggles.

1. People denied participation through one channel will seek (and usually find) another. Without consultation and incorporation of people's local knowledge and livelihood priorities *ex ante* in development processes, people are likely to organize protests;

seek redress in the courts; find supporters outside the jurisdiction to bring pressure on decisions; and/or influence investment decisions. To avoid community revolt and street protests, and economic decline fuelling long-term unrest, community engagement processes should

a. Involve all stakeholders and the local community in the advancement of a project-led vision and mission

b. Enlighten and edify the community about the logic and rationale of the intended or ongoing project, its expected impacts, its extent and duration, and other logistics that pertain to the project

c. Stimulate a sense of communal ownership and stewardship of the project through, for example, community benefits discussions and agreements

d. Generate enthusiasm and anticipation around prospects for the local people to benefit and raise their livelihood prospects so that they tend to embrace the project and safeguard it

e. Build robust networks, connections, and relationships between the project, the community, and other internal and external interests working for the project's success

f. Involve the young, women, people with disabilities, and less fortunate community members in various aspects of the project where they may work, contribute, and/or benefit

g. Appoint champions and facilitate an enlarging ring of support system/followers and defenders of the project for better synergy and joint ownership of the project

2. The findings established that Lamu Port construction has major impacts on fish and fishing activities, a mainstay of the local economy and of many community members' livelihoods. Discussion with various key informants in this study revealed that Lamu Port construction has already had negative impacts on the marine ecosystem and the community members who depend on it. The findings in this study are in agreement with findings by

Rodden (2014) that established that the LAPSSET project has the potential to severely affect the artisan fishing industry in Lamu. Our study shows that the fisherfolk of Lamu are most concerned about the lack of communication concerning the port's activities and the blocking of fishing areas during port construction and operation.

3. Participatory governance institutions are designed to build on local knowledge and assets, but they can be disrupted by incumbents (those who benefit from the status quo and fear change) as well as market pressures from outside the area. International solidarity for climate justice action can help to publicize and counter such regressive tendencies. From my many interactions with the community in Lamu, I learned that participatory governance largely depends on how well the project implementers are able to involve the local, regional, and international community and stakeholders at all levels of the project. The interactions with community and stakeholders add legitimacy and genuineness in the project's vision, mission, and intended goals. These interactions safeguard both interests: those of the community and also the interests of the implementers. In this way, the implementers are able to address the needs of the local community and help them build local support for the project. The local community ought to be involved from the initial stages of the project; this helps all stakeholders to have trust in each other.

4. This study revealed that Lamu Port development has incited rivalry between community members and investors and led to various land-related conflicts. Since port construction began, land uses have changed from the traditional farming and fishing to development of hotels and villas to support the population moving into the Lamu area, due to new economic prospects of the Lamu Port; these changes mostly benefit outsiders.

5. Local knowledge of ecosystems and how to protect them, and local people's commitment to do so when their livelihoods

depend on viable ecosystems, are significant assets and powerful forces for climate action. I also learned that there are a series of interrogations that one can use to assist in determining the features of a particular community ecosystem. The following axioms may help determine the needs of the community: socioeconomic and demographic features of the area; collective, financial, and ecological challenges or attributes of significance for the community; ongoing interactions and happenings that help determine neighbourhood improvement and identify emerging needs; livelihood dependency of the community in the region; and employability, sustainability, and developmental strategies to promote and uplift the lives of those in the community. These are some of the things community members are looking for when a project is broached within their vicinity, and mandatory public participation can help ensure that the project brings benefits for all.

NOTES

1   The Lokichar oil fields are being developed by the Canadian firm Africa Oil (which controls 25 per cent), along with Total (25 per cent) and the Anglo-Irish Tullow Oil (50 per cent) (Itayim, 2021).

2   Citizen participation theory (Horgan & Dimitrijević, 2019) provides governments, individuals, and stakeholders with a pragmatic rationale and platform for allowing all involved stakeholders to participate in decision making before a project is initiated (Bartoletti & Faccioli, 2016). The terms "citizen," "public," "involvement," and "participation" are frequently used synonymously (Thompson, 2007). Citizen participation refers to the direct involvement of constituents in decision-making (Hardina, 2003). Citizen involvement is a means to ensure that citizens have a direct voice (Richardson & Razzaque, 2006). In governance contexts, citizen engagement enhances social cohesion, responsibility, and stewardship (Bäckstrand, 2003). This includes bringing dissimilar and/or Indigenous coastal communities together, bringing "hard to reach" and "disadvantaged" groups into discussions, building relationships within and between different coastal communities and social groups (Beall & Ngonyama, 2009).

3   For details on this and other calculations, see Njenga, 2022.

4   Quantitative and statistical analysis of survey and questionnaire results using SPSS, and additional aspects of this study, are reported on and discussed in my dissertation: *Climate Change and Seaport Development on Ecosystem-Dependent Livelihoods through Climate Justice Lens at Lamu Port, Lamu County, Kenya* (Institute for Climate Change Adaptation, University of Nairobi, 2022).

5 As Irvin and Stansbury note, in most projects across the globe, this is not unusual; communities are rarely involved in public decision-making (Irvin & Stansbury, 2004).

6 Boni Forest is an indigenous open canopy forest found in coastal areas from Kenya to Mozambique—rich in biodiversity.

7 Overall in Lamu County, only 13 per cent of residents have a secondary level of education, and 33 per cent have no formal education (SID, 2013). I knew this fact and one of my mitigative approaches was to visit schools for referrals of those who had secondary education, so that I could select them for inclusion in the study. There are also gender implications to selecting more-educated participants since educational opportunities are more limited for women and girls.

Reference List

Aalders, J.T., Bachmann, J., Knutsson, P., & Musembi Kilaka, B. (2021). The making and unmaking of a megaproject: Contesting temporalities along the LAPSSET corridor in Kenya. *Antipode*, *53*(5), 1273–1293.

Bachmann, J., & Kilamba, B.M. (2021, May 20). Kenya launches Lamu Port. But its value remains an open question. *The Conversation*. https://theconversation.com/kenya-launches-lamu-port-but-its-value-remains-an-open-question-161301

Bäckstrand, K. (2003). Civic science for sustainability: Reframing the role of experts, policy-makers and citizens in environmental governance. *Global Environmental Politics*, *3*(4), 24–41.

Bartoletti, R., & Faccioli, F. (2016). Public engagement, local policies, and citizens' participation: An Italian case study of civic collaboration. *Social Media+ Society*, *2*(3), 2056305116662187.

Beall, J., & Ngonyama, M. (2009). Indigenous institutions, traditional leaders and elite coalitions for development: The case of Greater Durban, South Africa. *Crisis States Research Centre Working Paper No. 55*. https://www.lse.ac.uk/international-development/Assets/Documents/PDFs/csrc-working-papers-phase-two/wp55.2-indigenous-institutions-traditional-leaders-elite-coalitions.pdf

Benkenstein, A. (2018). Prospects for the Kenyan blue economy. *South African Institute of International Affairs (SAIIA) Policy Insights*, *62*. https://scholar.google.com/scholar?hl=en&as_sdt=0%2C5&q=Benkenstein%2C+A.+%282018%29.+Prospects+for+the+Kenyan+blue+economy.&btnG=

Callahan, K. (2007). Citizen participation: Models and methods. *International Journal of Public Administration*, *30*(11), 1179–1196.

CJI—Climate Justice International. (2022). https://climatejusticeinternational.org/about-us/who-we-are.html

Danielsen, F., Burgess, N.D., Balmford, A., Donald, P.F., Funder, M., Jones, J.P., Alviola, P., Balete, D.S., Blomley, T., Brashares, J., Child, B., Enghoff, M., Fjeldså, J., Holt, S., Hübertz, H., Jensen, A.E., Jensen, P.M., Massao, J., Mendoza, M.M., ...Yonten, D.

(2009). Local participation in natural resource monitoring: A characterization of approaches. *Conservation Biology, 23*(1), 31–42.

Gera, W. (2016). Public participation in environmental governance in the Philippines: The challenge of consolidation in engaging the state. *Land Use Policy, 52*, 501–510.

Glavovic, B.C. (2000). A new coastal policy for South Africa. *Coastal Management, 28*(3), 261–271.

Gusinsky, M., Lyrio, M., Lunkes, R., & Taliani, E. (2015). *Accountability through citizen participation and transparency in the public sector: An analysis in the City Hall of Florianópolis/Brazil* [Paper presentation]. Transparency Research 4th Global Conference, Lugano, Switzerland. Anais.

Hardina, D. (2003). Linking citizen participation to empowerment practice: A historical overview. *Journal of Community Practice, 11*(4), 11–38.

Horgan, D., & Dimitrijević, B. (2019). Frameworks for citizens participation in planning: From conversational to smart tools. *Sustainable Cities and Society, 48*, 101550.

Hoyle, B. (2001). Lamu: Waterfront revitalization in an East African port-city. *Cities, 18*(5), 297–313.

Irvin, R.A., & Stansbury, J. (2004). Citizen participation in decision making: Is it worth the effort? *Public Administration Review, 64*(1), 55–65.

Itayim, N. (2021, November 11). Kenya targets South Lokichar crude FID by mid-2022. Argus Media. https://www.argusmedia.com/en/news/2272784-kenya-targets-south-lokichar-crude-fid-by-mid2022

Jimoh, A. (2015). Maritime piracy and lethal violence offshore in Nigeria. *IFRA-Nigeria Working Papers Series* (51).

Johann, D. (2012). Specific political knowledge and citizens' participation: Evidence from Germany. *Acta Politica, 47*(1), 42–66.

Jordan, E.J., Vogt, C.A., Kruger, L.E., & Grewe, N. (2013). The interplay of governance, power and citizen participation in community tourism planning. *Journal of Policy Research in Tourism, Leisure and Events, 5*(3), 270–288.

Kearney, J., Berkes, F., Charles, A., Pinkerton, E., & Wiber, M. (2007). The role of participatory governance and community-based management in integrated coastal and ocean management in Canada. *Coastal Management, 35*(1), 79–104.

Kenya Law (2013). *The Constitution of Kenya: 2010*. National Council for Law Reporting with the Authority of the Attorney General. http://kenyalaw.org/kl/index.php?id=398

Kothari, C. (2013). *Quantitative techniques*. Vikas Publishing House Private Ltd.

Lawanson, T., & Agunbiade, M. (2018). Land governance and megacity projects in Lagos, Nigeria: The case of Lekki Free Trade Zone. *Area Development and Policy, 3*(1), 114–131.

Lesutis, G. (2022). Politics of disavowal: Megaprojects, infrastructural biopolitics, disavowed subjects. *Annals of the American Association of Geographers, 112*(8), 2436–2451.

Maritime Executive (2021, September 9). Kenya seeks $157M loan for Lamu Port Phase I Completion. https://maritime-executive.com/article/kenya-seeks-157m-loan-for-lamu-port-phase-i-completion

Mkutu, K., Müller-Koné, M., & Owino, E.A. (2021). Future visions, present conflicts: The ethnicized politics of anticipation surrounding an infrastructure corridor in northern Kenya. *Journal of Eastern African Studies, 15*(4), 1–21.

Mohamed, E., Jafari, P., & Hammad, A. (2021). Mixed qualitative-quantitative approach for bidding decisions in construction. *Engineering, Construction and Architectural Management, 29*(6), 2328–2357.

Mohamed, M., Abdel-Salam, H., & Bakr, A. (2021, September 7–10). *Developing cultural heritage plan in response to future challenges: Case of Lamu-Town world heritage site, Kenya* [Paper presentation]. CITIES 20.50–Creating Habitats for the 3rd Millennium: Smart–Sustainable–Climate Neutral. Proceedings of REAL CORP 2021, 26th International Conference on Urban Planning and Regional Development in the Information Society, Vienna, Austria.

Mwangi, O.G. (2022, February 14). Five reasons why militants are targeting Kenya's Lamu County. *The Conversation*. https://theconversation.com/five-reasons-why-militants-are-targeting-kenyas-lamu-county-176519

Njenga, S. (2022). *Climate change and seaport development on ecosystem-dependent livelihoods through climate justice lens at Lamu Port, Lamu County, Kenya* [Unpublished doctoral dissertation] University of Nairobi.

Njunge, J. (2019, August 30). Lamu Port Project impacts Kenyan fishermen's livelihoods. *The Maritime Executive*, translated and reprinted from *China Dialogue*. https://maritime-executive.com/editorials/lamu-port-project-impacts-kenyan-fishermens-livelihoods

Novaglio, C., Bax, N., Boschetti, F., Emad, G.R., Frusher, S., Fullbrook, L., Hemer, M., Jennings, S., van Putten, I., Robinson, L.M., Spain, E., Vince, J., Voyer, M., Wood, G., & Fulton, E.A. (2021). Deep aspirations: towards a sustainable offshore Blue Economy. *Reviews in Fish Biology and Fisheries, 32*, 209–230.

Nyangena, W. (2012). The Kenya Vision 2030 and the environment: Issues and challenges. *Environment for Development (EfD-Kenya)*, 45–56.

Obura, D. (2019, June 29). Court stops construction of Kenya's coal power plant. Here's why. *The Conversation*. https://theconversation.com/court-stops-construction-of-kenyas-coal-power-plant-heres-why-119550#:~:text=Kenyan%20judges%20have%20stopped%20plans,do%20a%20thorough%20environmental%20assessment

Owino, E.A. (2020). *The implications of large-scale infrastructure projects to the communities in Isiolo County: The case of Lamu Port South Sudan-Ethiopia transport corridor* [Unpublished master's thesis]. United States International University-Africa. https://www.researchgate.net/publication/340777844_The_

Implications_of_Large-Scale_Infrastructure_Projects_to_the_Communities_in_Isiolo_County_The_Case_of_Lamu_Port_South_Sudan_Ethiopia_Transport_Corridor

Pimentel, J.L. (2010). A note on the usage of Likert Scaling for research data analysis. *USM R&D Journal*, *18*(2), 109–112.

Praxides, C. (2021, October 21). Alarm over increased oil spills as more ships dock at Lamu Port. *The Star.* https://www.the-star.co.ke/business/kenya/2021-10-21-alarm-over-increased-oil-spills-as-more-ships-dock-at-lamu-port/

Republic of Kenya (2019). *2019 Kenya population and housing census, volume 1: Population by county and sub-county.* https://housingfinanceafrica.org/app/uploads/VOLUME-I-KPHC-2019.pdf

Richardson, B.J., & Razzaque, J. (2006). Public participation in environmental decision-making. *Environmental Law for Sustainability*, *6*, 165–194.

Rodden, V. (2014). *Analyzing the dynamics of the artisan fishing industry and LAPSSET Port in Lamu, Kenya.* Independent Study Project (ISP) Collection, 1765. SIT Digital Collections. https://digitalcollections.sit.edu/cgi/viewcontent.cgi?article=2781&context=isp_collection

Siami-Namini, S. (2018). Knowledge management challenges in public sectors. *Research Journal of Economics*, *2*(3), 1–9.

SID—Society for International Development. (2013). Lamu County—Education. *SID Forum.* http://inequalities.sidint.net/kenya/county/lamu/#education

Taljaard, S., Slinger, J., Arabi, S., Weerts, S.P., & Vreugdenhil, H. (2021). The natural environment in port development: A "green handbrake" or an equal partner? *Ocean & Coastal Management*, *199*, 105390.

Thompson, A.G. (2007). The meaning of patient involvement and participation in health care consultations: A taxonomy. *Social Science & Medicine*, *64*(6), 1297–1310.

Uku, J., Daudi, L., Alati, V., Nzioka, A., & Muthama, C. (2021). The status of seagrass beds in the coastal county of Lamu, Kenya. *Aquatic Ecosystem Health & Management*, *24*(1), 35–42.

Wanderi, H. (2019). Lamu Old Town: Balancing economic development with heritage conservation. *Journal of World Heritage Studies*, 16–22.

Yeri, T.M. (2018). *Determinants of successful implementation of infrastructure projects in devolved units in Kenya: A case of Kilifi county, Kenya* [Unpublished master's thesis]. University of Nairobi.

Yi, S. (2021, March 9). Kenyan cal project shows why Chinese investors need to take environmental risks seriously. *China Dialogue.* https://chinadialogue.net/en/energy/lamu-kenyan-coal-project-chinese-investors-take-environmental-risks-seriously/

Hydroelectricity, Water Rights, Community Mapping, and Indigenous Toponyms in the Queuco River Basin

Camila Bañales-Seguel

Introduction

The conceptualization of climate justice points to the fact that the negative impacts of the globe's warming climate are distributed unequally throughout the human population. It is the more vulnerable communities and individuals who feel the effects of climate change more adversely. Moreover, these key groups tend to have a disproportionately low responsibility for the human causes of climate change.

The supremacy of market-oriented policy ideas of dominant countries in the international environmental arena (Ciplet & Roberts, 2017; Newell & Taylor, 2020; Schlosberg & Collins, 2014) is part of a development model that has disregarded alternative forms of life-systems, including Indigenous livelihoods and non-human living beings (Lindenmayer & Laurance, 2016). These policy ideas are embodied in the UN's Sustainable Development Goals, which have in many cases served as a smokescreen for further environmental destruction (Zeng et al., 2020). One example is the international promotion of hydropower as a renewable source of energy (Lacey-Barnacle et al., 2020). Widespread support for renewable energies has often obscured the way in which energy generation projects are implemented, what their impacts are for local communities, and how "development"-driven inequities today are exacerbated by climate change.

This chapter invites the reader to learn about a place-based experience in an Andean river basin, located in the ancestral territory of Mapuche-Pewenche Indigenous People called Wallmapu (see Map 1, page 29). The principal objective of the research undertaken was to articulate local Indigenous knowledge and scientific knowledge about rivers to strengthen the river's resilience as a social-ecosystem. To achieve this goal, a participatory science outlook was adopted to implement strategies for co-production of knowledge that would lead to a more thorough understanding of the functioning of the Queuco River.

The Queuco River gathers the waters from diverse tributaries and flows freely among Andean mountains to meet the Biobío River—a river so large that its Indigenous name is *Butaleubü* (the Big River) and local Indigenous knowledge speaks of its sibling galactic river—the *Wenuleubü*—or Milky Way. The confluence between the Queuco and the Biobío rivers is named *Tratrawünko*, meaning encounter of big waters.

The Biobío is the second longest river in Chile (about 380 km long); it is located in the biogeographic transition between central and southern Chile, and forms part of the Valdivian Temperate Rainforest hotspot. This river network holds the highest diversity of native fish species in the country, including two endemic species, all of which are under threat of extinction (Vila & Habit, 2015). The river stands out for its enormous hydroelectric potential, which is estimated at around 2,902 MW (Ministerio de Energía, 2016).

Some decades ago, the Biobío used to be known worldwide by the white-water sports community as one of the most epic journeys from the mountains to the ocean. But since the mid 1990s the free-flowing waters of the Biobío have been interrupted by the construction of three mega-dams: Pangue (1996), Ralco (2004), and Angostura (2014). These dams use the river's high hydroelectric potential to produce energy for the Chilean Central Interconnected System.

The construction of these dams faced strong opposition from the local Indigenous Mapuche-Pewenche communities and diverse national and international conservation and activist groups. At the time of their evaluation and construction, the Chilean State had still not ratified International Labour Organization Convention n°169, the Indigenous and Tribal Peoples Convention, and there was no consultation or consideration for Indigenous communities' rights. The construction of the Ralco dam—the most controversial of them all—involved the relocation of eighty-one Mapuche-Pewenche families (approximately four hundred people) belonging to the Ralco Lepoy and Quepuca Ralco communities (Moraga, 2001; Namuncura, 1999). Their resistance was not enough to succeed against the country's development goals oriented towards economic growth and private profit.

With the filling of these dams, more than 4,600 ha of native forest and ancestral Pewenche land were flooded, including ceremonial sites and cemeteries. The State blatantly disregarded Indigenous communities and their livelihoods, incorporating forms of "participation" based on misinformation and division of the communities for the company's benefit, deepening the environmental injustice (Álvarez & Coolsaet, 2020). It is not just human communities living in this territory that have been affected: the aquatic communities of organisms that coexist within the river ecosystem have also been hurt by the construction of these projects. For example, native fish that were endemic (only occupying this river) have been extirpated from areas that are restricted upstream and downstream by dams (Habit et al., 2006; Valenzuela et al., 2019). The Biobío became a tragic example of how the dominant development model disregards alternative life systems, such as Indigenous cultures and lifestyles and non-human living beings (Lindgren, 2018). The forced displacement of Indigenous communities also showed how neo-liberal economies based on Nature's exploitation perpetuate colonialist logics and lead to what has been defined as "ecocide" (Crook et al., 2018; Higgins et al., 2013) on Indigenous lands.

Today, diverse environmental threats are still menacing the Biobío, its tributaries such as the Queuco, and their human and non-human inhabitants. In the middle of the COVID-19 pandemic in 2020, construction began for a fourth hydroelectric dam, which was approved by government authorities in 2015: Central Hidroeléctrica Rucalhue. And in early 2021, a new project proposal to construct a fifth dam (Central Hidroeléctrica Huequecura) was registered with the Environmental Impact Assessment Service and is currently undergoing the evaluation process within this state institution. These projects reflect a national policy to promote hydropower as clean energy (Pacheco, 2018); however, the compounding impact on ecosystems has not been adequately accounted for through the evaluation process.

Also, since 2018 a private corporation named "Reguemos Chile" has begun pushing for the construction of a so-called "hydric highway" to transport large volumes of water from rivers in southern Chile to central and northern Chile. The aim of this development project is to irrigate nearly a million hectares of export-oriented agricultural land and contribute to powering more hydro-power plants along the way. The justification from the powerful agriculture industry for this massive water transfer is that southern rivers have surplus water and that "water is lost to the sea." Another argument is "inter-regional solidarity," considering the harsh decade-long drought that central and northern Chile is facing. The corporation even claims that the project would help mitigate the effects of climate change.

These narratives mask strong economic interests, power imbalances in the Chilean water market model, and a profound scientific negationism that ignores evidence showing that negative social-ecological impacts largely outweigh the economic benefits of such projects (Vargas et al., 2020). Scientists from multiple universities and research centres have provided evidence against the water transfer project, indicating that it is a short-term solution that would only deepen climate injustices (Colin et al., 2021; Figueroa et al., 2020; Zúñiga & Ramos, 2021). Water scarcity is a reality at the national level, driven by climate change (Boisier et al., 2018; Garreaud et al., 2013; Rojas et al., 2019) and by the market-based water allocation model in Chile (Bauer, 2004; Budds, 2020). If all factors are considered, addressing water injustices would involve other "solutions" than further destruction of ecosystems.

One of the first rivers that the proposed water transfer project would affect is the Queuco River, where an average of 33 m^3/s (cubic metres per second) monthly between May and November are projected to be extracted. According to the company's documents, this projection was established based on water rights records for the river and its tributaries. However, these water rights have been allocated without any official river gauging station in the river or consideration of climatic projections on future water availability. Direct measurements of discharge and river levels show that this amount of water currently exceeds what actually flows through the Queuco during one of the highest discharge periods of the year (May–July). As climate change reduces precipitation and snow pack in this Andean basin, water scarcity will become more acute and future discharge is expected to decrease.

Through the analysis of water rights in the Queuco river basin, hidden threats to water bodies were unveiled—particularly the fact that an alarming 99.98 per cent of water rights in the basin are currently owned by people and companies from outside the basin, based on General Water Directorate data

(DGA, 2022). Many of these rights are registered for hydropower generation, so even if there are no projects developed or proposed yet, this constitutes a latent threat.

The Chilean Water Code, implemented mid-dictatorship in 1981, conceptualizes water as a natural resource over which property rights can be held. The underlying paradigm behind this law conceptualizes Nature as an object, as a commodity (Svampa, 2015). A development model that considers Nature a mere provider of natural resources is what has led to the current social-ecological crisis. At a global scale, it is precisely this paradigm that is being challenged by civil society activists, public demonstrations, and "side event" discussants at global climate change meetings, such as the United Nations Framework Convention on Climate Change annual "Conferences of the Parties" (COPs). From a legal angle, there are global advances in what has been called a rights revolution for Nature as a way to effectively protect nature. There are examples of rivers around the world being recognized as legal subjects (instead of as objects/resources), and it has been established through court rulings that their rights should be protected. In this sense, rivers have served as entities that reflect larger dynamics of environmental protection (Álvez-Marín et al., 2021).

Locally in Chile, many demands emerged during a massive popular uprising in 2019 that centered around environmental conflicts, particularly water conflicts. These uprisings led to a national plebiscite to change the 1980's constitution. Currently Chile is undergoing a constituent process and many proposals are being debated about how to relate to and with Nature in a less predatory manner. Different ways in which Nature's rights can be protected have been put forth by diverse members of the Constitutional Convention as well as citizens, interest groups, and organizations, through the participation platform implemented for this purpose.

In late 2021, two representatives of the Queuco and BioBío communities travelled to Glasgow, Scotland to participate in COP 26. As founding and leading members of the activist groups in their territories—Malen Leubü (a women's rafting team and collective from Alto Biobío), Red por la Defensa del Río Queuco (RDRQ—Queuco River Defense Network) and Grupo Juvenil SDL Rucalhue (Rucalhue Lirquén Seeds Youth Group; the Lirquén River is a tributary of the BioBío). They travelled as delegates of the non-governmental organization (NGO) Ríos to Rivers to deliver an important message at the United Nations climate forum: "Stop recognizing dams as clean energy." Ríos to Rivers is dedicated to youth exchanges from different endangered water-sheds, supporting youth leaders as the next generation of river guardians.

It was through one of these exchanges, held in the summer of 2019 in the upper Biobío River basin, that I initially met youth from local Pewenche communities who were members of the Malen Leubü, RDRQ, and Ríos to Rivers. I learned that their objective is to raise awareness about the new projects being pushed that threaten the Biobio and Queuco rivers. Our conversations led to finding common ground on the goal of learning about river well-being and constructing strategies for social-ecological resilience.

Decolonizing River Science

The study of rivers has traditionally been conducted from mono-disciplinary perspectives. Broad and deep research began to grow in disciplines such as hydrology, fluvial geomorphology, and ecology to improve human understanding of river systems. In recent decades, scholars from these disciplines have begun to collaborate and study rivers from more complex, inter-disciplinary approaches. This recognition that rivers are social-ecosystems (Dunham et al., 2018) accepts that natural and social dimensions are connected and interdependent. Such a framework considers humans as explicitly inseparable from Nature, and is especially appropriate to study territories with a focus on the feedbacks between social and ecological dimensions that lead to overall system resilience (Berkes & Folke, 1998; Folke, 2006). This also has opened a venue for dialogue between Western scientific disciplines and different knowledge systems such as local and Indigenous knowledge (Agrawal, 1995; Pretty, 2011).

My QES-funded research project, named "Keuko Leubü: Learning and Living with the Queuco River in Pewenche Territory," built a case study which was a key part of my doctoral research on river resilience and social-ecosystems in Indigenous territories of southern Chile. This research constituted an attempt to move the academic frontier one step closer towards a trans-disciplinary and decolonial approach to studying river systems.

The Queuco River is located in the upper Andean portion of the Biobío River basin, bordering the frontier between Chile and Argentina. Beyond this current political boundary, the basin is located in ancestral Indigenous Mapuche-Pewenche territory. This land has been historically inhabited by Pewenche communities and displaced Mapuche communities from both

Chile and Argentina. Currently, there are six established Mapuche-Pewenche communities in the basin: Callaqui, Pitril, Cauñicú, Malla Malla, Trapa Trapa and Butalelbún.

The main research activities I undertook in this project involved extensive field work during a six-month stay at Ralco, the only small town within the Pewenche territory of Alto Biobío. Moving to and living near the Queuco River and its riverside Pewenche communities was fundamental for two reasons: (1) to continue my work despite COVID-19 transportation restrictions and (2) to achieve a better cultural immersion and closeness to the experience of living in this territory and learning from the local inhabitants' perspectives. I conducted a series of semi-structured interviews, group interviews, or walking interviews (Evans & Jones, 2011; Guber, 2011) with diverse local inhabitants in the different communities living in the Queuco valley. The main focus was placed on *ngütram*, the local traditional format of extended conversation ranging across a series of topics.

In alignment with the goals of the partner organizations (the activist NGOs named above) and collaborating community members, this project sought to work on two main strategies of participatory science: river level monitoring and collective mapping of the river and its basin. Participatory science or open science encompasses a wide array of methodologies in which non-scientists are involved in scientific research. This involvement can happen in many ways, ranging from community members participating in the collection of data, to their being more deeply involved through the very development of the research questions being studied, and deciding the methodologies for carrying out the research. In this case, members of the local youth organization Red por la Defensa del Río Queuco (Queuco River Defense Network) were protagonists in the river monitoring as well as the mapping workshops.

Participatory River Monitoring

Community monitoring of the river was an example of participatory action research. With the help of local networks, we implemented two training workshops oriented towards youth and local community members. These workshops aimed to teach basic hydrological principles and to frame the importance of community participation in the context of commons governance of the watershed and external pressure for natural resource exploitation.

Engaging with these local organizations helped to continue and strengthen the participatory monitoring of the river's water levels.

Ecological research methods were used to better understand the functioning of the Queuco River. For the ecological dimension, in May 2021 we conducted the first of two field campaigns to sample aquatic communities. These campaigns focused on gathering habitat information and samples from the riparian habitat to characterize the fish communities as an indicator of river ecosystem health. The first four-day campaign included three technicians from the University of Concepción Fish Laboratory along with three members of the Pewenche riverside communities as participants. The data obtained from these field campaigns were analyzed so that the next field campaign could add to the results. An already meaningful outcome from these campaigns was testing a new, participatory manner of conducting biological field work in Indigenous territory. Bringing together local youth and community members with scientists and technicians from the university for the field campaigns is a first step towards getting to know each other better, building respect, and diminishing myths and mutual ignorance between these distant and quite different human groups.

Another research activity nourishing this strategy was the installation of a physical river gauge to measure the variations in river discharge (Figures 10.1 and 10.2). This river gauge helped to improve the precision of our community measurements, which were recorded using the citizen science cellphone app CrowdWater. At the monitoring site, in March 2020, we installed a digital river gauge with the support of an experienced technician from the Hydraulic Laboratory of the University of Concepción. The first data retrieval from this gauge was in late June 2020. The measurements from this digital gauge served to complement and validate the measurements from the community monitoring efforts (by taking cellphone photos, from a fixed location, of the river level in relation to a water gauge held at a fixed place in the river). Community monitoring already has yielded a year-round time series of river fluctuations, and is on-going. This monitoring initiative began at the start of the pandemic restrictions in May 2020. So far, we have implemented four monitoring sites in the Queuco River and one site in the Biobío River; we have collected 476 observations contributed by 21 community participants.

The main challenge related to this river monitoring strategy, as reported by the participants, was that they had to commit part of their personal time to travelling to the measuring site, located 8 km from Ralco, where most

Figs. 10.1 and **10.2** Top, bike ride to the Queuco River bridge, 8 km from the nearest town, where the trail to the measuring site starts. Bottom, installation of a physical river gauge for community monitoring.

of them live. This involved biking from their homes, walking down a short trail, and biking back. Still, this distance was not a deterrent for the most dedicated participants, who visited the site year-round, notwithstanding the weather. One participant said, "I loved doing it, I thought it was great because I couldn't go to the river very often as a Malen (member of the local female rafting team *Malen Leubü*, girls of the river), or participate in other activities, or meetings; I felt this was my way of contributing to the fight and defending the river. I also liked feeling this was my responsibility, giving myself the responsibility of contributing to the river defense."

Indigenous Counter-Mapping to Reclaim River *Kuibi Kimün (Ancestral Knowledge)*

Our project's second activity focused on implementing a series of participatory mapping workshops with the riverside communities of the Queuco basin. This required careful team building with members of local communities who helped as facilitators in these workshops and assisted with the systematization, translation, and validation of information. The main challenge had to do with the logistics and organization of the workshops. On the one hand, the work required the team members to commit part of their personal time. It also required prior organization to visit the participants in each workshop, who often lived in remote places and didn't have electronic means of communication. Finally, the context of the COVID-19 pandemic also made it difficult to carry out the workshops, as some meetings had to be suspended or postponed due to surges in infections in the communities.

With the help of four facilitators, one person in charge of documenting and systematizing the workshops, and one audio-visual professional, we carried out three workshops in the communities of Butalelbún, Cauñicú, and Callaqui (Figure 10.3). The plans and methodology, including the choice of workshop participants, was determined as a team along with community members.

Together we designed the structure, objectives and methodology for the mapping workshops. In each *lob* (area composed by a geographical territory and the Pewenche communities inhabiting it), a handful of selected people considered as *kimche* (wise, usually older, persons) in the community were invited to participate in an activity aimed at the recuperation of names of places (toponyms) in *Chedungün* (the Pewenche Indigenous language) and

Fig. 10.3 Collective mapping workshop with local *kimche* (traditional knowledge holders, usually elders).

stories related to water bodies. In the Mapuche-Pewenche worldview, bodies of waters are seen as living entities with whom inter-subjective relationships are established (Aigo et al., 2020). Local toponyms—place names, especially those derived from topographical features—can give us important clues about the way that communities have co-evolved with their environment and established intimate relationships with the places they inhabit (Salazar & Riquelme Maulen, 2020).

The workshop carried out to learn about local toponyms consisted of a presentation on scientific/geographic information on the Queuco River as currently available for the study area. This information was presented as, first, the IGM (Military Geographic Institute)'s 1:50,000 scale maps and, second, high-resolution satellite images of the local area. After sharing these "perspectives" on the territory, we invited participants collectively to draw a representation from the local perspective. In this sense, the work of the local facilitators was essential for the presentation of the workshop's objectives, information-sharing, evoking stories, the graphical representation of the river (actually drawing on blank paper!), and translating.

We worked on the systematization of information and digitalization using GIS (Geographic Information System mapping). In Butalelbún, the participants identified forty-two toponyms, forty-nine in Cauñicú, and forty in Callaqui. The information gathered aside from the toponyms was of different types: location of ceremonial sites, community meeting sites, historical foreign interventions, social-ecological conflicts, ecological or geomorphological landmarks, places with particular norms related to common water resources and their diverse uses.

Together with the meaning of Indigenous toponyms, we learned diverse *piam* (popular sayings, narratives, or stories) and *epeu* (traditional stories with a moral), which transmit tales and teachings associated with these places. Toponyms teach how people have co-evolved with their environment in time and how they have developed intimate relationships with different spaces through the experience of living there (Salazar & Riquelme Maulen, 2020). A map that gathers these Indigenous toponyms and their associated narratives allows people to get closer to knowing the territory through the lens of the traditional inhabitants' lived experience.

Next Turns in the River

At the heart of this research is a solid intention to achieve knowledge co-production and carry out the research process guided by values of respect, responsibility, reciprocity, and relevance, as inspired by endeavours such as the Decolonizing Water Project (https://decolonizingwater.ca/). In line with this, we hope to return the information gathered through this project in a format that is legible and beneficial for different audiences for community use. The GIS information was the scientific basis used to design illustrated maps for teaching purposes. These maps, printed on high-quality Size-A1 paper, will be shared with the rural schools where the workshops were hosted and with local NGO collaborators. Another way of sharing results with the community is via short videos on four topics: the participatory mapping workshops, the river monitoring experience, female leadership in environmental activism and science, and the role of universities as institutions dedicated to knowledge generation. These materials will also be hosted in an open online platform to make the research results widely accessible.

Historically, Indigenous knowledge systems have been seen through the binary lens of savage vs. civilized, undermining their value and making invisible the right of Indigenous peoples to self-determination. This superiority viewpoint has excluded Indigenous peoples from participating in knowledge creation, even about their own territories and common resources. Advancing beyond this lens requires explicit recognition of the legitimacy of multiple knowledge systems and the affirmation of the responsibilities that academic institutions have to foster horizontal dialogue with diverse knowledge holders. Participatory co-creation of knowledge, including understanding the stories behind the right words for naming places, is a strategy for strengthening opposition to ecologically damaging water diversions and dams through increased collaboration and allyship, political agency, and climate justice advocacy.

Reference List

Agrawal, A. (1995). Dismantling the divide between Indigenous and scientific knowledge. *Development and Change, 26*(3), 413–439. https://doi.org/10.1111/j.1467-7660.1995.tb00560.x

Aigo, J. del C., Skewes, J.C., Bañales-Seguel, C., Riquelme Maulén, W., Molares, S., Morales, D., Ibarra, M.I., & Guerra, D. (2020). Waterscapes in Wallmapu: Lessons from Mapuche perspective. *Geographical Review*, 1–19. https://doi.org/10.1080/00167428.2020.1800410

Álvarez, L., & Coolsaet, B. (2020). Decolonizing environmental justice studies: A Latin American perspective. *Capitalism, Nature, Socialism, 31*(2), 50–69. https://doi.org/10.1080/10455752.2018.1558272

Álvez-Marín, A., Bañales-Seguel, C., Castillo, R., Acuña-Molina, C., & Torres, P. (2021). Legal personhood of Latin American rivers: Time to shift constitutional paradigms? *Journal of Human Rights and the Environment, 12*(2), 147–176. https://doi.org/10.4337/jhre.2021.02.01

Bauer, C. (2004). *Siren song: Chilean water law as a model for international reform* (1st ed.). Routledge. https://doi.org/https://doi.org/10.4324/9781936331062

Berkes, F., & Folke, C. (1998). Linking social and ecological systems for resilience and sustainability. *Linking Social and Ecological Systems, 1*, 13–20.

Boisier, J.P., Alvarez-Garreton, C., Cordero, R.R., Damiani, A., Gallardo, L., Garreaud, R.D., Lambert, F., Ramallo, C., Rojas, M., & Rondanelli, R. (2018). Anthropogenic drying in central-southern Chile evidenced by long-term observations and climate model simulations. *Elementa, 6*. https://doi.org/10.1525/elementa.328

Budds, J. (2020). Securing the market: Water security and the internal contradictions of Chile's Water Code. *Geoforum, 113*(January), 165–175. https://doi.org/10.1016/j.geoforum.2018.09.027

Ciplet, D., & Roberts, J.T. (2017). Climate change and the transition to neoliberal environmental governance. *Global Environmental Change, 46*,148–156. https://doi.org/10.1016/j.gloenvcha.2017.09.003

Colin, N., Caputo Galarce, L., Fierro, P., & Górski, K. (2021, September). Carretera Hídrica, ¿solidaridad o egoísmo? *Ciper.* https://www.ciperchile.cl/2021/09/15/carretera-hidrica-solidaridad-o-egoismo/

Crook, M., Short, D., & South, N. (2018). Ecocide, genocide, capitalism and colonialism: Consequences for Indigenous Peoples and global ecosystems environments. *Theoretical Criminology, 22*(3), 298–317. https://doi.org/10.1177/1362480618787176

DGA (National Water Directorate/Dirección General de Aguas) (2022). https://dga.mop.gob.cl/Paginas/default.aspx

Dunham, J.B., Angermeier, P.L., Crausbay, S.D., Cravens, A.E., Gosnell, H., McEvoy, J., Moritz, M.A., Raheem, N., & Sanford, T. (2018). Rivers are social-ecological systems: Time to integrate human dimensions into riverscape ecology and management. *Wiley Interdisciplinary Reviews: Water, 5*(4), e1291. https://doi.org/10.1002/wat2.1291

Evans, J., & Jones, P. (2011). The walking interview: Methodology, mobility and place. *Applied Geography, 31*(2), 849–858. https://doi.org/10.1016/j.apgeog.2010.09.005

Figueroa, R., Rojas Hernández, J., Barra, R., Arumi, J.L., Delgado, V., Álvez, A., Parra, Ó., Torres, R., Urrutia, R., & Díaz, M.E. (2020, July). Por qué la carretera hídrica no es un proyecto sustentable. *Ciper Académico.* https://www.ciperchile.cl/2020/07/08/por-que-la-carretera-hidrica-no-es-un-proyecto-sustentable/

Folke, C. (2006). Resilience: The emergence of a perspective for social-ecological systems analyses. *Global Environmental Change, 16*(3), 253–267. https://doi.org/10.1016/j.gloenvcha.2006.04.002

Garreaud, R., Lopez, P., Minvielle, M., & Rojas, M. (2013). Large-scale control on the Patagonian climate. *Journal of Climate, 26*(1), 215–230. https://doi.org/10.1175/JCLI-D-12-00001.1

Guber, R. (2011). *La etnografía. Método, campo y reflexividad.* Siglo Veintiuno Editores.

Habit, E., Belk, M.C., Tuckfield, R.C., & Parra, O. (2006). Response of the fish community to human-induced changes in the Biobío River in Chile. *Freshwater Biology, 51*(1), 1–11. https://doi.org/10.1111/j.1365-2427.2005.01461.x

Higgins, P., Short, D., & South, N. (2013). Protecting the planet: a proposal for a law of ecocide. *Crime, Law and Social Change, 59*(3), 251–266. https://doi.org/10.1007/s10611-013-9413-6

Lacey-Barnacle, M., Robison, R., & Foulds, C. (2020). Energy justice in the developing world: a review of theoretical frameworks, key research themes and policy implications. *Energy for Sustainable Development, 55*, 122–138. https://doi.org/10.1016/j.esd.2020.01.010

Lindenmayer, D.B. & Laurance, W.F. (2016). The unique challenges of conserving large old trees. *Trends in Ecology and Evolution, 31*(6), 416–418. https://doi.org/10.1016/j.tree.2016.03.003

Lindgren, T. (2018). Ecocide, genocide and the disregard of alternative life-systems. *International Journal of Human Rights, 22*(4), 525–549. https://doi.org/10.1080/136 42987.2017.1397631

Ministerio de Energía. (2016). *Estudio de cuencas: Análisis de las condicionantes para el desarrollo hiroeléctrico en las cuencas del Maule, Biobío, Toltén, Valdivia, Bueno, Puelo, Yelcho, Palena, Cisnes, Aysén, Baker y Pascua.* https://cambioglobal.uc.cl/ images/proyectos/Documento_42_Estudio_de_cuencas_2.pdf

Moraga, J. (2001). *Aguas turbias: La central hidroeléctrica Ralco en el Alto Bío Bío.* LOM Edicione.

Namuncura, D. (1999). *Ralco: ¿Represa o pobreza?* LOM Ediciones.

Newell, P., & Taylor, O. (2020). Fiddling while the planet burns? COP25 in perspective. *Globalizations, 17*(4), 580–592. https://doi.org/10.1080/14747731.2020.1726127

Pacheco, M. (2018). Prólogo. In M. Pacheco (Ed.), *Revolución energética en Chile.* Ediciones UDP.

Pretty, J. (2011). Interdisciplinary progress in approaches to address social-ecological and ecocultural systems. *Environmental Conservation, 38*(2), 127–139. https://doi. org/10.1017/S0376892910000937

Rojas, M., Aldunce, P., Farias, L., González, H., Marque, J., Muñoz, C., Palma-Behnke, R., Stehr, A., & Vicuña, S. (Eds.). (2019). *Evidencia científica y cambio climático en Chile. Resumen para tomadores de decisiones.* Comité Científico COP 25; Ministerio de Ciencia, Tecnología, Conocimiento e Innovación. https://static.emol. cl/emol50/documentos/archivos/2019/12/04/file_20191204095012.pdf

Salazar, G., & Riquelme Maulen, W. (2020). The space-time compression of Indigenous toponymy: The case of Mapuche Toponymy in Chilean Norpatagonia. *Geographical Review, 112*(5), 1–26. https://doi.org/10.1080/00167428.2020.1839898

Schlosberg, D., & Collins, L.B. (2014). From environmental to climate justice: Climate change and the discourse of environmental justice. *Wiley Interdisciplinary Reviews: Climate Change, 5*(3), 359–374. https://doi.org/10.1002/wcc.275

Svampa, M. (2015). Commodities consensus: Neoextractivism and enclosure of the commons in Latin America. *South Atlantic Quarterly, 114*(1), 65–82. https://doi. org/10.1215/00382876-2831290

Valenzuela, F., McCracken, G.R., Manosalva, A., Habit, E., & Ruzzante, D.E. (2019). Human-induced habitat fragmentation effects on connectivity, diversity, and population persistence of an endemic fish, Percilia irwini, in the Biobío River basin (Chile). *Evolutionary Applications,* 1–14. https://doi.org/10.1111/eva.12901

Vargas, C.A., Garreaud, R., Barra, R., Vásquez-Lavin, F., Saldías, G.S., & Parra, O. (2020). Environmental costs of water transfers. *Nature Sustainability, 3,* 408–409. https:// doi.org/10.1038/s41893-020-0526-5

Vila, I., & Habit, E. (2015). Current situation of the fish fauna in the Mediterranean region of Andean river systems in Chile. *Fishes in Mediterranean Environments, 2,* 1–19.

Zeng, Y., Maxwell, S., Runting, R.K., Venter, O., Watson, J.E.M., & Carrasco, L.R. (2020). Environmental destruction not avoided with the Sustainable Development Goals. *Nature Sustainability, 3,* 795–798. https://doi.org/10.1038/s41893-020-0555-0

Zúñiga, D., & Ramos, M. (2021, March). "La carretera hídrica es una medida de corto plazo. Es mucho esfuerzo y dinero, solo para regar pobreza." *Ciper Académico.* https://www.ciperchile.cl/2021/03/20/la-carretera-hidrica-es-una-medida-de-corto-plazo-es-mucho-esfuerzo-y-dinero-solo-para-regar-pobreza/

Sentinels of Carelmapu: Participatory Community Monitoring to Protect Indigenous Marinescapes in Southern Chile

Francisco Araos, Florencia Diestre, Jaime Cursach,
Joaquin Almonacid, Wladimir Riquelme, Francisco Brañas,
Gonzalo Zamorano, José Molina-Hueichán, Darlys Vargas,
Manuel Lemus, Daniella Ruiz, and Claudio Oyarzún

Introduction

The Carelmapu "Community Sentinels" initiative in coastal Chile is a participatory citizen science experiment that aims to support local action to protect the territory and its inhabitants.

Protecting means establishing and sustaining caring relationships between people and all forms of life on the planet. Observation and awareness of the state of the territories is the first step to begin their care, through the communities' traditional knowledge, and/or through dialogue with the scientific knowledge provided by scientists.

Community monitoring can be carried out in any type of landscape or ecosystem: in marine-coastal areas, in the mountains, on lakes and rivers, and even in big cities. Observations can have ecological, environmental, social, historical, and cultural dimensions. Likewise, everyone can monitor, regardless of gender, age, ethnicity or other differences. The important thing is to have a motivated group that knows the territory or is interested in recognizing it, and is willing to learn the basic steps of the methodology.

Becoming a Sentinel also places us before a mirror, where the practices and actions of our communities become explicit, telling us about who we are and how we are inhabiting the place where we live. This is an act of consciousness that allows us to value our home and understand its development.

Community Sentinels is a low-cost initiative, open and adaptable to any situation and context, which allows monitoring of the multiple drivers and socio-environmental effects of climate change in the territories; citizens are in the front lines, facing it.

Know to Care, Care to Know

Knowing the ecosystems, cultures, and different ways of inhabiting territories and their socio-biodiversity allows us to identify the dangers that put life-networks at risk and alerts us to look after them.

Climate change, pollution, deforestation, and overfishing are producing transformations that afflict local communities, who perceive and call attention to them. These changes have revealed the importance of knowing the conditions of the territories to understand the pressures local communities are facing (Reyes-García et al., 2021). In this context, it is necessary to have tools to gather information about our environment and to help us to pose questions, generate alliances, and care for life networks (Iwama et al., 2021).

The exercise of "knowing through monitoring" is part of community practices where mutual care is an affective state between human and non-human beings. This is integrated with local cultural practices—for example, protection of threatened species' habitats, such as the kelp forest for marine otters. Habitat protection takes place through recognition of sacred or ritual practices in these places. Any organized community can carry out participatory monitoring, defining its objectives and adapting the methodology to its own contexts. Community monitoring includes recognizing and remembering our connections with the territories to which we belong, and also considering how we constantly interact and coexist with others on this one planet. In other words, what happens to other beings and territories affects us and has different consequences for all.

Central to community monitoring is that each Sentinel or group is in charge. Just as work in gardens or harvesting involves perceptions and functions involving the whole body, monitoring is also looking, smelling, touching, feeling, and being aware of the space and time of each process. This exercise helps to show changes, transformations, and movements. In addition, it makes it possible to show the transformational roles that other co-inhabitants play, such as insects, birds, mammals, pollinators, among others. In this way, the work of knowledge building in the monitoring process becomes collective and implies sharing care work, generating knowledge, and strengthening interpersonal relationships between human and non-human beings.

Many people of different ages, genders, origins, interests, occupations, and perspectives can participate, monitor, and implement this knowledge in other contexts. For example, school communities, environmental education centres, neighbourhood organizations, Indigenous communities, small producer organizations, among others, can all be involved. It also includes different objectives, such as: raising awareness and valuing nature in a school community, promoting the sustainable use of the territory in organizations, solving local problems such as water pollution, gathering data on species to motivate the conservation and protection of biodiversity, or making visible the various uses of resources such as a forest. Finally, multiple ecosystems and landscapes can be monitored: marine and coastal, archipelago, estuaries, mountains, lakes and rivers, valleys, forests, and wetlands.

Participatory community monitoring is a way to generate alliances and collaborate to protect territories and different life-ways. When the monitoring work becomes collective, the time and the different perspectives of the monitors are shared, contributing to dialogue and cultural exchanges, which promote the socio-ecological networks of the territories.

Citizen Science and Participatory Community Monitoring

The relationship between territories and their residents can be understood through citizen science, placing particular interest in monitoring how the components and processes of ecosystems are part of daily life. Although the monitoring exercises may be related to scientific research, the development of participatory community monitoring by citizens as a prevention strategy includes everything we do to maintain, monitor, and improve our world. It is a process that allows us to check old experiences, remember life-ways through stories and, above all, to recognize the threats to the territories and develop a better future.

Participatory monitoring is connected to science through behaviours and skills such as curiosity, questioning, observation, and feedback. In this way, Sentinels identify and monitor the present and future problems of the territory, motivating their inclusive governance (Araos and Ther, 2017).

Citizen science is a research approach where civil society participants collect, categorize, and/or analyze scientific data (Bonney et al., 2014), creating partnerships and collaborations between scientists and non-scientists (Jordan et al., 2012). Therefore, as an innovative method, citizen science can be helpful to promote individual and collective action on climate change, build social participation in environmental issues, and share experiences with the scientific process. Furthermore, it allows reconnecting with the natural world, motivates collective actions related to biodiversity protection, and strengthens commitments to participate and support place-based management (Groulx et al., 2017).

Community members' interest in protecting their living places against external dangers is expressed in a vision of the future related to the conservation and sustainable use of nature, integrating traditional practices and knowledge of elders, and promoting various ways of being connected to the environment (Brondízio et al., 2021). Thus, participatory community monitoring can contribute to commitments related to protecting vital processes and regenerating the places where people live.

This chapter presents a methodology for participatory and community monitoring to learn to understand territories and communities. Our case study illustrates its implementation. We have found that the recorded information is key to generating dialogue and solutions to the kinds of problems that can occur in many different regions.

Community Sentinels Methodology

This section presents a synthesis of the Sentinels methodology (LabC-ULAGOS, 2021). A full description is available for free download at: www. centinelascomunitarios.cl.

The process and stages of participatory community monitoring are simple. It begins with an organized group or community that is interested in

Table 11.1 The four stages of participatory community monitoring.

Stage	Description
1. Preparation	In this stage, the community agrees on their interests and monitoring objectives. They establish a list of key topics that allows them to decide where the attention of the Sentinels will be focused.
2. Organization	This stage begins with the organization of the group of Sentinels, considering the diversity of genders, ages, territories, and other characteristics that the organizing group deems essential. Then, the methodology is prepared according to the context in which it will be developed: identifying the time for data collection (days, weeks, months, etc.), defining the number of records to be made, creating key questions to support the Sentinels' tasks, and ensuring that all agreements reached are available to them. At the same time, they create the messaging group (e.g., WhatsApp or Telegram) where the recorded data is sent, and the technical team works on its review. Two people support the data collection. This stage ends with a monitoring test to check details and resolve unexpected occurrences.
3. Implementation	The third stage includes the implementation of monitoring and records transfer by the Sentinels group. The technical team works on data review, ensuring all reports have locations, visual support (photographs, videos, or drawings), and descriptions. In addition, the technical team goes with the Sentinels to help with the registration process, solving problems and collecting more specific data. The implementation stage ends when the monitoring time—previously defined—or the agreed number of records are completed. Finally, the transcriptions of the Sentinels' audios or narrations are copied, the logs containing all the monitoring materials are prepared, and the locations are identified and saved on maps (e.g., Google Maps, Open Street Maps, ArcGIS, or Google Earth).
4. Media campaign	The last stage consists of the dissemination of the material. The data is organized and shared on the platform www.centinelascomunitarios.cl, which allows it to be incorporated into the work of the community participatory monitoring network. From here, free and open access to the data and its networking with other territories is guaranteed.

Source: Developed by the authors.

collecting data about its local territory. The group chooses the elements and areas to be monitored. Then, one group (the Sentinels) carries out the information collection and recording, and a second group (the technical team) reviews, links, and shares the results.

Local People Watching Their Territories

The Sentinels work directly on the collection of participatory monitoring data. They oversee noticing, observing, and communicating about the environmental state of a territory or about some phenomenon of change or alteration.

In community participatory monitoring initiatives, local people are invited to be part of the Sentinel network in charge of observing the territories. Their experiences as inhabitants are expressed in daily experiences, such as telling stories they have heard throughout their lives that support their records. As a result, the Sentinels have knowledge associated with the landscapes and develop cultural and spiritual connections with the place where they live. Indeed, they are the local inhabitants who travel and experience the territory on a daily basis, who first and directly perceive the movements and changes that occur.

The Sentinels' vision of the state of their territories allows them to design and apply care practices associated with humans, the environment, climate change, and biodiversity. In addition, sharing these records is a way to participate, build relationships of affection and understanding with their environment and community.

Some of the elements that can be recorded are

- Biodiversity: components of the environment such as animals (birds, cetaceans, fish, rodents, amphibians, mollusks, insects, etc.); plants (native, introduced, edible, with or without flowers, etc.); algae; medicinal herbs; fungi; ecosystems (marine, wetlands, estuaries, lakes, rivers, forests, mountains, etc.); among others.

- Environmental risks: water, soil, and air pollution events; but also can be productive activities that are harmful or dangerous for the community and the environment, pollution situations, or risk events for health.

- Climate change: environmental events or phenomena related to climate change, ecosystem changes, population changes of some species, extreme weather events (droughts, floods, fires), impacts caused by these events (landslides, migration, etc.).

- Cultural landscape: Places or sites of importance to the inhabitants, spaces where they carry out their daily activities, sites related to cultural and spiritual practices, heritage and archaeological sites, and tourist places (natural and/or cultural).

- Activities: practices that are part of people's daily life, their communities, or groups of people, such as productive activities, fishing, diving, horticulture, harvesting, religious celebrations, recreational, tourism, educational activities, among others.

- Stories and memories: stories or narratives about past events, such as the origins or arrival of a group to the territory, past events that happened in the region, relationships between communities and their surroundings, memories, experiences, anecdotes, or stories.

Community Sentinels methodology offers an opportunity to generate and integrate multi-dimensional information, through the collaboration and participation of its inhabitants, based on traditional knowledge and the social and environmental situation of a place.

Case Study: Sentinels of Carelmapu

In Chile, since 2008, a door has been opened for the protection of the ocean for and by Indigenous people through the Lafkenche Law (No. 20.249), which created Indigenous Marine Areas (IMAs, or ECMPOs in Spanish). The IMAs are an institutional tool that protects the traditional uses of Indigenous communities (e.g., traditional fish practices, sacred places, rituals, cultural landscapes) and allows them to protect their way of life. In addition, through their implementation, Indigenous peoples contribute to biodiversity conservation and ecological restoration (Araos et al., 2020).

Carelmapu, or "green land" in the Indigenous language Mapudungun, is located on the western side of the Chacao Channel in south-central Chile (see Map 1, page 29) (Cursach, 2018). It has an area of 178.3 km² and 3,537 inhabitants (INE, 2019). Administratively, it is part of the Municipality of Maullín. Economic and subsistence activities are related to marine resources: boat or shore fishing, shellfish harvesting (mostly practiced by women), and finally, seaweed harvesting for consumption and commercialization (Rodriguez et al., 2014). In addition, Carelmapu has important breeding sites for marine birds and mammals, and the offshore Lenqui marine wetland is internationally considered an important site for the conservation of migratory shorebirds and marine-coastal biodiversity (Cursach, 2018).

The six Indigenous communities that belong to the Carelmapu Indigenous Communities Association (Associación de Comunidades Indígenas de Carelmapu) are: Encuramapu, Kalfu Lafken, Lafkenche Kupal, Lafken Mapu, Wetripantu, and Huerque Mapu Lafken. They formally requested the creation of the Carelmapu Indigenous Marine Area.

For the Indigenous communities, the IMA is an essential tool for the marine-coastal protection of Carelmapu, both from the dangers of environmental pollution connected to salmon farming and from local exploitation of natural resources (Cid and Araos, 2021). The requested area of the IMA covers 28,106 ha, from Astilleros to Amortajado, including the coast of Carelmapu Bay, Punta Chocoy, and including Doña Sebastiana Island and the Farellones de Carelmapu.

Traditional activities for protection were named in the IMA request and are as follows: fishing; boat fishing; shellfish harvesting for domestic consumption and seaweed gathering; religious and/or spiritual practices, recognizing ceremonial sites in the area of Los Corrales and Mar Brava Beach; the gathering of lawen (natural medicine) such as sargasso or huiro, limpia plata, seawater, and wolf oil; and recreational practices on sandy beaches such as Carelmapu Bay, Lenqui, and Mar Brava.

The importance of Doña Sebastiana Island is also recognized, as the island is an ancestral area for use of marine resources, and also known as a magical place with many stories and myths related to sunken ships, sightings of Caleuche (a mythical ghost-ship) and mermaids, which make the island a fundamental landmark in the cultural landscape of the IMA of Carelmapu (Cid and Araos, 2021). Finally, the Indigenous communities recognize the importance of the IMA for the conservation of biodiversity, the support of local livelihoods, and the sustainable development of the area.

In relation to the protection of the IMA, the Mapuche-Huilliche Association has expressed their interest in updated scientific information about the marinescape. Thus, our project entitled "Indigenous Marinescapes and Citizen Science: Enhancing Local Ecological Knowledge of Environmental Change in Southern Chile" was developed, based at the Citizen Science Laboratory of the University of Los Lagos, with the objective of providing Indigenous communities with a methodological strategy for participatory community monitoring that would allow them to collect, systematize, and share relevant data for traditional knowledge to support the protection and restoration of the Carelmapu IMA.

Participatory Monitoring Results

Using the Community Sentinels methodology described above, we carried out monitoring during the first half of 2021. It involved the participation of five Sentinels (two women and three men) who were in charge of observing and recording information about biodiversity, environmental dangers, and cultural landscapes in their territories, using their cellphones. The Indigenous communities selected the Sentinels based on their interest in participating in the project, while also considering gender, age, territorial diversity, etc. Thus, the methodology was adapted to the local context, taking into account the experiences of the Sentinels, their interests, and health conditions related to the context of the COVID-19 pandemic.

The geographical scale covered by this project was local, associated with the IMA area, with the possibility of covering a larger area with this initiative in the future. This first experience took about six months. Since the territory is protected by the IMA, it is possible to continue the monitoring exercise as long as necessary.

The information recorded by the Sentinels was organized into personal logbooks (see Figure 11.1 and Figure 11.2) and shared online and in a paperback book. In addition, the geospatial location was integrated into a GIS (geographic information system) platform, available on the Sentinels website.

Fig. 11.1 Manuel Lemus's personal logbook on biodiversity and environmental threats, and its edited version.

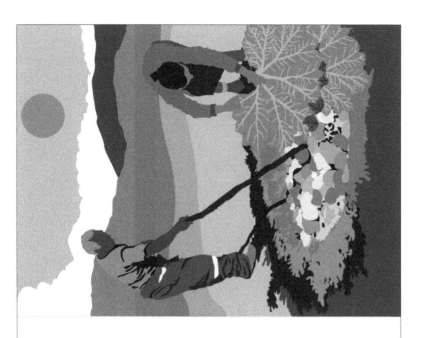

CHACAO CHANNEL

SHELLFISH AND SEAWEED GATHERING

📍 Don Juan Beach

Observations: Here we can see how the people gather luche. Upstream there is a lot of *luche* and some people go there to gather it, fill sacks with *luche*, and then rinse it with seawater. Then, they drain it and wrap it with *nalca* leaves, making small blocks. They cook those small blocks the "*curantiado*" way: they dig a hole in the ground, put firewood and stones inside it, then start a fire and let all the wood burn until there are almost no embers left. The small blocks of *luche* wrapped with *nalca* leaves are placed there and covered with a tarp or with a layer of soil. They cook it for an hour and a half or two, depending on the heat of the stones.

REPORTED BY
Darly Vargas [17-04-2021]

LUCHE GATHERING BY DARLY VARGAS

CENTINELAS COMUNITARIOS   82

Fig. 11.2.
Darly Vargas's personal logbook on biodiversity and cultural landscapes, and its edited version.

Manuel's photographs are included, and he notes in his observations that three years ago, this area was full of sea lions, but they are no longer seen. It's said that fishing boats hunt sea lions for bait. A speedboat came from Ancud to check the sea lion den and now nothing is there.

The Sentinelas report includes these observations, with an illustration, and also notes, "Sea Lion Colonies: One of the principal reproductive colonies of the common sea lion (*Otaria flavescens*) in the Los Lagos region is located on Sebastian Island. Besides the colony on Metalqui Island in Chiloë, it is the largest in southern Chile."

Source: Developed by the authors based on LabC-ULAGOS, 2021.

Darly's report includes photos of people collecting "luche" (sea lettuce) at Don Juan Beach, and recounts that it grows in abundance near the Sale River and some people go to search for it and bring sacks of it to wash in seawater on the beach. They wring it out and make little cakes wrapped in "nalca" (Chilean rhubarb) leaves. These cakes are cooked in "curantiado" fashion: by making a hole in the ground, and filling it with stones and wood which are burned until all the coals are gone. Then the wrapped luche cakes are put in and covered with canvas or turf. They are left for one and a half to two hours, depending on the heat of the stones.

This information is included, with an illustration, in the Sentinels report under "Chacao Channel: Seafood gathering activities."

Source: Developed by the authors based on LabC-ULAGOS, 2021.

Through community collaboration, the data collection process proceeded without difficulties. We also ensured that all data collected were available on a freely accessible GIS platform. In this case, the instruments used for data collection were cell phones, which supported the Sentinel network. The team of five Carelmapu Sentinels worked together on the project. During a day of training, they received technical support and methodological documents to help them easily record and share project information.

At the end of the project, the Sentinels received a certificate of participation in the initiative developed by the Citizen Science Laboratory of the University of Los Lagos, to attest to their experience, knowledge, and participation in the development of this methodology.

All monitoring information collected by community participants is shared through the Sentinels web platform. This allows the information that has been collected during the monitoring stage to be linked to new data as it emerges. Given its participatory nature and focus on territorial analysis, this project is part of the broader Community Sentinels initiative. This project can be replicated in other Indigenous communities who are applying for IMAs, such as other places in Los Lagos region, and include similar initiatives in both freshwater and marine zones with Indigenous communities and environmental activist organizations. In developing and using the Sentinels methodology, we even see the possibility of scaling up this work across the coastal zone of Chile in collaboration with local organizations who aim to gather information about their own marine areas in order to develop socio-environmental protection actions related to marine-coastal zone management.

Conclusion

The collaborative approach proposed by the Community Sentinels is based on the participation of local community members, and emphasizes the traditional knowledge of Indigenous peoples. This makes it possible to include the broad dimensions of IMAs and related territories, cultural landscapes, and the wide distribution of species, while also considering ecological, social, and economic areas.

Carelmapu participatory community monitoring contributes to integrating marine citizen science in the implementation of the IMAs, in order to challenge environmental changes and anthropogenic factors that endanger the marine zone. Participatory monitoring can be used as a tool for local people to contribute and share in scientific research and recording of data on biodiversity, cultural landscapes, and environmental risks. This methodology creates dialogues between local people and scientists and opens possibilities for developing individual and collective actions to address environmental changes in marinescapes.

We believe this has great value for marine-coastal conservation and management, by providing skills and information that allow real-time monitoring of the marinescape environment protected by the IMAs, and encouraging people to reconnect with nature, appreciate it, and protect it.

Acknowledgements

Thanks to the Mapuche-Huilliche Communities Association of Carelmapu for their willingness to co-create citizen science and for their hard work in the implementation of the pilot experience together with the Sentinels: José Molina, Darly Vargas, Manuel Lemus, Daniela Ruiz, and Claudio Oyarzún.

Thanks to the Queen Elizabeth II Diamond Jubilee Scholarships/Climate Justice, Commons Governance, and Ecological Economics project, York University (Canada), and ANID/FONDECYT Project 1220430 "La resurgencia de los comunes en el Antropoceno Azul en Chile" for making this initiative possible.

Thanks to Paulina Mansilla Haeger and Kathryn Wells for the English translation and review.

Reference List

Araos, F., Anbleyth-Evans, J., Riquelme, W., Hidalgo, C., Brañas, F., Catalán, E., Nuñez, D., Diestre, F. (2020). Marine Indigenous areas: Conservation assemblages for sustainability in southern Chile. *Coastal Management*, 48(4), 289–307. https://doi.org/10.1080/08920753.2020.1773212

Araos, F., & Ther, F. (2017). How to adopt an inclusive development perspective for marine conservation: Preliminary insights from Chile. *Current Opinion in Environmental Sustainability*, 24, 68–72. https://doi.org/10.1016/j.cosust.2017.02.008

Bonney, R., Shirk, J.L., Phillips, T.B., Wiggins, A., Ballard, H.L., Miller-Rushing, A.J., & Parrish, J.K. (2014). Next steps for citizen science. *Science*, 343(6178), 1436–1437.

Brondízio, E.S., Aumeeruddy-Thomas, Y., Bates, P., Carino, J., Fernández-Llamazares, Á., Ferrari, M.F., Galvin, K., Reyes-García, V., McElwee, P., Molnár, Samakov, A., & Shrestha, U.B. (2021). Locally Based, regionally manifested, and globally relevant:

Indigenous and local knowledge, values, and practices for nature. *Annual Review of Environment and Resources, 46*, 481–509.

Cid, D., & Araos, F. (2021). Las contribuciones del Espacio Costero Marino para Pueblos Originarios (ECMPO) al bienestar humano de las comunidades indígenas de Carelmapu, Sur de Chile. *CUHSO, 31*(2), 250–275.

Cursach, J. (2018). *Revisión bibliográfica sobre la biodiversidad marina del mar adyacente a Carelmapu, con especial énfasis en aves y mamíferos marinos* [Unpublished report].

Groulx, M., Brisbois, M.C., Lemieux, C.J., Winegardner, A., & Fishback, L. (2017). A role for nature-based citizen science in promoting individual and collective climate change action? A systematic review of learning outcomes. *Science Communication, 39*(1), 45–76. https://doi.org/10.1177/1075547016688324

INE (Instituto Nacional de Estadística). (2019). *División político administrativa y censal*. Departamento de Geografía.

Iwama, A.Y.; Araos, F.; Anbleyth-Evans, J.; Marchezini, V.; Ruiz-Luna, A.; Ther-Ríos, F.; Bacigalupe, G., & Perkins, P.E. (2021). Multiple knowledge systems and participatory actions in slow-onset effects of climate change: Insights and perspectives in Latin America and the Caribbean. *Current Opinion in Environmental Sustainability, 50*, 31–42. https://doi.org/10.1016/j.cosust.2021.01.010

Jordan, R., Ballard, H., & Phillips, T. (2012). Key issues and new approaches for evaluating citizen-science learning outcomes. *Frontiers in Ecology and the Environment, 10*(6), 307–309. https://doi.org/10.1890/110280

LabC-ULAGOS (2021). *Centinelas comunitarios. Guía metodológica para realizar monitoreos participativos comunitarios*. Laboratorio de Ciencia Ciudadana de la Universidad de Los Lagos, Osorno, Chile. https://www.researchgate.net/publication/359962150_centinelas_comunitarios_guia_metodologica_para_realizar_monitoreos_participativos_comunitarios

Reyes-García, V., Fernández-Llamazares, Á., García-del-Amo, D., & Cabeza, M. (2020). Operationalizing local ecological knowledge in climate change research: Challenges and opportunities of citizen science. In M. Welch-Devine, A. Sourdril, B. Burke (Eds.), *Changing climate, changing worlds* (pp. 183–197). Springer, Cham.

Rodríguez, D., Gajardo, C., & Ther, F. (2014). *Chile litoral 2025: Modelo de gestión territorial para asentamientos de pescadores artesanales*. Carelmapu, Provincia de Llanquihue, Región de Los Lagos: Serie etnografías Litorales. Universidad de Los Lagos, Osorno.

Inequality in Water Access for South Africa's Small-Scale Farmers Amid a Climate Crisis: Past and Present Injustices in a Legal Context

Patience Mukuyu and Mary Galvin

Introduction: *Access to Water, Food Security, and Climate Change*

In South Africa, as in most developing countries, small-scale farming is central to achieving food security, particularly in communal areas (Khalil et al., 2017)). These communal areas (formerly called homelands, imposed by the apartheid regime) are characterised by high poverty levels alongside widespread unemployment (von Fintel & Pienaar, 2016). Although several government policies aim to improve the agricultural productivity of historically disadvantaged farmers on communal land, their limited access to water is a critical constraint. Access is marred by two interrelated factors: climatic variations and a divisive history surrounding water allocations and access.

South Africa is generally described as water stressed (Denby et al., 2016), meaning that water is needed for many uses but the available water resources are too limited—particularly due to frequent droughts—to meet the high and growing demand. This water stress is compounded by highly unequal distribution of water resources between Black historically disadvantaged individuals (HDI) and white historically advantaged individuals (HAI), skewed infrastructure distribution, and limited and weak water-use rights among the vulnerable.

The 2021 Intergovernmental Panel on Climate Change (IPCC) report projected with high confidence increasing temperatures, a decrease in mean rainfall and frequent drought occurrences (IPCC, 2021). In line with these projections, parts of South Africa will face drought or biophysical water scarcity, which will be even more acute with already insufficient water resources. Yet the challenges faced by farmers are not only biophysical, but also hydropolitical. In 2006, a United Nations Development report boldly asserted that

> there is more than enough water in the world for domestic purposes, for agriculture and for industry. The problem is that some people— notably the poor—are systematically excluded from access by their poverty, by their limited legal rights or by public policies that limit access to the infrastructures that provide water for life and for livelihoods. In short, scarcity is manufactured through political processes and institutions that disadvantage the poor (UNDP, 2006, 3).

The recognition of manufactured scarcity means that "water equity—fair shares in access and entitlements to water, and benefits from water use— should form a central ambition in the decades to come" (Calow & Mason, 2014, p. 2).

The allocation of water amongst those with competing needs is often influenced by water tenure, defined by the Food and Agricultural Organization (FAO, 2020, p. 3) as

> The relationships, whether legally or customarily defined, between people, as individuals or groups, with respect to water resources.

The definition of water tenure captures the recognition of customary norms, reflecting social relationships, or to embody legal rights. Yet what is most important is that the concept of water tenure is closely aligned to security of tenure. Water tenure security allows for the realisation of both water and food security, which is especially critical in the context of climate change. Water tenure security depends on the legal recognition and enforcement under formal, statutory law, and there are debates about how legal systems can best achieve this.

The focus of the global debate among practitioners and academics on water tenure security is on the needs of vulnerable, small-scale farmers in the face of resource scarcity and competing water needs. In other words: How does the implementation of statutory water law help to reduce small-scale farmers' experiences of water insecurity? Reporting on a case study that is one aspect of a wider research project, this chapter discusses the legal framework for water resource allocation in view of both customary and statutory water laws, in the context of South Africa's discriminatory colonial history. This provides the context for understanding how small-scale farmers' access to water is affected by statutory and customary law and governance institutions.

Our participatory research approach, which included interviews with small-scale farmers directly affected by water access challenges, gave us insights into how various forms of water rights are intertwined with social inequities, exacerbated by climate-related rainfall changes, and how small-scale farmers are taking action to protect their livelihoods. The chapter concludes by indicating the direct relevance of our findings for policy concerned with achieving equity in the allocation of water resources. In particular, we explore how water tenure arrangements are (and could be) formulated to ensure equitable water access and to narrow the divide between policy, science, and implementation.

Research Design and Methodology

This study focuses on the Inkomati Catchment (or watershed; see Map 6; see Map 6, page 242), located in a semiarid region with variable rainfall and frequent droughts. In particular, it focuses on the communal land within the Sabie-Sand sub-catchment, one of the three sub-catchments in the Inkomati Catchment (alongside the Crocodile and Inkomati sub-catchments). Formerly part of apartheid homelands of Gazankulu and Lebowa, the area exemplifies the complexity of implementing both customary and statutory legal frameworks on communal land.

Water on communal land areas is often interlinked with access to water for both domestic and productive use, with communities relying on water from various sources (e.g., wells, rivers, and streams) for multiple uses (e.g., domestic use, livestock watering, and crop irrigation) (van Koppen et al.,

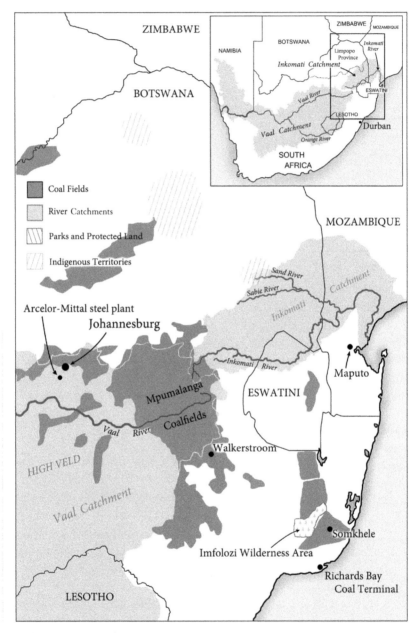

Map 6 South Africa—Vaal and Inkomati Catchments, and Coalfields

Fig. 12.1 The first author interviews a small-scale farmer in the Inkomati Catchment.

2020). The typical separation of domestic and agricultural uses of water is not clear cut in rural areas, where communities have such multiple water-use systems (Mukuyu et al., 2021; van Koppen et al., 2020; Hofstetter et al., 2021). Clearer recognition by government and local authorities of community efforts to supply their own water needs under customary water tenure can further enhance water access for vulnerable small-scale farming communities where their access is protected through statutory law (van Koppen et al., 2021).

Using a social constructivist methodology, based on the theory/awareness that reality and knowledge are constructed through shared discourse and social conventions, it is possible to generate knowledge about water access and water tenure arrangements using a ground-up approach to build and share understanding. This approach parallels a participatory research methodology where systematic inquiry is conducted in direct collaboration with those affected by the research, and provides a learning experience for both the researcher, local participants, and stakeholders (Couto, 1987; Vaughn and Jacquez, 2020).

This chapter is based on a review of secondary literature as well as twenty-five in-depth interviews conducted by the first author with small-scale farmers between April and August 2021, as well as her involvement in a related research project being conducted by the International Water Management Institute (IWMI).

Interviews focused on small-scale farmers' experiences and perceptions of water access and entitlement to water (Figure 12.1), and were guided by an interview questionnaire. A local non-profit organisation that has worked in the area for decades, the Association for Water and Rural Development (AWARD), helped introduce the researchers and gain access to villages. Households were selected through purposive sampling to target those involved in farming and irrigation activities. The local AWARD community facilitator who is fluent in the local languages of SiSwati and Shangaan assisted with translation, and interviews were recorded with the permission of interviewees.

Our work in the area is ongoing, and the next stages of this study will include observation of catchment forum meetings, secondary data analysis for technical contextual background in water-use and availability projections, as well as key informant interviews to understand relationships between farmers and local catchment agency and government departments.

Water Dispossession in the Colonial and Apartheid Era

The history of land dispossession set the foundation for inequitable water access in democratic South Africa. In particular, the 1913 Natives Land Act had a devastating impact on the majority Black population, who were relegated to ethnically defined "homelands" that comprised only 13 per cent of the country's land. The apartheid regime forcibly removed people from their land, controlling its natural resources such as minerals, forests, and water. The white minority consolidated power, amassing the most strategic and favourable land and accompanying natural resources—including water resources.

Paradoxically, South Africa was a legally pluralistic country with clear lines of legal jurisdictions. In other words, while customary laws were upheld in the homelands, statutory laws applied in the exclusively white areas. The main impetus of the early 1900s irrigation development period, supported by the 1912 Irrigation and Conservation Act, was to ensure adequate water supply to support agricultural development by the growing settler community (Tempelhof, 2017). As such, white farmers received considerable government support for their irrigation activities, such as subsidized infrastructure and water rights recognised under the prevailing laws. Infrastructure developed to meet these needs was concentrated in areas dominated by white people, creating further disparity with Black people (Tempelhoff, 2017; Klug, 2021). During the 1950s, the government also developed irrigation schemes within homelands, created as concessions by the apartheid government to keep Black people from migrating into the cities (Tempelhoff, 2017).

Riparian rights were applied to water appurtenant to land; in other words, land owners had rights over the water that flowed over their land. Over time other water uses began to materialise due to growing demands from industry, mining, and urban development. In response, the 1956 Water Act (which repealed the narrowly focused 1912 Irrigation Act) broadened the scope of governance by creating government control areas alongside riparian rights. Enforced mainly in the "white only areas," it aimed to ensure that irrigation development was balanced with providing water for other activities such as mining and industry. Through this Act, the State consolidated control over public water resources, alongside private water[1] rights (including riparian rights) that also applied to groundwater (Tempelhoff, 2017; Klug, 2021).

Statutory law was not applied in former homelands or communal areas, except in Government Water Control areas and government-owned irrigation schemes. On communal land, customary water tenure was the dominant legal system presided over by traditional authorities, and it seemingly played no role in the development of statutory water law in South Africa (Thompson, 2006).

Water Management and Tenure in the Former Homelands

Close to 70,000 ha of communal areas of Limpopo Province are informally irrigated (van Koppen et al., 2017), including using a hose pipe to tap into a nearby stream, shallow well, or wetland (Figure 12.2). Such informal irrigation is largely self-financed with no direct government involvement and is thus invisible to government when identifying areas under irrigation and formulating policy (van Koppen et al., 2017). Definitions of irrigation often exclude certain irrigation activities such as carrying water with a bucket, while conventional irrigation technologies are typically the focus of policy discussions (Venot et al., 2021). During this research, small-scale farmers were using this type of irrigation on plots ranging in size from 0.5 ha to 4 ha.

While customary tenure is prevalent in communal areas, customary water law is not explicitly recognised within statutory law (Murombo, 2021). This means that the water rights of farmers who have been irrigating their land are not adequately provided for under the National Water Act (NWA)

Fig. 12.2 Hose irrigation on a small-scale farm near a tributary of the Sand River in the Inkomati Catchment.

of 1998. The invisibility of these water uses within regulatory frameworks to provide secure entitlements to water renders them vulnerable to exploitation and increases small-scale farmers' water insecurity. Water uses in communal areas therefore face the risk of not being accurately considered in water allocation discourses, perpetuating a dismissal of Black small-scale farmers' capabilities in terms of productively using water resources (Dube, 2020).

Of course, even given this lack of consideration, customary norms continue to evolve and people living on communal land legitimise these customary norms and laws through acknowledgement and adherence to their local norms (van Koppen, 2022). Most recently, authority over water resources management has switched from traditional authorities to municipal authorities and the catchment management agency. The impact that this move has on shaping community-based water tenure systems is an ongoing area of inquiry of this research effort.

It is important to recognise important differences among the practices of small-scale farmers, including the institutional framework applicable to their context. During fieldwork conducted in the Inkomati Catchment, three types of small-scale farmers were encountered, each with different relations to statutory water rights. First, small-scale farmers in communal areas invest in their own access to water through pumps and storage tanks. Here customary water tenure applies. Second, there are small-scale farmers in government-owned irrigation schemes where water rights (conferred during the apartheid era) are still held by the government departments. Finally, there are "emerging farmers" on land restituted (bought, or legally reclaimed) from white owners. Here water rights are either linked to the restituted land or are at times separated from the land during restitution by former white owners and traded separately, thereby prejudicing the new Black owners. This chapter considers the first two types of water rights in this complex context, which we will continue to explore in future research.

An overview of the evolution of statutes relating to water rights provides some important background.

Current Legal Framework for Water Allocation: Righting the Wrongs of the Past?

South Africa's progressive constitution was formulated to address apartheid injustices and to ensure that specific rights are protected, such as the rights to food and water, equality and non-discrimination, administrative justice and redress (RSA, 1996). These constitutional foundations have formed the basis of legislation including the 1998 NWA, which repealed earlier legislation to advance equitable access to water resources and to allow for redistribution. Overall, the success of post-apartheid South African law—including water law—in righting the wrongs of the past has been criticized, despite its widely claimed progressive nature. This is mainly because some of the provisions have not been implemented as envisaged, particularly at the local level, and, in practice, have even perpetuated inequalities in water resource allocations between HDI and HAI. The following sections explore why and show how western legal systems imposed on communal land have served to perpetuate inequalities in water access: in particular, the impact of water-use licensing and the slow implementation of compulsory licensing in entrenching administrative injustices and hampering water access for small-scale farmers.

The Impact of Compulsory Licensing

As a transitional measure, the 1998 NWA included clauses for the recognition of existing lawful uses (ELU) in the two-year period before its enactment. This allowed for water entitlements issued during previous dispensations to remain valid, which was perhaps a reasonable transition from one Act to another and may have prevented legal and administrative upheaval. However, this "sunset" provision has largely benefitted the HAIs and has remained in place almost twenty-five years after the 1998 NWA was passed.

Compulsory licensing is a process that allows the government to review all water use in a catchment and to reallocate water if necessary. It is the only tool available to legally abolish ELU in the 1998 NWA. Supporting compulsory licensing are two other processes: verification of lawfulness of the ELU and validation of the extent of the ELU. The highly consultative nature of the compulsory licensing process results in a democratised process (which arguably is in line with Constitutional rights for all), but which in practice may not yield the expected redress outcomes due to long and drawn-out contestations. In the Inkomati Catchment where this study is based, the process is only about 60 per cent complete, even though it has been ongoing since 2010 (IUCMA, 2020).

The Act also authorizes the Department of Water and Sanitation to issue water-use licences to potential water users, entitling them to access specified amounts of water. The licensing process itself is resource intensive and can take years (Kidd, 2016). The majority of Black small-scale farmers do not have the administrative capacity to engage in this process and have largely been excluded.

Regarding the continued recognition of historical water rights (i.e. ELU) the State has at its disposal the power to either "deprive" (reduce) or expropriate rights—both of which should be implemented following due process, according to the Constitution. Constitutional provisions on the limitation of rights require that rights may only be limited "to the extent that the limitation is reasonable and justifiable in an open and democratic society based on human dignity, equality and freedom, taking into account all relevant factors" (RSA, 1996, s36). Due to water scarcity, in other words, water use should be rationally and fairly regulated.

According to comments from an unnamed Department of Water and Sanitation (DWS) national government official (Water Research Commission

[WRC], Project reference group meeting minutes, 19 May 2021), the role of compulsory licensing and indeed the National Water Act, is not to expropriate water rights but rather to limit or deprive those rights:

> The whole National Water Act is not couched on any expropriation but is couched on deprivation which is different from expropriation if you want to analyse what section 25^2 of the Constitution says.... We are implementing the National Water Act based on the limitation of rights and not on the total elimination of rights.

The use of the word "deprivation" is associated with the limitation of rights (for example, a reduced assurance of supply). Marais (2018, p. 2) defines deprivation as referring to the "state's police power to regulate the use, enjoyment and exploitation of property in the public interest, mostly without compensation." In other words, holders of water-use rights can retain their entitlements, while these rights are curtailed; for example, limits can be placed on the duration and place of exercising an entitlement. Given the historical imbalances in water access between the Black majority and white minority, expropriating (if the government budget allows) or deprivation of rights would seem the most effective means of redressing past inequities.

Licensing and Administrative Injustices

The fact that the compulsory licensing process, which entails reviewing and reallocating water resources, is still incomplete or has not been initiated in most of the country's catchments is a clear indication that it has not been a government priority. Moreover, water-use licences (WUL) as they were introduced in the 1998 NWA, were meant to equitably authorise water use post-1998. In practice, however, the licensing process itself has only perpetuated a skewed distribution of water towards white applicants. Further, national data show that the ratio of ELU to WUL is 4:1 (Hydrosoft Institute, 2021) indicating a slow progression in the conversion of ELU to WUL and a lower number of WUL applicants compared to ELU authorisations post-1998.

The water-use licensing process had hitherto faced delays and backlogs resulting in applicants lodging complaints against the DWS due to the lengthiness of the process, which impacted economic productivity. While the new administrative speedup of the licensing process[3] is a commendable move and likely to benefit many have who been stuck with an un-adjudicated licence application, from an equity perspective care must be taken to ensure that due process is not compromised to benefit "economic productivity" while prejudicing the small water users. It is up to the state department to decide within its powers what constitutes high impact and for whom.

A potential pitfall for swift adjudication of licence applications means allocable water may be quickly allocated to HAIs at the expense of current and future uses of HDI. This highlights the urgent need to render HDIs' water uses visible and protected, and realise how today's allocations shape future allocation for this vulnerable group. According to Dube (2020), there is a widely held perception among the elites that Black users do not use water productively and need only a very small volume of water. She terms this "deficit thinking" where it is assumed that Black people do not need large volumes of water anyway and therefore the water can be allocated to the supposedly more economically productive, largely HAIs.

Water-use licences have been presented as an economic enabling tool as opposed to a redistributive tool and as such have been used by HAIs to amass water supposedly with the intention of creating employment and sustainable development.

Legal Attempts to Ensure Water for Small-Scale Farmers

The majority of Black small-scale water users are provided for through legal exemptions for what are termed *de minimis* or minimal uses. Van Koppen (2007, p. 56) describes these uses as "second class," which are given a "status of being negligible and invisible by design for the mere reason—not their own fault—of not being administrable." These *de minimis* exemptions are gazetted as schedule 1 uses, which include basic domestic and household non-commercial uses. If a water use exceeds schedule 1 yet is below licensing requirements, a general authorization (GA) licensing exemption applies.

Farmers using more than the set GA threshold, which varies according to catchment conditions, need to obtain a licence. In "water stressed" catchments this GA threshold can be as low as 0.3 ha equivalent (or 2000 m^3),[4] an area which in some cases is smaller than the average areas cultivated by small-scale farmers. However, because licensing is largely administratively inaccessible to the majority Black water users, this renders all water use above the GA threshold illegal. The administrative process is prohibitively bureaucratic, with several technical requirements and assessments demanded from the applicant. The majority of Black small-scale water users are therefore administratively discriminated against as they do not fully grasp or have the means to fulfill these requirements.

The South African Government indeed recognizes how the licensing system excludes the majority of South Africans. The National Water Resources Strategy 2 (NWRS, of 2013) states that "mechanisms that reduce the administrative burden of authorising water use must be implemented. Current processes are often costly, very lengthy, bureaucratic and inaccessible to many South Africans" (DWA, 2013, p. 48). This concession by the DWS thus constitutes an administrative injustice according to s33 of the South African Constitution on just administrative action. Van Koppen (2007) has criticized the adoption of permit systems in unequal environments such as the Inkomati. She argues that permit systems boil down to the formal dispossession of water for the rural, informal water users who manage their water under community-based arrangements.

Another provision in the NWA of 1998 establishes the Ecological and Basic Human Needs Reserve, which is the only specific right to water included within the NWA. It is central in the legally binding South African NWRS, which establishes the Reserve as the country's first and utmost water allocation priority. This means that before any other allocations are made, ecological and people's basic water needs must be met. It is important to note that the Basic Human Needs Reserve is based on a minimum of 25 litres of water per person per day, which may meet the World Health Organisation (WHO) minimum quantities for domestic use but is insufficient for other equally essential uses such as irrigation and livestock watering, vital in rural contexts.

Despite the potential that the Basic Human Needs Reserve may have in redressing water access inequalities, its practical implementation has fallen short—even at the minimal quantities. In the Sabie-Sand sub-catchment of the Inkomati, it is unclear which authority or agency has responsibility for implementing the Basic Human Needs Reserve within the rivers.

However, it is precisely this human-rights-based approach for prioritisation of water use that is meant to protect those in communal areas as well as other marginalised groups, such as farm workers (Marcatelli, 2018). Moreover, this equity orientation is supported further by the NWRS 2 prioritization of poverty eradication as a basis for water allocation (DWA, 2013). The aim of poverty eradication is to improve livelihoods and advance racial equity in areas where poverty is endemic, such as former homelands. While the NWRS is a legally binding document according to the NWA, there has not yet been significant change in practice with respect to how water allocations are distributed.

Field Observations and Findings

Policies are only as effective as their implementation. Based on our participatory research with small farmers—HDIs—in the Inkomati Catchment, this section provides field insights into how policy efforts for redress and legislative tools have translated in local practice. In the villages discussed here, small-scale farmers with plot sizes ranging from 0.2 ha to 4 ha were interviewed by the first author, most living in the upstream and downstream reaches of the catchment. The bulk of the farmers fall within the GA category (no more than 2000m³/annum of surface water). During interviews, only one farmer acknowledged having a form of authorisation for his water use of more than 2000m³, likely a GA; while he knew the name of the official who told him about it, he was not well informed about its details.

In the upstream parts of the catchment where the Sand River originates, one community has taken water supply matters into their own hands and collectively brought water through gravity-fed pipes into their village and homes. Here, water is used for both domestic and irrigation water supply, illustrating the inapplicability of separating domestic water from other water uses in the rural setting. This water is not regulated by the municipality and villagers maintain their pipe system collectively through a committee that they set up for this and other community related issues. This water use is, however, not licensed and the local water resource management agency (IUCMA) has neither interfered nor regularised this use.

Other villages in the upstream areas have not mobilised in a similar manner to address their water issues due to distance and resource constraints. One village in particular was concerned about not being able to use the water from a nearby government-owned dam. Farmers did not know who to approach for authorisation to access this water. The dam is within their community and yet they have no access to the resource. Nonetheless, the dam supplies irrigation water to government-owned irrigation schemes in the midstream and downstream reaches of the catchment. Water rights for this water use are held by the Department of Agriculture, which has absolute control over these water rights as conferred during the apartheid government. Farmers in these irrigation schemes are largely concerned with maintaining irrigation infrastructure and less about ownership of water rights—since the water is provided to them anyway through the government irrigation project.

Preliminary interviews in the upstream and downstream parts of the catchment raised questions about the role of traditional authorities and their legitimacy in managing water allocations. Communities were dissatisfied with how these authorities have shifted focus from their traditional role of protecting the community's natural resources to one of making money off selling land parcels. Nonetheless, there are instances where their involvement has been useful such as in conflict resolution between HDIs over water uses. As a result, taking into account how they now function in democratic South Africa, it seems that the role of traditional authorities in equitable water allocation needs further exploration.

None of the farmers interviewed were registered water users, as per the statutory requirement. Farmers had no knowledge of the registration process or what was needed. Their water use thus remains invisible in government water allocation and planning processes. In a context where customary water laws are not explicitly recognized within statutory law, these water uses are overlooked and become vulnerable to having their water rights usurped by third party users, for example holders of a higher ranked water-use licence. Redistributing more water through, for example, expropriation of apartheid-era water rights, will free up much needed water to allocate to small-scale users and perhaps also warrant a raising of the GA threshold.

Discussion and Conclusion

South Africa's history of dispossession has shaped post-apartheid legislative reform in three ways. First, the South African legal terrain has always been pluralistic in nature due to the numerous cultural and religious influences that define the political landscape. However, even within such a pluralistic environment, a dominant legal system is widely implemented. Van Koppen and Schreiner (2019) show how water permitting systems—implemented through statutory water law with colonial foundations—have worsened water insecurity for small-scale users in five African countries including South Africa. They advocate a "hybrid" legal system that considers customary norms and is tailored to specific users.

Van Koppen and Schreiner (2019) argue that permit systems adopted from the Western ways of managing water do not translate well to local contexts in Africa where customary laws are upheld. As such, permit systems have been used as a tool to continue the disempowerment of Black water users by colonialists. As water permits or licences are largely inaccessible to the rural Black majority, they have been used by the white elite to amass water rights at the expense of future and current water use for the vulnerable Black majority in the communal areas. If more water was to be made available for uptake by this majority through more efficient water use and management (reduced wastage and losses, new dams, aquifer protection, etc.), equity considerations and due diligence in licensing, then some equity in allocation could be achieved.

Secondly, while customary laws are recognised in the South African Constitution, customary water laws are not explicitly recognised in the NWA of 1998. This results in a weaker recognition of customary water laws, since they often derive legitimacy from existing statutory laws (Murombo, 2021).

Third, the skewed distribution of infrastructure, and thus entitlement to stored or piped water, is an important aspect in understanding water access inequalities. In the Inkomati there are four dams in the Sabie-Sand sub-catchment that were constructed to serve government-owned irrigation schemes for Black farmers. However, other farmers who live in proximity to this state-owned infrastructure but outside of the irrigation schemes do not have a right to the water. Water management planning practices implemented during the apartheid era continue to be upheld, which disempowers small-scale farmers in the communal areas who are not operating within the formal structures of government irrigation schemes and creates tiers of inequity. Reopening discussion of government water rights along with improved implementation of the NWA and NWRS would help to address these inequitable inefficiencies.

In conclusion, the critical role of legislation and water allocation regulations in empowering small-scale farmers to attain equitable access to water is irrefutable. As our research shows in the unequal communal areas of the Sabie-Sand sub-catchment, water security for HDI communities requires integrating communal and customary governance systems with property rights and greater legislated water access.

NOTES

1   In the 1956 National Water Act, private water was defined as "water which rises or falls naturally on any land or naturally drains or is led on to one or more pieces of land which are the subject of separate original grants, but is not capable of common use for irrigation purposes" (RSA, 1956).

2   Section 25 of the 1996 South African Constitution deals with the property clause as regards deprivation, expropriation, and compensation. Water rights are considered as property rights in this respect.

3   A turn-around time of 90 days for water use licensing was emphasized by the president in 2022 during a State of the Nation address. https://www.gov.za/speeches/president-cyril-ramaphosa-2022-state-nation-address-10-feb-2022-0000

4   Irrigation water use is measured in volume per hectare equivalent, especially under Existing Lawful Uses.

Reference List

Calow, R., & Mason, N. (2014). *The real water crisis: Inequality in a fast-changing world.* Overseas Development Institute.

Couto, R.A. (1987). Participatory research: Methodology and critique. *Clinical Sociology Review, 5*(1), 9.

Denby, K., Movik, S., Mehta, L., & van Koppen, B. (2016). The "trickle down" of IWRM: A case study of local-level realities in the Inkomati Water Management Area, South Africa. *Water Alternatives, 9*(3), 473–492.

Dube, B. (2020). Deficit thinking in South Africa's water allocation reform discourses: A cultural discourse perspective. *Journal of Multicultural Discourses, 16*(4),293–312. https://doi.org/10.1080/17447143.2020.1835926

DWA (2013). *National water resources strategy 2.* Pretoria, South Africa Department of Water Affairs.

FAO (Food and Agriculture Organization). (2020). *Unpacking water tenure for improved food security and sustainable development.* Land and Water Discussion Paper 15s. Rome. https://doi.org/10.4060/cb1230en

Hofstetter, M., van Koppen, B., & Bolding, A. (2021). The emergence of collectively owned self-supply water supply systems in rural South Africa–what can we learn from the Tshakhuma case in Limpopo? *Water SA, 47*(2), 253–263.

Hydrosoft Institute (2021). *Decolonising water access and allocation: A renewed effort to address persistent inequalities in the water sector.* Water Research Commission Report, Project No. K5/2858.

IPCC (Intergovernmental Panel on Climate Change). (2021). *Regional fact sheet—Africa: Sixth assessment report: Working Group 1—The physical science basis.* Retrieved on 20 January 2023, from https://www.ipcc.ch/report/ar6/wg1/resources/factsheets/x

IUCMA. (2020). *Inkomati-Usuthu Catchment Management Agency annual performance plan 2020/2021*. 1 April 2020–31 March 2021.

Khalil, C.A., Conforti, P., Ergin, I., & Gennari, P. (2017). *Defining small-scale food producers to monitor Target 2.3 of the 2030 agenda for sustainable development*. Food and Agricultural Organization of the United Nations (FAO).

Kidd, M. (2016). Compulsory licensing under South Africa's National Water Act. *Water International, 41*(6), 916–927.

Klug, H. (2021). Between principles & power: Water law principles & the governance of water in post-apartheid South Africa. *Dædalus, 150*(4), 220–239.

Marais, E.J. (2018). Narrowing the meaning of "deprivation" under the property clause? A critical analysis of the implications of the Constitutional Court's Diamond Producers judgment for constitutional property protection. *South African Journal on Human Rights, 34*(2), 167–190.

Marcatelli, M. (2018). The land-water nexus: A critical perspective from South Africa. *Review of African Political Economy, 45*(157), 393–407. https://doi.org/10.1080/0305 6244.2018.1451318

Mukuyu, P., van Koppen, B., & Jacobs-Mata, I. (2021). *Inventory and analysis of water use practices and regulations in Inkomati: Deliverable 3 to the Water Research Commission on the Operationalizing Hybrid Water Law for Historical Justice Project* [Unpublished report].

Murombo, T. (assisted by Seme, N.). (2021). *Recognition of customary water rights in South African legislation* [Draft working paper prepared for International Water Management Institute (IWMI) as part of PIM 5.1.1 project "Recognizing customary water tenure in hybrid water law: legislative and local perspectives" and paper presentation]. Water Institute of Southern Africa's (WISA) Biennial Conference and Exhibition 2020, online.

RSA (Republic of South Africa). (1996). The Constitution of the Republic of South Africa.

RSA (Republic of South Africa). (1956.) Water Act, No. 54 of 1956. Repealed.

Tempelhoff, J. (2017). The Water Act, No. 54 of 1956 and the first phase of apartheid in South Africa (1948–1960). *Water History, 9*(2), 189–213.

Thompson, H. (2006). *Water law: A practical approach to resource management and the provision of services*. Juta and Company Ltd.

UNDP (United Nations Development Programme). (2006). *Human development report 2006: Beyond scarcity: Power, poverty and the global water crisis*. Palgrave Macmillan.

Van Koppen, B. (2007). Dispossession at the interface of community-based water law and permit systems. In B. van Koppen, M. Giordano, and J. Butterworth (Eds.), *Community-based water law and water resource management reform in developing countries* (pp. 46–64). CABI International.

Van Koppen, B. 2022. *Living customary water tenure in rights-based water management in Sub-Saharan Africa*. IWMI (International Water Management Institute) Research Report 183. https://doi.org/10.5337/2022.214

Van Koppen, B., & Schreiner, B. (2019). A hybrid approach to statutory water law to support small-scale farmer-led irrigation development (FLID) in Sub-Saharan Africa. *Water Alternatives, 12*(1), 146–155.

Van Koppen, B., Nhamo, L., Cai, X., Gabriel, M.J., Sekgala, M., Shikwambana, S., Tshikolomo, K., Nevhutanda, S., Matlala, B., & Manyama, D. (2017). *Smallholder irrigation schemes in the Limpopo Province, South Africa*. IWMI (International Water Management Institute) Working Paper 174. https://doi.org/10.5337/2017.206

Van Koppen, B., Molose, V., Phasha, K., Bophela, T., Modiba, I., White, M., Magombeyi, M.S., & Jacobs-Mata, I. (2020). *Guidelines for community-led multiple use water services: Evidence from rural South Africa*. IWMI (International Water Management Institute) Working Paper 194. https://doi.org/10.5337/2020.213

Van Koppen, B., Schreiner, B., & Mukuyu, P. (2021). Redressing legal pluralism in South Africa's water law. *The Journal of Legal Pluralism and Unofficial Law, 53*(3), 383–396.

Vaughn, L.M., & Jacquez, F. (2020). Participatory research methods—Choice points in the research process. *Journal of Participatory Research Methods, 1*(1). https://doi.org/10.35844/001c.13244

Venot, J.P., Bowers, S., Brockington, D., Komakech, H., Ryan, C.M., Veldwisch, G.J., & Woodhouse, P. (2021). Below the radar: Data, narratives and the politics of irrigation in sub-Saharan Africa. *Water Alternatives, 14*(2), 546–572.

Von Fintel, D., & Pienaar, L. (2016). *Small-scale farming and food security: The enabling role of cash transfers in South Africa's former homelands*. Institute for the Study of Labor, Discussion Paper Series Report no. IZA DP No. 10377.

Activist Citizen Science: Building Water Justice in South Africa

Ferrial Adam[1]

Introduction

This chapter builds on my doctoral research with water justice activists in the Vaal region of South Africa (Map 6, page 242). The Vaal area, situated to the south of Johannesburg, is a rustbelt where shrinking industrial activity since the 1980s has left densely populated communities with significant legacies of pollution. The history of the Vaal has its roots in apartheid, which has influenced its social, political, environmental, and economic realities.

Drawing on my prior knowledge and experience as an activist, I carried out participatory research with the Vaal Environmental Justice Alliance (VEJA), a coalition of five environmental justice organisations focusing on air quality and health, water quality, waste, energy, and climate change in the Vaal area. The intention was to document the contexts of the region's water struggles and the processes of activist knowledge building, decision-making and broader strategy development being used in local communities.

The chapter explores how activist citizen science (ACS) can build water justice through processes co-created with community members, and shows why communities need such an approach. This perspective is grounded in the work of Paulo Freire and Aziz Choudry on activism, as well as Alan Irwin and Melissa Leach on citizen science (CS).

To start, the chapter briefly discusses the science of water, what informs that science, new understandings and challenges, and how this all affects the way water is seen and managed. I then summarize South Africa's water realities, highlighting the associated challenges of climate change, prolonged droughts, pollution and poor infrastructure, coupled with extreme water scarcity.

The next section overviews how water resources are being managed by government agencies and policies in South Africa. These policies are based on the principle of water as a basic human right, a key focus of the South African Human Rights Commission (SAHRC). But the picture that emerges is one in which those policies and the governance that flows from them may be failing people on the ground. People and communities are turning to developing local solutions to manage their water resources—namely, "water justice from below."

CS—depending on the type, and how it is used—can be key to building water justice from below. I document how local people, organized through VEJA, have built ACS, building on their prior experience with "bucket brigade" methods for local monitoring of air pollution; they have seen the advantage of using science to further their aims. VEJA has worked with traditional health practitioners (THPs[2]) to win recognition as key users of water in the Vaal area, using ACS to protect healers and their patients against polluted water and supporting THPs' access to rivers.

At the end of the chapter, I give examples of the practical application in South Africa of the "contributory" and "collaborative" types of CS, and then reflect on VEJA's experiences, discussing additional examples of how ACS can fit the needs of activists who are challenging environmental, climate, and water injustice in South Africa.

The Science of Water

The importance of water to life has been recognised throughout history and time. It has influenced where people live and how people live. The earliest towns and cities were located near rivers and seas; traditional and Indigenous healers have long identified the spiritual and physical health associated with water. In short, an understanding of water has always been imperative for life and local knowledge.

The science of water, thus, is not a new science. It can be traced to Indigenous knowledge and people's science rooted in traditional practices and experiences which, as shared wisdom, are centuries old. The early history of science has shown that historically, contributions were made by ordinary people who used science as a way of challenging the prevalent ideologies and cultures of the societies within which they lived. But in the nineteenth century, science was institutionalised in a way that divided scientists from other people (Irwin, 1995; Dyson, 2006).

Science became a field of professional expertise and knowledge. It has been used to leverage power and limit basic information that people need about their worlds. In the environmental sphere, people at the frontline of pollution and environmental damage have found it difficult to challenge polluting industries, as science has been used to counter their arguments and refute their claims. Science has thus been used to sustain injustices and perpetuate power differentials.

More recently, what kinds of knowledge are needed to inform bodies of science are being debated, alongside the power of science itself. Understandably, this constitutes a profound challenge to the closed spaces that frame mainstream science. Martinez-Alier et al. (2014) speak of "popular epidemiology" which emphasizes the validity of "lay" knowledge and raises the importance of local knowledge. Here we see the emergence of a theoretical call for reframing science that takes into account the cultural boundaries that shape its perspectives and does not dismiss other ways of knowing. Making science more accessible and including local knowledge leads to the democratisation of science which can, in turn, lead to the empowerment of the marginalised and those whose voices have been ignored (Irwin, 1995; Leach et al., 2005; Munnik, 2015; Visvanathan, 2016).

Creating such a new type of relationship between scientific and other knowledge is part of a broader movement towards the decolonisation of science and cognitive justice. Shiv Visvanathan defines cognitive justice as "the constitutional right of different systems of knowledge to exist as part of dialogue and debate" (Leach et al., 2005, p. 92). The decolonisation of science can create respect and acknowledge contributions of both Western science knowledge and local and Indigenous knowledge (Visvanathan, 2016). Respecting other forms of knowledge relieves the powerlessness felt by communities (Martinez-Alier et al., 2014). Boaventura de Sousa Santos (2007) states that

Fig. 13.1 Women farmers in the Vaal.

"the struggle for global social justice must be a struggle for global cognitive justice" (de Sousa Santos, 2007, p. 53).

In this respect, those with Indigenous and local knowledge[3] in areas of farming, fishing, and traditional healing have been important players in identifying and spotlighting injustices—for example, related to corporate control of seeds and fishing permits, access to rivers, and access to plants. Their knowledge and experience have been used to map out the challenges and impacts of climate change. In the context of water, a healthy relationship between scientific knowledge and traditional or Indigenous knowledge is desirable, especially in developing countries where technologies for prediction and modelling are least developed. Sharing and exchanging information and knowledge can foster better responsibility and care of water resources. Farmers, for example, can enhance their skills in soil and water management, while sharing their knowledge of micro-climates (Levidow et al., 2014). In other words, these groups conduct CS as part of their expertise or profession, and they are doing this by claiming a space that may also represent a way of slowly reclaiming the commons—shared use of land, water, and other requisites of sustainable livelihoods for all.

The State of South Africa's Water

Water scarcity is fast becoming one of the most serious concerns facing the planet. It is estimated that more than one-third of the world's population lives in water-stressed regions, with 663 million people facing a daily struggle to access clean and safe water. This is predicted to worsen, as suggested by the United Nations estimations that the world will only have 60 per cent of the water it needs by 2030 (DWA, 2012; Stewart, 2017).

South Africa, as the thirtieth driest country in the world, is not immune to this scarcity. With an average annual rainfall of 490 mm (well below the world average of 860 mm a year), it is estimated that less than 9 per cent of the precipitation eventually finds its way into South Africa's river systems. Even then, much of it is lost to erratic runoff and high levels of evaporation (CSIR, 2010; DWA, 2012; WWF-SA, 2016).

The country's water ecosystems are not in a healthy state. Of the 223 river ecosystem types, 60 per cent are threatened, with 25 per cent of these critically endangered by a changing climate and human activities. Less than 15 per cent of river ecosystems are located within protected areas, and 65 per cent of the wetland ecosystems have been identified as threatened, while 48 per cent are critically endangered (Department of Water and Sanitation, 2018).

To make matters worse, the little water available is being polluted and wasted. The main factors contributing to the deterioration of water quality in South Africa are mining, manufacturing industries, agriculture, crumbling infrastructure, and poor wastewater treatment (CSIR, 2010). It is estimated that 37 per cent of South Africa's clean, potable water is lost and wasted through poor infrastructure such as leaking pipes (News 24, 2014). As the responsibility for supplying water lies with the local municipalities, there is a clear problem with the management of water at a local level.

There is quite a big gap between those who do and those who don't have access to water. Powerful interest groups including agricultural, industrial, and mining sectors are prioritised by government due to their contribution to the country's gross domestic product (GDP), as evident in the National Water Resources Strategy 2 (NWRS2) (DWA, 2012). These powerful groups are in a position to influence the allocation of South Africa's scarce supply. Almost 98 per cent of the country's fresh water is already allocated. Nearly 61 per cent is used by the agricultural sector, while 27 per cent is for domestic and urban use, 8 per cent for mining and energy, and 3 per cent for forestry (Department of Water and Sanitation, 2018; Mkhonza, 2017). Marginalised and poor rural and urban communities experience high levels of water insecurity and many do not have access to a reliable potable water supply.

Climate change is making water scarcity even worse (Dwortzan, 2021; Isaacman et al., 2021). According to the World Wildlife Foundation, eight out of nine provinces were declared disaster areas in 2016 due to the ongoing drought. The more vulnerable and poor are hardest hit. Not only do they lack the material means to protect themselves against the impacts of climate change, but they also rely directly on polluted water (Cock, 2006; Munnik, 2007).

The dire state of South Africa's water resources has resulted in a myriad of responses—from government policies and programmes, to communities finding local solutions, to businesses using expensive technologies.

Government's Water Policy as a Solution?

In South Africa, water as a basic human right is enshrined in the South African Constitution's Bill of Rights and associated legislation.[4] It stipulates that the state holds the environment and the water resources in public trust for the people, the principle being that both are public "goods" (commons) and should be enjoyed equally by all. National government is thus responsible for the regulation and allocation of water, and local government is responsible for supplying water.

South Africa has four core pieces of water legislation and policy that govern water resources in the country: the National Water Policy (1997), the National Water Act (1998), the Water Services Act (1998), and the National Water Resource Strategy (CSIR, 2010). The National Water Act and the Water Services Act together provide for the establishment of institutions for management and distribution of water. The National Water Policy rests on the concept of Integrated Water Resource Management (IWRM) on a catchment basis, and the National Water Resource Strategy is centred around a recognition of water as a basic human need and its critical role in equitable socio-economic development (CSIR, 2010; Goldin, 2010; DWA, 2012). One significant element is the incorporation of a free basic water allowance of 25 litres per person per day, although there was a government proposal to scrap this in 2018 and it is not fully applied (OECD, 2021).

Not surprisingly, there is a gap between water policy and its implementation. While the policies constitute an attempt to redress historical inequalities of the past, constraints determined by racial, economic, and social structures retain and reproduce dominant power relations. Empowering communities, and facilitating transfer of local knowledge via CS, can help to change this. Goldin (2010) suggests that people can learn about basic science, such as the water cycle and the effects of patterns of water consumption on people and the environment, empowering them to be able to make choices and be active in decisions concerning the institutions that are set up to manage water. More importantly, Goldin (2010) stresses that it is also important that those with "scientific" knowledge and expertise in these areas gain knowledge about the living conditions of the poor so that there is an exchange of different types of knowledge, all important for good water management.

Privatisation of Water: A Response to Failing Government

Managing the gap between water policy goals and implementation can be an enormous task. The trend in some countries is to privatise the country's water, either by selling resources to an investor or by developing public-private partnerships. While there are arguments that privatisation can result in improvements in the efficiency and quality of service, there is ample evidence to show that privatisation has not worked for the majority of people and is incompatible with ensuring a human right to water, both in terms of access and affordability. Amongst other examples, privatisation plans in Bolivia and Tanzania were aborted (Public Citizen, 2003; Brown, 2010).

In South Africa, privatisation has taken many forms and has been met with varying responses. As early as 1996, municipalities involved the private sector in water and sanitation service provision, mostly through public-private partnerships. One privately owned contract is the thirty-year deal that was awarded to Siza Water Company for providing water and sanitation services to the iLembe District Municipality, along the KwaZulu Natal northern coastline (Food and Water Watch, 2015).

There is growing anger and frustration by communities that have no or limited access to water, with an increase in protests over poor or privatised service delivery (water, sanitation, and electricity in particular), social marginalization, and unequal access to water (Harrington, 2014). In 2014, just less than half of all households in South Africa obtained their water from a tap inside their home, while a further 27 per cent had a tap on their property, and 12 per cent walked up to 200 m to get water. Approximately 6 per cent of the population accessed piped water at a distance greater than 200 m (the target for basic services) and around 9 per cent of the population did not have access to piped water, relying on springs, rivers, and wetlands (WWF-SA, 2016).

Water as a Human Right

While government interventions have been important, they have not managed to reduce the challenges people face regarding water as a basic human right. The policies and institutions that have been developed are good on paper and come with good intentions, but the implementation has been poor. An example is the SAHRC, which is established under Chapter 9 of the Constitution. One of the areas it covers is service delivery for housing, water, and sanitation. On the issue of water, the SAHRC in 2018 held hearings with various groups and stakeholders in the Vaal region on pollution in the Vaal River, and declared that government was responsible for the pollution and must respond urgently. In addition, in 2019 the SAHRC ordered six education members of the executive councils to address the lack of sanitation and water and the continued use of pit toilets at schools in the Eastern Cape, KwaZulu-Natal, North West, Mpumalanga, Limpopo, and Free State (SABC News, 2021).

The SAHRC has been active and has regularly tabled reports on the state of socio-economic rights in Parliament, but these interventions are not achieving the desired results as government has been slow to react. The SAHRC has very little power to hold relevant government departments and municipalities accountable.

The reality is that people on the ground have lost faith in government bodies and institutions, and increasingly people and organisations at the local level are creating change from below, seeing no other option.

Water Justice from Below

There is a growing global movement for water justice, as people and communities are coming together to fight for access to clean water, to end pollution of water resources, to end privatisation, and to engage in increased efforts to manage scarcity and find solutions to combat the impacts of climate change.

In South Africa, for example, the civil society alliance Tshintsha Amakhaya has embarked on a Water Justice Campaign, while organisations like the Centre for Environmental Rights, VEJA, groundWork, Environmental Monitoring Group, and the South African Water Caucus are but a few of the structures and movements that are taking up the water justice fight.

Many of the measures government has taken to address water challenges are not easily available to communities, most of whom do not have the funds or knowledge to implement and sustain such measures. These include the use of high-level technical strategies, improving water use efficiency, development of new infrastructure, re-use and recycling, desalination, and the removal of water hungry alien invasive plants.

But communities are finding their own innovative ways to manage water resources. These responses can take many forms, such as simple technology solutions, Indigenous knowledge, water use efficiency, and CS. In some cases, local strategies offer cheaper alternatives: decentralised projects are more effective than the large-scale, centralised approaches that have dominated in the past.

Simple technology solutions include such practices as rainwater harvesting as used in India, fog catching/cloud harvesting in Nepal, and wastewater treatment in Cambodia (Asian Development Bank, 2006). In Bangladesh, where there is an extreme lack of safe drinking water, rickshaw pullers are using a pedal-powered water filter that provides clean and safe water. In India, a group called the Bengaluru "water warriors" challenge citizens in the city to be "water kanjoos"[5] using a WhatsApp group to promote water conservation (Pinto, 2017). In some communities in South Africa, a simple mechanism using recycled tires to collect water is being used. The tires are cut and buried just beneath the soil where they act as a trap to collect excess water that has trickled down, which then lies available for plants/crops to consume at a later stage. The traps also make the soil warmer, which means that crops can grow faster and they yield more produce.

The bottom line is that communities are using traditional knowledge to cope with water shortages and climate change, especially in rural areas. Traditional knowledge has been gained over time and across generations, with communities that live close to natural resources often observing and closely understanding the environment around them. As a result, they are able to easily identify any changes and adapt accordingly, using CS.

Citizen Science

The participation of the general public in the generation of new scientific knowledge goes back a long way. What is newer is the term "citizen science," which is now used to describe ordinary people participating in activities that involve science. The phrase was first used in the 1990s by Rick Bonney to describe the volunteer birdwatchers who shared their data with the Cornell University ornithology lab. CS activity (Irwin, 1995), which can include various fields such as astronomy, nature conservation, nuclear science,[6] and environmental protection, gained traction in the 2000s and has been described by many as the participation of the general public or volunteers in gathering and collecting information and data over large geographic areas (Kearns et al., 2016; Kullenberg, 2015; Buytaert et al., 2014; Cooper et al., 2007). Large networks of citizen scientists have been established in the US, Europe, and Australia.

The Oxford English Dictionary added CS in 2014 and defined it as "scientific work undertaken by members of the general public, often in collaboration with or under the direction of professional scientists and scientific institutions." However, this definition fails to take into account the complexity and nuances of CS. One of the more inclusive definitions is by Ceccaroni et al. (2017): "work undertaken with citizen communities to advance science, foster a broad scientific mentality, and/or encourage democratic engagement, which helps society address complex modern problems" (Ceccaroni et al., 2017, as cited in Eitzel et al., 2017, 6). This definition concurs with the view that CS refers to ordinary people using science, whether as volunteers gathering data or as a partnership between a community and scientists, to act on issues of concern (Vayena & Tasioulas, 2015).

There are three general models of CS. The first model is contributory. It involves volunteers only in data collection—for example, counting types of trees, identifying birds or capturing rainfall data. The second model is collaborative. Volunteers are more widely involved in data collection and they assist with the design of the research. In both models, the interpretation and analysis are conducted by professional scientists and the contribution made by other people is mostly towards gathering data (Goodwin, 1998). The third type of CS is co-created, where people are involved in all aspects of the scientific process and associated research. Although there is consensus on the potentially positive impact that CS can have on the environment, it is questionable whether all models can empower affected communities, particularly where people only collect data. Buytaert (2014) views CS, where community members work with professional researchers to seek solutions for problems at the community level, as participatory action research (PAR) and not as CS. However, the two are not mutually exclusive. PAR is a collaborative process that can be and often is a key component of CS. The more closely community members are engaged with the research design and research questions, the more opportunities arise for knowledge exchange and the more empowering the process is for community members.

CS is not new to South Africa. Vallabh (2021) has documented over sixty projects that utilise CS, ranging from identifying birds and plant species to river health actions. There are also good examples in other streams of knowledge such as health. For example, in the late 1990s and early 2000s, activists in the Treatment Action Campaign used CS to educate people about HIV and AIDS, antiretrovirals, and healthy living. In this example, activists used CS as an awareness-raising and education tool that was important to counter the dangerous and harmful views of people in high positions of power, such as the HIV denialism from then-president Thabo Mbeki (Cullinan, 2016).

An example of how knowledge can be used to fight environmental injustice is evidenced in the work of environmental organisations such as groundWork and the South Durban Community Environmental Alliance (SDCEA) who have used CS to monitor air pollution in refinery-impacted communities around South Africa. The activists in these organisations used the "bucket brigade" air monitoring system, in which an air quality sample is drawn into a bucket and then taken for laboratory analysis. This form of monitoring empowered communities to act on air pollution themselves, building empirical evidence to take both government and private companies to task and providing communities with a weapon to question and critique the environments within which they live. Crucially, this simple method managed to demystify science (Hallowes, 2014). VEJA also strategically uses science as a way to participate in spaces such as the Water Catchment Management Forums and to expose and challenge understatements about the levels of pollution.

Another way that CS has been used is in legal challenges, as demonstrated by the Centre for Environmental Rights (CER)[7] in response to South Africa's 2020 emission standards. CER analyses environmental impact assessments (EIAs) for new mining or industrial developments and asks trained activists to observe and report on their lived experiences, in order to expose

Fig. 13.2 Spilling sewage.

discrepancies in industries' arguments and what scientific data they include. This is an example of how CS can be used as a tool to enhance people's resistance in the course of environmental and water struggles.

All types of CS can be regarded as forms of knowledge production, but it can be used to sustain the status quo or to challenge it, depending on how that knowledge is developed. This could result in what Epstein (1996) refers to as "the scientization of politics that simultaneously brings about a politicization of science such that political disputes could become technical disputes," which could exacerbate the levels of exclusion of communities with less power (cited in Leach & Scoones, 2007, 16). Likewise, Weiler (2011) suggests that "areas of life are scientized and taken out of reach of participatory politics to be handed over to experts" (Weiler, 2011, p. 211). In this respect, it could be argued that the first two types of CS, in which active participation by the less powerful is not foundational, would also likely result in less action to resolve community problems. Here, citizen scientists could believe that they are privileged to participate in a science "endeavour" and that might foster or sustain an uncritical attitude towards how science is used and for what purposes. In turn, this could mean that there is less critique of the systemic causes of the environmental injustices and thus less challenging of top-down power.

CS that is only focused on collecting data for use within science models is not likely to challenge environmental injustice. According to Leach and Scoones (2007), defining a problem as a technical one will result in the solution being a technical and short-term one, when in fact it may require tackling root causes and systemic issues like environmental injustice and exclusion of marginalised communities. For example, burst pipes and overflowing manholes could be seen as an infrastructural issue that need technical expertise to fix—but on its own a technical fix could be a short-term solution. A deeper analysis could reveal a failing local government due to corruption and maladministration that neglects the provision of basic services to poor racialized communities. This, then, would require more than a technical fix—it requires a political approach as well.

Activist Citizen Science

As indicated earlier, in general there are different types of CS with varying levels of participation in activities that use science for awareness, education, monitoring the environment, and fun. The purpose of describing and discussing the applications and associated limitations of the options is not to identify the "best" example of CS, but rather to show that CS can be moulded to fit different situations and purposes. The kind of CS that is useful to grassroots activists dealing with injustices—which shares the general characteristics of the co-created type but also adds to and expands an understanding of what the co-created can entail—is ACS in which both critical reflection and action are necessary, by definition.

This concept of ACS emerges from a combination of the relevant literature, analysis of the types of CS, and the development of environmental justice activism in South Africa. The following three components are proposed as key to formulating and using ACS: challenging and building knowledge production, building networks and social movements, and shifting power relations.

ACS is a knowledge tool that can be used to work for justice through participation and conscientization. Paulo Freire (2000) not only argued that power and knowledge are inextricably linked, but that the type of education and the way one learns are relevant to shifting power relations. Instead of filling people with information, they can be conscientized to analyse and be critical of the world around them. Further, authors such as Choudry (2015) and Irwin (1995) argue that activists and people on the ground also create and produce skills and knowledge through their lived experiences and local knowledge. Within these frames, ACS can therefore be defined as an approach in which activists use science as a galvanising tool to redirect power to people and to challenge injustice.

ACS can be the nexus that brings together the science, the lived experiences of activists and communities, and an understanding of the power relations and politics at play. Leonard and Lidskog (2021) have found that the integration of knowledge between science experts and people's lived experience can build an alternative expertise with the potential to increase levels of trust and interventions between communities and industries. Irwin (1995, p.144) convincingly argues that public education, like CS, must "include the wider social, economic and political aspects" as people on the ground are more interested in safety and health than the technical details. Having the science without the ability to advocate for change may result in CS becoming just another hobby or, as described by Leach and Scoones (2007), a case of "responsibilised citizens who come together to articulate individual rights in relation to public goods" (Leach & Scoones, 2007, p. 27).

Nonetheless, even all the learning and knowledge creation may not be enough to shift power relations because the learning is done in a way that supports the status quo, or the balance of forces is strongly in favour of those with power. Choudry's (2014) view closely aligns to that of Freire: knowledge in itself does not lead to empowerment. The power held by government and the private sector could be strong enough that they can ignore activists, even if the activists have evidence and knowledge to back up their claims. That is why the building of collective voice and action, as carried out by social movements and civil society networks—through an ACS approach—is a way for activists to build counter knowledge (and thereby counter power) to government and the private sector. ACS thus presents itself as a way to build a movement of water warriors, who get involved in local or regional water justice issues and understand the link between water and broader issues of social justice.

The Vaal Environmental Justice Alliance and Activist Citizen Science

To learn more about ACS in action, I chose the VEJA as a case study because water conflicts in the Vaal were familiar to me from my own activism, and I knew about VEJA's use of science as a tool to fight their struggles. In addition, VEJA works on water and environmental issues in poor and marginalised communities, central to the key research question—how does CS contribute to environmental justice?

VEJA uses detailed scientific knowledge as well as locally shared knowledge and observation to support participation in water governance. They built their ACS model through their earlier experience with the bucket brigades that monitored air pollution, as described above.

VEJA has built knowledge among activists to mobilise and create spaces for engagement. I attended community meetings, discussed my research goals with VEJA organizers, took advice from them about how to engage with community members, and collaborated with them on related projects during dozens of visits, meetings, workshops, and community events. I have observed four ways that they have made use of ACS in their water struggles. The first is that they try to give people a basic understanding of water science and politics through workshops, training, and site visits. The second aspect relates to participation in the catchment management forums[8] that are otherwise mostly made up of farmers and industry representatives. The third involves the practical use of data they have collected to influence policy and debates. VEJA has used their own observations to challenge "scientific statements" made in the catchment forum meetings, for example the claim that ArcelorMittal Steel was a zero-effluent facility. VEJA had documented evidence (such as photographs with the time, date, and place) and monitored the effluent-laden water that came out of the facility over a period of several months. The fourth aspect is mobilisation, which uses all of the above to challenge and shift power relations.

VEJA has practised ACS by using scientific information and language to include marginalised people and to create public awareness on water pollution. The key result was local communities' inclusion and recognition in various forums and structures. This is a reflection of a shift in power which, even if relatively minor, can be regarded as an important achievement.

Conclusion

The challenges involved in tackling water scarcity and injustice are complex and no one sector can simply "solve" the problem—whether it is government, business, or civil society. However, there are many examples globally, and in South Africa, of people taking control of their situations to make things better. These people are playing an important role in the sustainable management of natural resources and more specifically, water—the key to life.

It is becoming increasingly important for ordinary people, especially in a country like South Africa, to become more active on water issues. In a drought-prone country with deep inequalities, it is essential that water be democratised through people's power. The laudable right to water in the South African Constitution can only be practically realised if people are an integral part of managing and controlling water resources through water sovereignty.

While the plethora of proposed solutions to water challenges should not be disregarded, there are fundamental limitations when the dominant focus is at a technical level. Technology can only go so far. It removes all blame from the system and those with political and economic power, and thus undermines community action for substantive change.

The key focus of ACS is not on how individuals can reduce their water consumption so that industry and government can keep polluting and using most of the resources, but rather to inspire activism within communities that directs water resources away from industry and into the hands of the people, to promote water justice and the health of common water systems.

Learning the basic science does not necessarily change the system of water management, but it is increasingly being called on to provide consensus in political disputes. Science experts are often summoned and empowered to settle political and social disputes, such as those over polluted water or a lack of water supply. This politics is influenced by the science, and at the same time the science is influenced by the politics.

It is within this context and these realities that ACS has the potential to ensure that water is seen, both conceptually and in practice, as a common good for all. ACS not only offers a better means to increase people's knowledge and skills to monitor water resources and act as a vehicle to democratise water science, but it can also radically shift power dynamics that can lead to systemic change.

NOTES

1. Part of this chapter draws from my PhD Thesis and appeared in a booklet produced by the Rosa Luxemburg Stiftung Southern Africa, "No Easy Walk to Water," 2021. My fieldwork was made possible through the QES program, funded by the International Development Research Centre and the Social Sciences and Humanities Research Council of Canada.

2. THPs have been campaigning for recognition as healers. They have been marginalised as their Indigenous methods are often disregarded by the medical establishment (See Louw & Duvenhage, 2017).

3. There is another form of CS in South Africa that is not covered by the three types mentioned in this chapter and could be linked to the result of epistemicide mentioned here. This form of CS links to the issues of popular epidemiology and cognitive justice. It revolves around the recognition of Indigenous and local knowledge in areas of farming, fishing, and traditional healers, where people use community organised seed banks in farming, monitoring of fish stocks, and recognition of plants for healing with traditional healers.

4. Some of the key water laws and policies that govern water in South Africa include the National Water Policy (1997), the National Water Act (1998), the Water Services Act (1998) and the National Water Resource Strategy 2 (2012)—all founded on the government's vision to redress past inequalities and build a sustainable water future.

5. *Kanjoos* is an Urdu word that means stingy.

6. Greenpeace trains activists on the basics of nuclear science, enabling them to measure radiation levels in food, water, etc. I was trained by Greenpeace and spent a week in Fukushima with Japanese activists measuring the radiation levels in various parts of the city.

7. Such as the "Deadly Air: groundWork's section 24 challenge." For a full list of all CER's litigation, visit https://cer.org.za/programmes/pollution-climate-change/litigation.

8. The purpose of the Catchment Management Agencies (CMAs) and catchment management forums is to involve various stakeholders and local communities in regional or catchment level water resource management. See https://www.citizen.org/wp-content/uploads/migration/waterprivatizationfiascos.pdf (visited on 19 September 2020).

Reference List

Asian Development Bank. (2006, March). *Fostering participation: Water management by local communities.* https://www.adb.org/publications/fostering-participation-water-management-local-communities

Brown, R. (2010). Unequal burden: Water privatisation and women's human rights in Tanzania. *Gender & Development, 18*(1), 59–68.

Buytaert, W., Zulkafli, Z., Grainger, S., Acosta, L., Alemie, T.C., Bastiaensen, J., De Bièvre, B. Bhusal, J., Clark, J., Dewulf, A., Foggin, M., Hannah, D.M., Hergarten,

C., Isaeva, A., Karpouzoglou, T., Pandeya, B., Paudel, D., Sharma, K., Steenhuis, T., ... Zhumanova, M. (2014, October 22). Citizen science in hydrology and water resources: Opportunities for knowledge generation, ecosystem service management, and sustainable development. *Frontiers in Earth Science*, 1–21. https://doi.org/10.3389/feart.2014.00026

Choudry, A.A. (2015). *Learning activism: The intellectual life of contemporary social movements*. University of Toronto Press.

Cock, J. (2006). Connecting the red, brown and green: The environmental justice movement in South Africa. In R. Ballard, A. Habib, & I. Valodia (Eds.), *Voices of protest: Social movements in post-apartheid South Africa* (pp. 179–201). University of KwaZulu-Natal Press.

Cooper, C.B., Dickinson, J., Phillips. T., Bonney, R. (2007). Citizen science as a tool for conservation in residential ecosystems. *Ecology and Society, 12*(2), 11. https://www.ecologyandsociety.org/vol12/iss2/art11/

CSIR (Council for Science and Industrial Research). (2010). *A CSIR perspective on water in South Africa*. CSIR Report No. CSIR/NRE/PW/IR/2011/0012/A. https://mpaforum.org.za/wp-content/uploads/2016/08/CSIR-Perspective-on-Water_2010.pdf

Cullinan, K. (2016, March 7). Mbeki still believes his own AIDS propaganda. *Health-e News*. https://health-e.org.za/2016/03/07/mbeki-letter-believes-aids-denialism/

Department of Water and Sanitation. (2018). *National water & sanitation master plan: Volume 1:call to action (draft)*. http://www.dwa.gov.za/National%20Water%20and%20Sanitation%20Master%20Plan/Documents/NWSMP%20Call%20to%20Action%20Final%20Draft%20PDF.pdf

DWA (Department of Water Affairs). (2012, July). *Proposed National Water Resource Strategy 2 [NWRS2]: Summary*—Managing Water for an Equitable and Sustainable Future. https://static.pmg.org.za/docs/120911proposed_0.pdf

Dwortzan, M. (2021, November 5). Scientists project inreased risk to water supplies in South Africa this century. *MIT News*. https://news.mit.edu/2021/scientists-project-increased-risk-water-supplies-south-africa-this-century-1105

Dyson, F. (2006). The scientist as rebel. *The New York Review of Books*.

Eitzel, M.V., Cappadonna, J.L., Santos-Lang, C., Duerr, R.E., Virapongse, A., West, S.E., Kyba, C.C.M., Bowser, A., Cooper, C.B., Sforzi, A., Metcalfe, A.N., Harris, E.S., Thiel, M., Haklay, M. Ponciano, L., Roche, J., Ceccaroni, L., Shilling, F.M., Dörler, D., ... Jiang, Q. (2017). Citizen science terminology matters: Exploring key terms. *Citizen Science: Theory and Practice, 2*(1), 1. https://doi.org/10.5334/cstp.96

Epstein, S. (1996). *Impure science: AIDS, activism, and the politics of knowledge*. University of California Press.

Food and Water Watch. (2015). *Water privatisation: Facts and figures*. https://www.foodandwaterwatch.org/insight/water-privatization-facts-and-figures

Freire, P. (2000). Pedagogy of the oppressed—30th anniversary edition (30th ed.). Continuum.

Goldin, J.A. (2010). Water policy in South Africa: Trust and knowledge as obstacles to reform. *Review of Radical Political Economics, 42*(2), 195–212. https://doi.org/10.1177/0486613410368496

Goodwin, P. (1998). "Hired hands" or "local voice": Understandings and experience of local participation in conservation. *Transactions of the Institute of British Geographers, 23*(4), 481–499. https://doi.org/10.1111/j.0020-2754.1998.00481.x

Hallowes, D. (Ed.). (2014). *Slow poison: Air pollution, public health and failing governance.* groundWork. https://static.pmg.org.za/141028slow_poison_2014_groundwork.pdf

Harrington, C. (2014, May 2). *Water security in South Africa: The need to build social and ecological resilience.* Sustainable Security. https://sustainablesecurity.org/2014/05/02/water-security-in-south-africa/

Irwin, A. (1995). *Citizen science—A study of people expertise and sustainable development.* Routledge.

Isaacman, A.F., Musemwa, M., & Verhoeven, H. (2021). Water security in frica in the age of global climate change. *Daedalus, 150*(4), 7–26.

Kearns, G., Fegler, M., & Schreiber, M. (2016, April 1). *Citizen science programs, parks & recreation.* National Recreation and Park Assocation. https://www.questia.com/magazine/1P3-4042304311/citizen-science-programs

Kullenberg, C. (2015). Citizen science as resistance: Crossing the boundary between reference and representation. *Journal of Resistance Studies, 1*(1), 50–76. https://scientometrics.flov.gu.se/files/Kullenberg-Citizen-Science-Journal-of-Resistance-Studies-issue-1.pdf

Leach, M., & Scoones, I. (2007). *Mobilising citizens: Social movements and the politics of knowledge.* Institute of Development Studies. https://opendocs.ids.ac.uk/opendocs/handle/123456789/4030

Leach, M., Scoones, I., & Wynne, B. (Eds.). (2005). *Science and citizens: Globalization and the challenge of engagement.* Zed Books.

Leonard, L. & Lidskog, R. (2021). Industrial scientific expertise and civil society engagement: Reflexive scientisation in the South Durban Industrial Basin, South Africa. *Journal of Risk Research, 24*(9), 1127–1140. https://doi.org/10.1080/13669877.2020.1805638

Levidow, L., Zaccaria, D., Maia, R., Vivas, E., Todorovic, M., & Scardigno, A. (2014). Improving water-efficient irrigation: Prospects and difficulties of innovative practices. *Agricultural Water Management, 146*, 84–94. https://doi.org/10.1016/J.AGWAT.2014.07.012

Louw, G., & Duvenhage, A. (2017). Does the traditional healer have a modern medical identity in South Africa? *Australasian Medical Journal, 10*(2), 72–77. https://doi.org/10.21767/amj.2017.2730

Martinez-Alier, J., Anguelovski, I., Bond, P., Del Bene, D., Demaria, F., Gerber, J.-F., Greyl, L., Haas, W., Healy, H., Marín-Burgos, V., Ojo, G., Porto, M. Rijnhout, L., Rodríguez-Labajos, B., Spangenberg, J., Temper, L., Warlenius, R., & Yánez, I.

(2014) Between activism and science: Grassroots concepts for sustainability coined by environmental justice organizations. *Journal of Political Ecology, 21*(1), 19–60. https://doi.org/10.2458/v21i1.21124

Mkhonza, A. (2017). Presentation by the Centre for Environmental Rights (CER) at a Vaal Environmental Justice Alliance workshop on 20 May, Vanderbijlpark, South Africa.

Munnik, V. (2007, November). *Solidarity for environmental justice in Southern Africa*. groundWork. https://groundwork.org.za/wp-content/uploads/2022/07/Solidarity-for-enviroment.pdf

Munnik, V. (2015, November 4–7). *Making space for cognitive justice: Creating a level playing field for catchment forums in South Africa* [Paper presentation]. Action Learning, Action Research Association (ALARA) World Congress, Pretoria, South Africa.

News 24. (2014, November 3). South Africa's looming water disaster. https://www.news24.com/News24/South-Africas-looming-water-disaster-20141103

OECD (Organisation for Economic Co-operation and Development). (2021). *Water governance in Cape Town, South Africa*. OECD Studies on Water, chapter 1. https://doi.org/10.1787/a804bd7b-en

Pinto, N. (2017, April. 9). Bengaluru "water warriors" challenge citizens in city to be "water kanjoos" this summer. *India Today*. http://indiatoday.intoday.in/story/bengaluru-water-shortage-challenge-kanjoos-summer/1/924530.html

Public Citizen. (2003). *Water privatization fiascos: Broken promises and social turmoil*. https://www.citizen.org/wp-content/uploads/migration/waterprivatizationfiascos.pdf

SABC News. (2021, October 1). *SAHRC orders six education MECs to address lack of sanitation and water at schools*. https://www.sabcnews.com/sabcnews/sahrc-orders-six-education-mecs-to-address-lack-of-sanitation-and-water-at-schools/

de Sousa Santos, B. (2007). Beyond abyssal thinking: From global lines to ecologies of knowledges. *Review (Fernand Braudel Center), 30*(1), 45–89. http://www.jstor.org/stable/40241677

Stewart, R. (2017). *Wild water: The state of the world's water*. www.wateraid.org.

Vallabh, P. (2021). *For the common good: A counter-hegemonic (re)stor(y)ing of the citizen sciences*. Rhodes University.

Vayena, E., & Tasioulas, J. (2015). We the scientists: A human right to citizen science. *Philosophy and Technology, 28*(3), 479–485.

Visvanathan, S. (2006). Alternative science. *Theory, Culture & Society, 23*(2–3), 164–169.

Weiler, H.N. (2011). Knowledge and power: The new politics of higher education. *Journal of Educational Planning and Adminstration, 25*(3), 205–221.

WWF-SA. (2016). *Water: Facts and futures—rethinking South Africa's water future*. https://wwfafrica.awsassets.panda.org/downloads/wwf009_waterfactsandfutures_report_web__lowres_.pdf

PART IV

Collective Resilience for Climate Justice

Conflicting Perspectives in the Global South Just Transition Movement: A Case Study of the Mpumalanga Coal Region in South Africa

Andries Motau

Introduction: Coal As a Problematic Commodity in South Africa

South Africa's coal mining sector has had significant social, economic, and environmental impacts on the country. Coal is the source of over 90 per cent of the nation's electricity, roughly 30 per cent of the liquid fuel, and approximately 70 per cent of total energy needs. Coal mining in South Africa has been associated with negative legacies, especially in the province of Mpumalanga, where most coalfields are located (see Map 6, page 242). Benefits throughout coal mining regions in South Africa have not been equitably distributed, and coal mining communities are characterised by high levels of poverty, socio-economic inequalities, and environmental degradation.

In chapter five of the National Development Plan (South African National Planning Commission, 2013), South Africa's government has proposed Vision 2030, which addresses environmental sustainability and the equitable transition to a low-carbon climate-resilient economy and society. However, the proposal does not articulate clear pathways to a Just Transition.[1] There has also been public outcry, especially from local and international civil society organizations. These groups argue that South Africa's coal industry must not expand any further; rather the country must transition away from coal and create a shared vision of a different future, and this must be accelerated.

A Just Transition in the context of a coal-dependent developing country such as South Africa presents many inter-related challenges, both on a national and local scale. Stakeholders have different, and often conflicting, priorities and perspectives on how the energy transition will impact communities, workers, the environment, and the economy. These complexities have created difficulties in reaching a consensus on how various trade-offs can be managed and how an inclusive and equitable transition can be achieved, from both local and national perspectives.

Given the many differences in information access, power, and agency (among other factors) across affected stakeholders, climate justice requires that a wide range of information and opinion sources be shared and weighed as part of decision processes, both formal and informal.

As in other such complex situations, participatory and multi-method research approaches are useful to help investigate and analyse different perspectives, discourses, and synergies towards realising a Just Transition in the Mpumalanga region of South Africa.

Coal plays a major role in many economies around the world, like in South Africa, and there is a growing realisation that coal-based growth is not sustainable. According to Kretschmann (2020) there is increasing disinvestment in coal mining in many coal regions, due to consumer and investor pressure, and as a result many entities are implementing decarbonization strategies across their portfolios. These disinvestment decisions are influenced by the impacts of coal at both global and local levels. These impacts include greenhouse gas (GHG) emissions, which contribute to climate change. In South Africa some of the impacts are a result of historic injustices from the apartheid era and inconsistent legal compliance by the coal sector. These have exacerbated levels of poverty and systemic inequality, as well as corruption and nepotism in mining-affected communities, and have reinforced the notion that mining operations disproportionately favour mining companies and the State (Fine & Rustomjee, 2018; Marais, 2013; McCarthy & Humphries, 2013; Baker et al., 2014; South African Human Rights Commission, 2017; Shongwe, 2018; Mandel, 2019).

At local levels, the impacts of coal mining include water contamination affecting the water quality, physical and chemical land degradation, air pollution through dust fall-out and emissions of particulate matter (PM) and toxic gases (Shongwe, 2018). Despite the negative impacts posed by coal production and utilisation, it needs to be recognised that coal mining plays a critical role in the mineral economy, as it contributes to energy needs, employment, exports, local communities' livelihoods, and gross domestic product (GDP) (Minerals Council South Africa, 2020). According to Keles and Yilmaz (2020) the negative repercussions overshadow the beneficial consequences of coal mining and coal consumption. Thus, there are calls and plans for the globe to move away from coal mining and production, with countries like Germany setting a phase-out by 2035. There have been many charges that the current trajectory of South Africa's coal dependency and growth is not sustainable, and that South Africa needs to transition to a low-carbon economy (Hallowes & Munnik, 2019). Whilst this need is undisputed, it is equally important that the transition be done in a just (morally justifiable) and fair manner, especially in a developing country such as South Africa (World Bank, 2018). The Just Transition movement is gaining momentum in South Africa, with campaigns such as "Life After Coal," which involves various non-governmental organisations, and "A Green New Eskom" (South Africa's public electricity company), led by the Climate Justice Coalition.

While South Africa's National Development Plan includes mention of a national Just Transition pathway (South African National Planning Commission, 2019), showing recognition of the need to collectively plan toward a transition, the government remains committed to coal and is driving a political agenda to invest in coal due to its economic implications and benefit. (Hallowes & Munnik, 2019). Nalule (2020) argues that though there is a realisation in many developing countries of the need to transition, countries such as South Africa are still conflicted, with the government having plans to transition whilst also seeking developments in the coal sector. Montmasson-Clair (2017) maintains that although there are good policies in place for South Africa's sustainability, there has been a lack of government initiative in implementation.

This conflict of interest escalates already existing tensions between different stakeholders with different priorities. For example, environmental justice groups are pushing for a coal phase-out due to its negative externalities, whilst trade unions are concerned about job losses for coal workers (Cock,

2019). These conflicting perspectives have resulted in a lack of coordination, and concerns about the lack of inclusivity in the Just Transition (Swilling and Annecke, 2012; Baker and Wlokas; 2014; Bond, 2019; Cock, 2019; Reinouad, 2019). These concerns are driven by South Africa's previous economic growth trajectories, where only a few mining companies benefited whilst mining communities were left worse off, as elaborated below (Fine & Rustomjee, 2018; Marais, 2013; McCarthy & Humphries, 2013; Baker et al., 2014). The minerals-energy complex centred on coal represents a continuation of such an economic path (Froestad et al., 2018).

Mpumalanga is one of the provinces where negatively affected mining communities are found, especially in the Emalahleni area where the economy is mainly dependent upon coal mining. After many years of coal mining in the area, civil society came together to raise concerns around mining. According to Munnik (2019), the discourse on climate change and anti-coal activism in Mpumalanga gained momentum in 2014 when a group of communities affected by coal came together with activists and non-governmental organizations (NGOs) in a forum called "push back coal," the aims of which were to coordinate and share knowledge and build resistance against coal and put in place plans to transition away from fossil fuels. Furthermore, the push-back forum continued to do community work on environmental and climate justice with NGOs such as groundWork, Earthlife Africa, and the Centre for Environmental Rights, establishing a core alliance with the "Life After Coal" campaign, which emphasises a move from resistance against coal to a Just Transition (Munnik, 2019).

The resistance to coal has been mostly from environmental groups, but when it comes to the Just Transition the situation is more complex, as there are several different actors across various levels involved with different and sometimes conflicting priorities, perspectives, and goals. Mining communities are often not involved in debates, but rather represented by third parties with their own agenda, whilst government policies are not consistent or well aligned. These complexities and gaps are preventing a coordinated and inclusive Just Transition approach in South Africa. An in-depth understanding and analysis of the actors, their discourses and shifting power relations is required to inform transformative Just Transition policy and planning.

Local Environmental and Social Impacts of Coal Production and Combustion

South Africa is ranked amongst the world's top eight countries for coal production and consumption and, due to the country's large supply of coal and its pressing development priorities, advocating lower coal consumption is politically difficult (Burke & Nishitateno, 2013). Coal is South Africa's principal source of energy for fuel and petrochemical production, and a significant contributor to the country's GDP and to its socio-economic development (Mathu & Chinomona, 2013; Zhao & Alexandroff, 2019). Although coal mining contributes to the economy, coal mining and combustion have significant adverse impacts on local environments and communities, in South Africa as in other countries across the globe.

Studies on Just Transition have tended to focus on economic impacts of fossil fuel dependence, and this has resulted in a growing gap in literature on Just Transition coordination, especially when it comes to actors and their various vested interests (Cahill & Allen, 2020). In Mpumalanga province, studies have not only focused on the economic contributions of coal mining and production but they have also looked at the social and environmental impacts. For instance, a study by Aneja et al. (2012) found that activities such as surface coal mining are a source of air pollution through the blasting and wind erosion of exposed areas, and these emissions also occur during transportation, handling of coal at the mines, and during coal processing. A Greenpeace (2019) investigation of a full year of TROPOMI (Tropospheric Monitoring Instrument) satellite monitoring of nitrogen dioxide emissions and other scientific datasets declared the coal-fired power plant and industrial cluster in Mpumalanga to be the world's worst hotspot for nitrogen dioxide (NO_2) and sulphur dioxide (SO_2) emissions, and the area overall ranks fourth for NO_2 and third for SO_2 emissions in the world (Greenpeace, 2019). These toxic pollutants can result in increased risk of respiratory infections, increased risk of stroke, and increased risk of death from diabetes. Due to the high levels of gaseous pollutants, the Emalahleni area was declared a Priority Area, in terms of section 18(1) of the South African National Environmental Management: Air Quality Act, 2004 (Act No. 39 of 2004) (Department of Environmental Affairs, 2018; Gray, 2019).

Studies have also investigated impacts of coal mining and production on water pollution; it was found that there are severe strains on the quality of water in Mpumalanga as a result of coal mining, due largely to acid mine drainage which has contaminated and disturbed vast amounts of land, destroyed wetlands, rerouted streams, and contaminated ground and surface water (Geldenhuis & Bell, 1998; McCarthy, 2011; Nzimande & Chauke, 2012; Mhlongo et al., 2018; Gupta & Nikhil, 2016; Kholod et al., 2020).

Research has also been conducted on the negative social impacts of coal mining in local communities (Munnik, 2019). Coal mining communities are often faced with detrimental effects that include damage to homes due to blasting, damage to roads and infrastructure due to large vehicles, forced and unplanned removals, and negative impacts on health and well-being (Centre for Environmental Rights, 2016). Coal mining has also had negative impacts on crop and livestock farming in Mpumalanga, and these negative impacts have contributed to rising tension and conflicts within communities between farmers and miners, resulting in activism, protests, and litigation (Shongwe, 2018).

From the different arguments about the environmental, social, and economic impacts of coal mining and production, it is still unclear how social justice and environmental sustainability can be balanced in achieving a Just Transition in Mpumalanga, or in South Africa overall. The reality is that the issues of social justice and environmental sustainability are extremely complex when translated into practice, as there are many trade-offs that must be considered and mediated (Ciplet & Harrison, 2020; Culwick & Patel, 2020). An understanding of the complexity of issues of social justice and environmental sustainability requires consideration of how justice and sustainability interact at different temporal, social, and spatial scales, as these issues are both conceptual and practical (Fatti et al., 2021). Also involved in each particular context are social trust, institutional structures for mediating conflicts and deciding benefit-sharing, and the overall governance system in which power and agency are distributed and exercised.

Coal Mining and Global Climate Change

Like other fossil fuels, mining and utilisation of coal give rise to the GHG emissions responsible for global warming. According to Pandey et al. (2018), coal mining activities are responsible for direct emissions of GHGs such as methane and carbon dioxide (CO_2), as well as indirect emissions through the consumption of fossil-fuel derived electricity and other materials. Coal mining also emits increased GHGs through spontaneous combustion (Carras et al., 2009; Mohalik et al., 2016), which can occur when coal is stored in bulk. Coal-fired power plants are the largest emitters of GHGs in the coal-to-power value chain in South Africa.

According to Strambo et al. (2019), 50 per cent of South Africa's GHG emissions are accounted for by public electricity utility Eskom and chemical firm Sasol, and they are responsible for 85 per cent of the coal used in the local market by volume. This has resulted in South Africa's being ranked the world's fourteenth-largest emitter of GHGs (Climate Transparency, 2020). Emissions remain high even as areas of South Africa such as the Highveld plateau, which includes part of Mpumalanga, are warming at double the global rate (groundWork, 2018). Changing climatic conditions resulted in one of the worst historic droughts in the Highveld in 2015/16 that withered the maize crop and sent prices spiralling; this affected poor people negatively due to a reduction in the availability and quality of food (groundWork, 2018).

Seen through justice and environmental sustainability lenses, such issues as pollution and food security drive marginalisation and inequalities within communities. Decades after the end of apartheid, areas such as the Highveld are still characterised by prolonged inequalities and environmental degradation, despite being resource rich. Davis (2010) argues that climate injustice derives from both the causes and effects of climate change; those who are likely to be most adversely affected have not only contributed the least to and benefited the least from the development and consumption of resources that have caused climate change, but have also had limited influence over decisions that affect future impacts. Thus, increased agency for marginalized people must be part of the Just Transition agenda.

Just Transition: Drivers and Barriers

The concept of a Just Transition emphasises the need to phase out industries that pose harms for workers, community health, and the planet, whilst at the same time creating opportunities and pathways for workers to transition to other jobs (Smith, 2017; Climate Justice Alliance, 2018). An analysis of the literature indicates that research on the Just Transition has been mainly focused on developed countries rather than the Global South. Furthermore, much existing research emphasises the economy rather than the above priorities (harm for workers, community health, and the planet) and tensions between different actors (Snell, 2018; Cock, 2016). According to Smith (2017), the global definition of the Just Transition has changed from a focus on shifting coal and other fossil fuel workers to "green jobs," to embracing broader issues across economic, social, and environmental dimensions of sustainability. According to Cock (2016), transitioning to green jobs is a more moderate version of shifting to a low-carbon economy as this involves a shallow, reformist transformation focused narrowly on building a new energy regime with "green" jobs, new technology, social protection, and consultation. Cock (2016) argues that a Just Transition approach of this nature is somewhat defensive as it is often manipulated to act as if it is representing and protecting the interests of the most vulnerable. Ward (2018) holds that a Just Transitions discourse has progressed beyond the jobs-versus-environment argument. Through interaction with the environmental justice and climate justice movements, a Just Transition has evolved as a broad framing that supports an expanded scale of reflections across economic, social, and environmental aspects of sustainability (Ward, 2018).

Whatever the evolution and different understandings of the term, Just Transition remains a challenge in South Africa, and this is because the current political economic systems that are in place in South Africa need to change for South Africa to have a transformative Just Transition. Ward (2018) argues that the approaches to South Africa's sustainability, especially looking at the green economy accord signed at the South African National Parliament in 2011, have not been explicit about a Just Transition but have referred to it only in a passive and minimalist sense. This poses a threat in achieving a transformative Just Transition if it is expected to occur within a capitalist economy that has been enabled by current policies and unjust political systems, both old and new (Ward, 2018). Avelino and Wittmayer (2016) argue that transformative approaches are key to sustainability transitions; however, transformation cannot be perceived from a single perspective, as there are often shifts in power from different actors and sectors who play crucial roles in transition debates. Montmasson-Clair (2017) holds that although there are multiple plans and strategies in place for a transition to development that is sustainable in South Africa, there are great challenges from a policy and institutional perspective, and this is due to the inconsistencies and the misalignments in strategies and plans. Efforts to achieve transformations towards sustainability will always be contested due to the highly political nature of transformations, which can result in different actors being affected in different ways with both gains and losses (Meadowcroft, 2011, van den Bergh et al., 2011).

A major barrier to achieving a Just Transition in line with South Africa's sustainability principles is the large number of actors with diverse and sometimes competing priorities, and the tensions between ecological and socio-economic imperatives. There are often competing social and ecological concerns in a Just Transition; for instance, social adjustments take time while climate change requires immediate action (Snell, 2018). This is particularly the case in South Africa, where there is considerable conflict between environmental activists and trade unions as they have deeply held but differing perspectives on social and ecological concerns (Snell, 2018). This conflict has revealed fault lines within the Just Transition discourse as there are tensions between labour and environmental movements on Just Transition priorities.

Global South Anti-Coal Campaigns

There have been mass protests and campaigns in South Africa against coal mining and production. These campaigns have been catalysed by global calls about the changing climate, pollution from coal mining, threats to water security, and human rights (Baker et al., 2014, Shongwe, 2018). Some of the more prominent campaigns are "Life After Coal," which is a joint campaign by Earthlife Africa Johannesburg, groundWork, and the Centre for Environmental Rights. This campaign aims to discourage the development of new coal-fired power stations and coal mines and support the reduction of emissions from existing coal infrastructure while encouraging a coal phase-out in response to both climate change and South Africa's previous exploitive trajectories of economic growth (Munnik, 2019).

"Life After Coal," along with eight civil society organizations, challenged the authorization of a proposed coal mine in the Mabola Protected Environment, a strategic water source area near Wakkerstroom in Mpumalanga Province. The opposition was considered successful, because the constitutional court ruled in favour of the civil society organisations (Life After Coal, 2019). Another climate justice campaign that has gained prominence is in Somkhele, Kwazulu-Natal, which aims to unify opposition to the Fuleni Coal Mine on the border of the iMfolozi Wilderness area. It is led by the civil society organizations Global Environmental Trust (GET) and the Mfolozi Environmental Justice Organisation (MCEJO).

Bond (2019) argues that environmental groups in South Africa are still not yet as militant and effective as in other places, comparing South African movements with Germany's Ende Gelände annual movement. Ende Gelände is an anti-nuclear and anti-coal movement that has been in existence since 2015, organizing non-violent direct-action protests against lignite and coal mining and coal-fired plants in Germany. According to Bond (2019), despite local opposition to fossil fuels' environmental impacts in several places, and the emergence of anti-coal activist networks, South Africa's climate justice groups remained fragmented and unable to focus on broad climate justice priorities: the need for ecologically constructive employment via Just Transition programs spanning gender, race, geographic, and intergenerational equity.

Some South African movements oppose the short-term impacts of coal mining and production, which include local and direct environmental impacts related to mine pollution, and community health and well-being, while others focus on the long-term impacts of coal such as GHG emissions and global warming, which also affect the quality of life and livelihoods of local communities while including equity-based critiques of the economic systems that cause climate change. This broader understanding of climate justice creates the possibility of linking labour unions, women's movements, environmentalists, low-income community activists, and youth climate movements in unified or networked decarbonization campaigns.

Activist researchers can assist this broadening of perspectives by working with local organizations to document and share the nuanced information that emerges from engaged participatory research with organisations such as groundWork, including their involvement with smaller organisations such as the Vukani Environmental Justice Movement in Action (VEM, Vukani Environmental Movement, for short), one of the grassroots movements in Emalahleni, which to date have proven to be the most successful push-back measures on coal.

My doctoral research on climate justice in South Africa's coal industry involves collaboration with groundWork and the VEM, guided by a participatory action research approach. Debates and planning for a Just Transition have presented many viewpoints from different stakeholders, and some of these stakeholder voices, especially those of the most marginalized, tend to be lost or ignored. Participatory research creates opportunities to engage and reflect the arguments of all active stakeholders within the Just Transition debates, even those not regarded as influential actors. For example, VEM is a community-led climate-justice movement that advocates against coal's impacts, especially on community health, pollution, worker safety, racial and gender equity. VEM as a grassroots organisation plays a critical role in obtaining and sharing needed information that can inform community members and also influence planning decisions for a better Just Transition at the grassroots and local level. If grassroots movements such as VEM are disregarded because they are not seen as having much influence, unlike more established union-linked movements such as groundWork, this can result in their messaging not making it all the way to be included in decision-making processes, which is a real loss for climate justice.

Participatory research highlights issues of local concern regarding coal-mining impacts, and identifies how grassroots organisations such as VEM can partner better with other established organisations in influencing democratic decision-making. A Just Transition involves negotiating many contested perspectives, and in the context of Mpumalanga participatory research plays a critical role in understanding and influencing the ever-changing landscape, debates, and power dynamics.

Conclusion

Coal mining and production have had significant impacts in South Africa and many of these are still playing out, as South Africa remains a country highly reliant on coal. The challenges to a Just Transition in South Africa are multi-faceted, with the main barrier being how to interpret and implement the country's sustainability principles; there are a large number of actors with diverse and sometimes competing priorities, so tensions emerge between eco-logical and socio-economic imperatives. Thus, coordination of a transition or phase-out of coal in South Africa has been hampered by the presence of these diverse viewpoints. To achieve consensus on how to move ahead and create a transformative economy, and to build institutions that allow this, many differing perspectives need to be understood, even when most stakeholders agree that a new and different trajectory is needed in contrast to previous environmentally damaging and socially exploitative systems.

Civil society remains key in pressing for any sort of Just Transition, and this can be seen from civil society organizations' involvement in opposing coal mining and utilization. The organisation groundWork has been at the fore-front and has built strong credibility with labour unions, including miners' unions. Through groundWork's growing involvement with other organisations, it is evident that the narrative about a Just Transition has moved far beyond the "jobs versus environmental protection" debate to include how those who previously have been marginalised can begin to benefit economically. The slow progress in policy development and implementation, and the South African government's lack of efforts to fast-track decisions on green-movement initiatives, have been a great challenge that has held back some of the victories of civil society in the fight against human rights violations. Civil society activism on Just Transition issues should be seen as a starting point in conversations with key stakeholders who are moving South African politics on this pressing issue. A transformative and progressive Just Transition is needed, and can be achieved; its success relies on communication among many stakeholders within frameworks that fairly weigh the priorities of the environment, the people, and the economy.

NOTE

1   Just Transition means a framework of social processes, interventions and practices to
     secure workers' rights and livelihoods, and place-based community agency and power,
     in shifting from extractive to sustainable, regenerative economies.

Reference List

Aneja, V.P., Isherwood, A., & Morgan, P. (2012). Characterisation of particulate matter (PM10) related to surface coal mining operations in Appalachia. *Atmospheric Environment, 54,* 496–501.

Avelino, F., Wittmayer, J.M., Pel, B., Weaver, P., Dumitru, A., Haxeltine, A., Kemp, R., Jørgensen, M.S., Bauler, T., Ruijsink, S., & O'Riordan, T. (2016). Transformative social innovation and (dis)empowerment. *Technological Forecasting and Social Change, 145,* 195–206.

Avelino, F., & Wittmayer, J.M. (2019). The transformative potential of plural social enterprise: A multi-actor perspective. In P. Eynaud, J.-L. Laville, L. dos Santos, S. Banerjee, F. Avelino, L. Hulgård (Eds.), *Theory of social enterprise and pluralism* (pp. 193–221). Routledge.

Baker, L., Newell, P., & Phillips, J. (2014). The political economy of energy transitions: the case of South Africa. *New Political Economy, 19*(6), 791–818.

Baker, L., & Wlokas, H.L. (2014). *South Africa's renewable energy procurement: A new frontier?* Energy Research Centre, University of Cape Town.

Bond, P. (2019, September 4). Fighting fossil fuels in South Africa: Campaigners invoke specters of climate chaos. *Counterpunch*. https://www.counterpunch. org/2019/09/04/fighting-fossil-fuels-in-south-africa-campaigners-invoke-specters- of-climate-chaos/

Burke, P.J., & Nishitateno, S. (2013). Gasoline prices, gasoline consumption, and new- vehicle fuel economy: Evidence for a large sample of countries. *Energy Economics, 36*, 363–370.

Cock, J. (2016). Alternative conceptions of a "just transition" from fossil fuel capitalism. In A. Bieler, R. O'Brien, & K. Pampallis (Eds.), *Challenging corporate capital: Creating an alternative to neo-liberalism* (pp. 55–66). Rosa Luxemburg Foundation.

Cock, J. (2019). Resistance to coal inequalities and the possibilities of a just transition in South Africa. *Development Southern Africa, 36*(6), 860–873.

Cahill, B., & Allen, M.M. (2020). *Just transition concepts and relevance for climate action: A preliminary framework.* Center for Strategic and International Studies.

Carras, J.N., Day, S.J., Saghafi, A., & Williams, D.J. (2009). Greenhouse gas emissions from low-temperature oxidation and spontaneous combustion at open-cut coal mines in Australia. *International Journal of Coal Geology, 78*(2), 161–168.

Centre for Environmental Rights. (2016). *Zero hour: Poor governance of mining and the violation of environmental rights in Mpumalanga.* https://cer.org.za/reports/zero- hour

Ciplet, D., & Harrison, J.L. (2020). Transition tensions: Mapping conflicts in movements for a just and sustainable transition. *Environmental Politics, 29*(3), 435–456.

Climate Justice Alliance. (2018). *Just transition principles.* https://climatejusticealliance. org/wp-content/uploads/2018/06/CJA_JustTransition_Principles_final_hi-rez.pdf

Climate Transparency. (2020). *Climate transparency report: Comparing G20 climate action and responses to the COVID-19 crisis.* https://www.climate-transparency.org/g20- climate-performance/the-climate-transparency-report-2020

Culwick, C., & Patel, Z. (2020). Building just and sustainable cities through government housing developments. *Environment and Urbanization, 32*(1), 133–154. https://doi. org/10.1177/0956247820902661

Davis, M. (2010). Who will build the Ark? *New Left Review, 61*, 29–46.

Department of Environmental Affairs. (2018, October 26). National Environmental Management: Air Quality Act, 2004 (Act No. 39 of 2004) and the 2017 National Framework for Air Quality Management in the Republic of South Africa.

Fatti, C.C., Cohen, B., Jennings, G., Kane, L., Rubin, M., & Simpson, E.T. (2021). *In pursuit of just sustainability.* Gauteng City-Region Observatory. https://www.gcro.ac.za/ outputs/research-reports/detail/pursuit-just-sustainability/

Fine, B., & Rustomjee, Z. (2018). *The political economy of South Africa: From minerals– energy complex to industrialisation.* Routledge.

Froestad, Jan, Nøkleberg, M., Shearing, C., & Trollip, H. (2018). South Africa's minerals-energy complex: Flows, regulation, governance, and policing. In Y. Omorogbe & A. Ordor (Eds.), *Ending Africa's energy deficit and the law: Achieving sustainable energy for all in Africa*. Oxford Scholarship Online. https://doi.org/10.1093/oso/9780198819837.003.0016

Geldenhuis, S., & Bell, F.G. (1998). Acid mine drainage at a coal mine in the eastern Transvaal, South Africa. *Environmental Geology, 34*(2–3), 234–242.

Gemmill, B., & Bamidele-Izu, A. (2002). The role of NGOs and civil society in global environmental governance. In D.C. Esty & M.H. Ivanova (Eds.), *Global environmental governance: Options and opportunities* (pp.77–100). Yale School of Forestry and Environmental Studies.

Gray, H.A. (2019). *Air quality impacts and health effects due to large stationary source emissions in and around South Africa's Mpumalanga Highveld priority area (HPA)*. Gray Sky Solutions.

groundWork. (2018). *Coal kills: Research and dialogue for a just transition*. https://groundwork.org.za/wp-content/uploads/2022/07/Coal-Kills.pdf

Gupta, S.K. & Nikhil, K. (2016). Ground water contamination in coal mining areas: A critical review. *International Journal of Engineering and Applied Sciences, 3*(2), 69–74.

Greenpeace. (2019). Latest satellite data reveals Mpumalanga is the world's largest power plant emission hotspot, ranked fourth overall. https://www.greenpeace.org/africa/en/press/6600/latest-satellite-data-reveals-mpumalanga-is-the-worlds-largest-power-plant-emission-hotspot-ranked-fourth-overall/

Hallowes, D., & Munnik, V. (2019). *Down to zero: The politics of just transition*. groundWork.

Keles, D., & Yilmaz, H.Ü. (2020). Decarbonisation through coal phase-out in Germany and Europe—Impact on emissions, electricity prices and power production. *Energy Policy, 141*, 111472.

Kholod, N., Evans, M., Pilcher, R.C., Roshchanka, V., Ruiz, F., Coté, M., & Collings, R. (2020). Global methane emissions from coal mining to continue growing even with declining coal production. *Journal of Cleaner Production, 256*, 120489.

Kretschmann, J. (2020). Sustainable change of coal-mining regions. *Mining, Metallurgy & Exploration, 37*, 167–178. https://doi.org/10.1007/s42461-019-00151-2

Life After Coal (2019, 18 November). Constitutional Court rules against coal mining in Mpumalanga Protected Area. https://lifeaftercoal.org.za/media/news/constitutional-court-rules-against-coal-mining-in-mpumalanga-protected-area

Mandel, S. (2019). Green hydrogen and the future of sustainable energy use in South Africa. *Scilight*. https://doi.org/10.1063/1.5118335

Marais, H. (2013). South Africa pushed to the limit: The political economy of change. Zed Books Ltd.

Mathu, K., & Chinomona, R., (2013). South African coal mining industry: Socio-economic attributes. *Mediterranean Journal of Social Sciences, 4*(14), 347–347.

McCarthy, T.S. (2011). The impact of acid mine drainage in South Africa. *South African Journal of Science, 107*(5–6), 1–7.

McCarthy, T.S., & Humphries, M.S. (2013). Contamination of the water supply to the town of Carolina, Mpumalanga, January 2012. *South African Journal of Science, 109*(9/10), 1–10.

Meadowcroft, J. (2011). Engaging with the politics of sustainability transitions. *Environmental Innovation and Societal Transitions, 1*(1), 70–75.

Mhlogo, S., Mativenga, P.T., & Marnewick, A. (2018). Water quality in a mining and water-stressed region. *Journal of Cleaner Production, 171*, 446–456.

Minerals Council South Africa. (2020). *Facts and figures pocketbook 2020.*

Mohalik, N.K., Lester, E., Lowndes, I.S., & Singh, V.K. (2016). Estimation of greenhouse gas emissions from spontaneous combustion/fire of coal in opencast mines–Indian context. *Carbon Management, 7*(5–6), 317–332.

Montmasson-Clair, G. (2017). *Governance for South Africa's sustainability transition: A critical review.* Trade & Industrial Policy Strategies (TIPS), Green Economy Coalition (GEC).

Munnik, V. (2019). *"COAL KILLS": An analytical framework to support a move away from coal and towards a just transition in South Africa.* Society, Work & Politics Institute University of Witwatersrand.

Nalule, V.R. (2020). Transitioning to a low carbon economy: Is Africa ready to bid farewell to fossil fuels? In G. Wood & K. Baker (Eds.), *The Palgrave handbook of managing fossil fuels and energy transitions* (pp. 261–286). Palgrave Macmillan Cham.

Nzimande, Z., & Chauke, H. (2012). Sustainability through responsible environmental mining. *Journal of the Southern African Institute of Mining and Metallurgy, 112*(2), 135–139.

Pandey, B., Gautam, M., & Agrawal, M. (2018). Greenhouse gas emissions from coal mining activities and their possible mitigation strategies. In S.S. Muthu (Ed.), *Environmental carbon footprints: Industrial case studies* (pp. 259–292). Butterworth-Heinemann. https://doi.org/10.1016/B978-0-12-812849-7.00010-6

Reinouad, C. (2019). *Localising the "just transition": A case study in Mpumalanga, South Africa* [Unpublished master's thesis]. Oxford University.

Shongwe, B.N. (2018). *The impact of coal mining on the environment and community quality of life: A case study investigation of the impacts and conflicts associated with coal mining in the Mpumalanga Province, South Africa* [Unpublished doctoral dissertation]. University of Cape Town.

Smith, S. (2017). *Just transition: A report for the OECD.* Just Transition Centre.

Snell, D. (2018). "Just transition"? Conceptual challenges meet stark reality in a "transitioning" coal region in Australia. *Globalizations, 15*(4), 550–564.

South African Human Rights Commission. (2017). *National hearing on the underlying socio-economic challenges of mining-affected communities in South Africa.* https://www.sahrc.org.za/home/21/files/SAHRC%20Mining%20communities%20report%20FINAL.pdf

South African National Planning Commission (2013). *National development plan 2030: Our future—Make it work.* http://www.dac.gov.za/sites/default/files/NDP%20 2030%20-%20Our%20future%20-%20make%20it%20work_0.pdf

South African National Planning Commission (2019). *Social partner dialogue for a just transition, May 2018 to June 2019.* https://www.nationalplanningcommission. org.za/assets/Documents/Vision%20and%20Pathways%20for%20a%20Just%20 Transition%20to%20a%20low%20carbon%20climate.pdf

Strambo, C., Burton, J., & Atteridge, A. (2019). *The end of coal? Planning a "just transition" in South Africa.* Stockholm Environment Institute.

Swilling, M., & Annecke, E. (2012). *Just transitions: Explorations of sustainability in an unfair world.* UCT Press/Juta and Company (Pty) Ltd./United Nations University Press.

Van den Bergh, J.C., Truffer, B., & Kallis, G. (2011). Environmental innovation and societal transitions: Introduction and overview. *Environmental Innovation and Societal Transitions, 1*(1), 1–23.

Ward, M. (2018). Just transitions and the green economy: Navigating the fault lines [Working paper]. University of Witwatersrand.

World Bank. (2018). *Managing coal mine closure: Achieving a just transition for all.* https:// www.worldbank.org/en/topic/extractiveindustries/publication/managing-coal-mine-closure

Zhao, S., & Alexandroff, A. (2019). Current and future struggles to eliminate coal. *Energy Policy, 129,* 511–520.

Saving Our "Common Home": A Critical Analysis of the "For Our Common Home" Campaign in Alberta

Chrislain Eric Kenfack

Introduction

The recent years have witnessed a fast-growing wave of social collaborative mobilizations, demands for a more aggressive fight against climate change, for climate justice, and for a Just Transition to a post-carbon society around the world. This chapter is a case analysis of one of these collaborative and solidarity struggles, the "For Our Common Home" campaign in Edmonton, Alberta. Led by Development and Peace—Caritas, the official international development organization of the Catholic Church in Canada, the "For Our Common Home" campaign is a multi-year, faith-inspired, climate-justice campaign aimed at pushing Canadian companies operating in the Amazon to be more environmentally responsible in their activities and to respect the voices of local Indigenous environmental activists. With a third of Canadians (about 12.8 million) citing Catholicism as their religion, and with churches' considerable social influence worldwide, their role in motivating climate action and cultural transformation is receiving attention (Müller & Ozyürek, 2021; Jenkins et al., 2018). Moreover, as Jenkins et al. note (pp. 85, 101), "Responses to climate change by Indigenous people challenge the categories of religion and of climate change in ways that illuminate reflexive stresses between the two cultural concepts.... (R)adically new religious formations and imaginations may be under development in the many cultural spaces of climate change."

I explored this broad issue by means of the following question: How can a faith-inspired movement like Development and Peace—Caritas Canada, through a religious environmental campaign involving Indigenous communities in Canada and Brazil, participate in the development of social cohesion and the advancement of social justice?

As Pope Francis stated in his 2015 encyclical *Laudato Si'*, though it is true that the Amazon region is facing an ecological disaster, it also has to be made clear that "a true ecological approach always becomes a social approach; it must integrate questions of justice in debates on the environment, so as to hear both the cry of the earth and the cry of the poor" (Pope Francis, 2015). The endemic persecution of environmental activists in the Amazon makes it almost impossible to raise up that combined voice of the earth and of the poor. It is therefore in a solidarity effort for an environmentalism "that is concerned for the biome but [does not ignore] the Amazonian peoples" (*Instrumentum Laboris*, quoted in Pope Francis, 2020, p. 7), that the "For Our Common Home" campaign was developed to support environmental struggles in the Amazon with a focus, among others, on advocating for Canadian mining companies operating in the Amazon region to take their environmental responsibilities seriously and be held accountable and liable for their environmentally destructive activities.

After describing the campaign's context and its supporters, methods and goals, along with my research methodology, I explore the implications of this form of Catholic climate action.

Care for Our Common Home: When Indigenous and Christian Environmental Demands Meet

The key demands of the care "For Our Common Home" campaign fall within the scope of the ecological teachings of Pope Francis, developed in his encyclical letter *Laudato Si'*, namely his call for *integral ecology, ecological conversion*, and the *culture of care*. In Pope Francis' words, "it cannot be emphasized enough how everything is interconnected. Time and space are not independent of one another, and not even atoms or subatomic particles can be considered in isolation. Just as the different aspects of the planet—physical, chemical and biological—are interrelated, so too living species are part of a network which we will never fully explore and understand. A good part of our genetic code is shared by many living beings. It follows that the fragmentation of knowledge and the isolation of bits of information can actually become a form of ignorance, unless they are integrated into a broader vision of reality (Pope Francis, 2015: Number 138). From this perspective, integral ecology points at the nature and level of interconnectedness that exists between the living and non-living, and that defines the very essence of our common nature as fundamentally relational beings. In other words, nobody and nothing in the order of creation can live in isolation; our very existence is essentially relations and connections to other living and non-living things as well as everything surrounding us. The understanding of nature as integrality and interconnectedness among all its constituencies, including humans, leads us to the need for ecological conversion: simultaneously a call and a responsibility. It is a call for a complete paradigm shift from the way we look at nature, and its human and non-human inhabitants, not as means and objects of profit and capital accumulation; they must instead be contemplated as manifestations of the creator and as sisters and brothers in the order of creation. Ecological conversion is a call to shift away from the culture and practices of overconsumption, and the belief that humans are masters of the earth, as patriarchal colonialist imperialism has accustomed us to, towards a culture of care and sustainability-partnership with Mother Nature and with other creatures. Such visions, which also recall the Indigenous concepts of *Rematriation*, *interconnectedness*, and *stewardship*, support my analysis in this chapter.

It should be noted that the visionary calls of Pope Francis are not unique among world religious leaders. In fact, on November 10, 2016, on the eve of the global climate change conference called COP (Conference of the Parties) 22, some 304 religious leaders from 58 countries issued a joint declaration on climate change. Their declaration calls world leaders to "ground their [climate] decisions in a humble and compassionate reverence for the interconnectedness of all life," and invites believers and their communities to reduce emissions, divest from fossil fuels, and reinvest in low carbon solutions (Interfaith Statement on Climate Change, 2016). This call represents a confluence with the one made by Pope Francis in mid-June 2015 in advance of the previous year's climate change conference, COP 21, when both as religious authority and head of state he issued a radical encyclical entitled *Laudato Si': On care of our common home*, with the aim of influencing the then-forthcoming Paris summit on building a legally binding post-Kyoto climate agreement. In his encyclical, Pope Francis developed the concepts of integral ecology and ecological conversion, which together allow for a focused analysis of the current climate crisis from a holistic perspective. For Pope Francis, climate change is symptomatic of socially unjust neo-liberal capitalist models that oppress the poor and workers for the sake of profits and capital accumulation (Pope Francis, 2015). As such, to address the resulting challenges, we should: 1) respect nature, its laws and equilibrium; and 2) go beyond partisan interests to put the well-being of current and future generations, particularly the most vulnerable, at the centre of political preoccupations (Pope Francis, 2015; Pope Francis, 2016).

While the Pope uses a religious vocabulary, concepts similar to what he names as integral ecology have long been incorporated into other worldviews. According to many Indigenous cosmologies, nature is perceived as a whole, and Indigenous cultures are often based in worldviews that do not put humans at the centre of creation, as ultimate masters, but situate them "within a web of life in which all entities, be they inanimate, plant, animal or natural, possessed a spiritual dimension of their own" (Stonechild 2005, p. 2). "Species of animals and plants are siblings or close relatives of human communities among many Indigenous peoples and thus must be treated respectfully as they too have rights and needs" (Kapyrka & Dockstator 2012, p. 101). Saint Francis of Assisi, far from any anthropocentric or anthropomorphic representation of nature, embraces such relationship views when, in his famous Canticle of the Creatures, he praises "Brother Sun," "Sister Moon and the stars," "Brother Wind," "Sister Water," "Brother Fire," "Sister Mother Earth," "and Sister Bodily Death" (Saint Francis quoted in Gatlif, 2012).

The two sets of visions above (Christian and Indigenous) stress a conception that goes beyond the normative Western understanding of nature as an externality, or a "commodity to be exploited or owned," to include a spiritual relationship (Richardson, 2008; Cardinal, 2001; Verney, 2004) and an inalienable dimension of mutual respect (Steinhauer, 2002; Alfred, 2010; Kovach, 2013). The holistic, spiritual, and reciprocal respect dimensions are key to Indigenous worldviews and therefore, from Indigenous perspectives, defending nature is not simply a matter of protecting an externality, but it is a matter of defending an identity, maintaining relationships, and protecting survival.

From a non-Indigenous perspective, religious ecological conversion is not about adapting production and consumption patterns (within the existing neo-liberal capitalist system) through multiple market mechanisms, techno-fixes, and patches, but rather ecological conversion involves a systematic and systemic change, in order to adopt models that respect nature, workers, and the specificities of affected populations (Kenfack, 2018). In short, "what is required is an act of re-orientation away from unsustainable practices. This act is part of a larger process that can be named 'essential recovery,' which needs to occur both on the level of worldview and in terms of bringing forward past sustainable practices" (Hrynkow, 2014, p. 119). This environmental model can only become possible if societies and individuals learn to live according to sustainable modes of resource use, consumption, and care through acts of transformative learning—so transformative that it will lead to changes in our worldviews, to make them more holistic (O'Sullivan; 1999; Goodman, 2002; Hrynkow, 2014; Hrynkow & Creamer, 2015) and accelerate the ecological transition based on ideas and projects of essential recovery and ecological Rematriation. In fact, "the Indigenous concept of Rematriation refers to reclaiming of ancestral remains, spirituality, culture, knowledge and resources.... It simply means back to Mother Earth, a return to our origins, to life and co-creation, rather than Patriarchal destruction and colonization" (Muthien, 2021, Rematriation section). As such, Rematriation mostly appears as a counter-narrative, countermovement, and an alternate sustainability lifestyle. From such a perspective, and applying this concept to the specific case of ecological crisis, I understand Rematriation as an Indigenous reaction to the current dominant colonial, paternalistic, and capitalist-inspired view that has turned nature into fragmented simple commodities. Beside that oppositional stand, ecological remediation puts forward a feminist-based view that claims a humble return to Mother Nature, understood as an inexhaustible network of relations among human and non-human as well as living and non-living beings, in a continuity that involves past, present, and future generations from a horizontal perspective, and the divine from a vertical perspective. In fact, the return to the ancestral teaching and approaches of Indigenous people offers the possibility of learning from those who, around the world, have always and continue to be uncontestable stewards of nature. As Pope Francis, taking the example of Indigenous people of the Amazon, states:

If the care of people and the care of ecosystems are inseparable, this becomes especially important in places where "the forest is not a resource to be exploited; it is a being, or various beings, with which we have to relate".... The wisdom of the original peoples of the Amazon region "inspires care and respect for creation, with a clear consciousness of its limits, and prohibits its abuse. To abuse nature is to abuse our ancestors, our brothers and sisters, creation and the Creator, and to mortgage the future".... When the indigenous peoples "remain on their land, they themselves care for it best," ... provided that they do not let themselves be taken in by the siren songs and the self-serving proposals of power groups. The harm done to nature affects those peoples in a very direct and verifiable way, since, in their words, "we are water, air, earth and life of the environment created by God. For this reason, we demand an end to the mistreatment and destruction of Mother Earth. The land has blood, and it is bleeding; the multinationals have cut the veins of our Mother Earth" (Pope Francis, 2020, p. 42).

Sustainability can be fostered through acts of essential recovery in which past practices are re-contextualized to meet present challenges. In this scenario, historically sustainable ways of life can be rediscovered, not simply to clone or appropriate past practices, but rather as a renewing *ressourcement*, a return to the sources, in the spirit of Vatican II (O'Malley, 2008). With regard to the importance of *ressourcement* in the Christian tradition, it should be emphasized that, a few years after his election, Pope John XXIII launched the idea of the second Vatican Council with a statement that became emblematic in the history of the Roman Catholic Church. In the conservative-dominated context of the bi-millenary institution, he called Catholics to "throw open the windows of the church and let the fresh air of the spirit blow through." These words of the pope gave a new strength and breadth to the progressive wing of the Church that was already and is still divided into two movements: the partisans of *aggiornamento*, and the partisans of *ressourcement*. *Aggiornamento* refers to the radical progressive movement demanding a complete adjustment, adaptation, and accommodation of the Church to the standards and demands of the modern world. The disciples of *ressourcement*, on the other hand, endorse a more balanced position: *ressourcement* implies "a return to the authoritative sources of Christian faith, for the purpose of rediscovering their truth and meaning in order to meet the critical challenges of our time" (Echeverria, 2014, p. 1). In other words, *ressourcement* demands a creative and fruitful dialogue between the past, the present, and the future. In my use of this concept, I insist on the interconnectedness it implies among various epochs and the importance it gives to the past as an inspirational source for current human-nature-divine relations. Thus, an ecological conversion that is based on a return to the "authoritative sources of Christian faith" implies a renewing *ressourcement*, or rather a continuous reliance on the biblical, traditional, and hierarchical sources of the Church that all advocate respect for nature, or, better said, a genuine spiritual environmentalism. In this regard, our shared humanity with its common survival needs and its shared environmental concern, in the sense of care for life-sustaining ecosystems, can become potential sources of wisdom for living out proper human-earth-divine relationships (Hrynkow, 2016a), as opposed to something to be discarded in favour of narrowly understood manifestations of progress and development. In line with the eco-ethical imperatives laid out in Pope Francis's *Laudato Si'*, this energizing movement is about rediscovering green roots that can be cultivated to branch out in a contemporary context (Pope Francis, 2015; Hrynkow, 2016b). Such a transition to ecological sustainability could gain inspiration from and respond to an Indigenous holistic worldview and traditional knowledges, as "Indigenous peoples interpret and react to the impacts of climate change in creative ways, drawing on traditional knowledge and other technologies to find solutions which may help society at large to cope with impending changes" (United Nations Permanent Forum on Indigenous Issues, 2008, p. 2; see also Kuhn and Duerden, 1996; McGregor, 2004). It is also inspired by Pope Francis's integral ecology, motivating an ecological conversion and the achievement of adaptive and resilient eco-ethical living.

The rediscovery of the interconnectedness of all living and non-living beings and the return to the sources through acts of Rematriation are aimed at bringing humanity to develop a deeper sense of ecological conversion based on responsible stewardship (Kenfack, 2020). The call for responsible stewardship is an invitation to a radical shift in the way we view nature and our relationship with nature. Nature, from this perspective, is not just a provider of goods and services, and we are not "masters of creation"; we are part of nature and, at the same time guardians (not owners) of nature, because the whole creation belongs to a supreme and transcendent entity called God, Great Spirit, Great Mystery, the Great One, the Mighty Spirit, the Divine, the

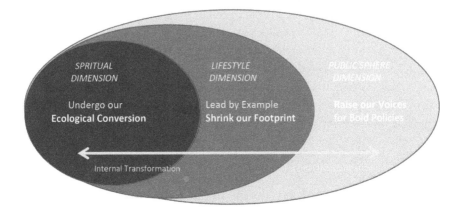

Fig. 15.1 Diagram from *Laudato Si'* Movement's online training webinar for Catholic climate activists, explaining the steps involved to "turn *Laudato Si'* into action and tackle the climate crisis." **Source**: *Laudato Si'* Movement, 2021. https://drive.google.com/file/d/1ep7tIWtf1KR5s uvoicv3P7e4A6M0dzX6/view.

Transcendent, the One who lives above, or Creator, among others, depending on spiritual traditions. From the Biblical perspective, we are called to have dominion over the earth, but this does not imply the kind of dominion involving majestic royal power of modern-time kings and leaders, a domination and mastery that lead to unsustainable exploitation. Dominion, from the Biblical perspective, refers to the stewardship-dominion of the ancient Kings of Israel who were chosen by God to serve the people and "to exercise care and responsibility for God's domain particularly in the interest of those who were poor and marginalized" (Butkus, 2002, p. 1). "The steward is one who has been given the responsibility for the management and service of something belonging to another, and his office presupposes a particular kind of trust on the part of the owner or master" (Hall, 1994, p. 32); that is why, in the context of human-induced unsustainable exploitation and destruction of our common home, Pope Francis, like Indigenous communities around the world, calls for a radical change of patterns and the adoption of those that are more respectful of living and non-living beings. Such a conversion, following in the footsteps of the *Laudato Si'* Movement, is to be undertaken at three important levels: the spiritual, the lifestyle, and the public sphere dimensions (Figure 15.1).

Briefly, the spiritual dimension that entails ecological conversion calls for a complete mindset change, a renewed look at nature from a holistic, relational, and interconnectedness perspective. The lifestyle dimension calls for change in our exploitation, consumption, and disposal patterns to adopt those that are sustainable, less consumerist, less polluting, and more environmentally friendly. The public sphere dimension basically calls for mobilizations, advocacy, and solidarity. For the purposes of this chapter, even though the "For Our Common Home" campaign integrates all three dimensions, I focus mainly on the public sphere dimension. Decentralized and participatory governance, as increasingly advocated by climate and religious institutions, gives more opportunities to non-sovereign actors to raise their voices, and the "For Our Common Home" campaign aims at making the voices of the formerly voiceless heard by decision-makers. In fact, unlike the former decision-making approach in global climate organizations where negotiation texts were unilaterally prepared by the United Nations Framework Convention on Climate Change (UNFCCC) secretariat and discussed by states during COP meetings, in the current approach the main responsibility in building climate policies is transferred to states through institutionalization of their Intended Nationally Determined Contributions (INDCs) (UNFCCC, 2015). In such contexts, the international community mandates the future of climate politics and actions to states under the coordination of the COP, and only has recourse to "naming and shaming" tactics to encourage countries to action (Busby, 2016; Ivanova, 2016; Falkner, 2016; Morgan, 2016). In this scenario, climate justice movements, Indigenous environmental defenders, and faith-inspired environmentalists who had little to no impact in the former political context are able to influence the conception of intra-national and national climate policies, and the elaboration of national reduction pledges. Given this new context, it is interesting to study the metamorphosis that faith-inspired movements are going through, and the mechanisms they are putting in place to induce decision-makers to take their views and perspectives into consideration.

Deep "reflexive stresses" (Jenkins et al., 2018,p. 1), related to the meaning of spirituality and religion, are also entangled with the many people's growing resistance against environmentally destructive practices, investments, and policies, which have huge impacts on Indigenous lands, resources, and health in Canada in general and Alberta in particular. Faith-inspired environmentalism and the solidarity of that form of environmentalism with Indigenous struggles for the protection of their lands, livelihood, and the environment, have spiritual as well as political implications. The "For Our Common Home" campaign, as a faith-based initiative that tries to emphasize the importance of fighting climate change, ensuring social-climate justice ideals, and building solidarity with Indigenous environmental and climate leaders, offers an opportunity to examine these questions in detail.

Methodological Considerations

This chapter discusses the results of a participatory research approach involving both direct participant observation and participation in virtual meetings and events as part of the "For Our Common Home" campaign, which carried on its activities despite the restrictions imposed as a consequence of the global COVID-19 pandemic. I attended events held at the Saint Thomas Aquinas French Catholic Church of Edmonton and at the Sacred Heart Church of the First Peoples in Edmonton, which is a "unique Catholic community of Indigenous peoples and settlers who pray together using symbols, music, and ritual which are meaningful to our People and to our ... Native and Métis culture" (Sacred Heart, 2022). Taking into consideration the Indigenous backgrounds of the members of one of these communities, this research was governed by the guidelines for "Research Involving First Nations, Inuit and Metis Peoples of Canada" (University of Alberta, 2018). Upon establishing firm connections with Indigenous communities through my earlier involvement and work with Development and Peace—Caritas Canada, and subsequently receiving other participants' free prior informed consent at the first meeting, I made sure the basic principles of ethical research with Indigenous people were respected at every level and step of the research initiative. Abiding by such principles was aimed at minimizing risks and maximizing benefits to participants and communities. Far from being a simple arms-length observer, I participated in the development and implementation of "For Our Common Home" campaign.

The "For Our Common Home" Campaign in Alberta

The "For Our Common Home" campaign was launched by Development and Peace—Caritas Canada in 2020. It deals with ecological justice and Indigenous rights issues, with a specific focus on solidarity with Indigenous peoples in the Amazon region. The campaign, officially called "For Our

Common Home: A Future for the Amazon, a Future for All," is a call to reflection, solidarity, and action with Amazonian Indigenous peoples who are continuously battling against 1) deforestation (with new highways and railroads opening the forests to cattle ranching and industrial agriculture, the Amazon is losing one to three soccer fields' worth of forest cover every minute!); 2) resource extraction (essentially driven by insatiable consumer demand, oil extraction and mining that are polluting the Amazon's land, air, and waterways); 3) life-threatening and livelihoods-threatening risks (from megaprojects like hydroelectricity dams that are uprooting communities, and where those who defend their territories are being threatened, criminalized, and killed). The campaign was built around activities such as education and training on integral ecology, ecological conversion, and calls for solidarity with Indigenous communities fighting for the protection of their land and of the Amazon Forest, as an expression of the culture of care by Canadian Catholics. Education and training activities mostly took place during Sunday celebrations during which campaigners elaborated on the meaning of those concepts developed by Pope Francis, and on their implications for the life of Catholic communities and believers. Stories of lives, struggles, and per-secutions of Indigenous communities and some environmental activists in the Amazon were presented. The culmination of this activity was the col-lection of solidarity signatures and messages from Canadian Indigenous and non-Indigenous people. Those signatures were initially supposed to be handed over to two Amazonian Indigenous communities (the seringueiros of Machadinho d'Oeste and the Mura people of Manaus) in solidarity and sup-port of their struggles during a visit that ended up being cancelled because of travel restrictions due to the COVID-19 pandemic. As a result, those signa-tures were given to representatives of the two Amazonian Indigenous com-munities, who received them with great satisfaction and gratitude on behalf of their respective communities, during a webinar organized by Development and Peace and entitled "From Canada, with love, to the defenders of the Amazon," on October 4, 2020. Other activities included advocacy, reflections, and pledges to reduce carbon footprints through individual or community actions to be taken as participation in the global fight for the protection of the environment. Pledges were "invitations [to] people of all ages to commit to at least one lifestyle change for the sake of the environment. Examples in-clude reducing meat consumption and using public transport" (Development and Peace, 2019a, p. 5), as acts of communities' and individuals' ecological

Fields marked * are required

For our Common Home, I pledge to (check one or more boxes): *

☐ Choose more plant-based meals
☐ Choose eco-friendly transport and reduce my emissions
☐ Consume wisely (buy less, organic or fair trade, reduce waste)
☐ Reconnect with nature (gardening, hiking, walking or playing outdoors, etc.)
☐ My own idea

First Name *

Last Name *

Email

Your age group *

∨ - Select -

If the signatory does not have an email
address (e.g., a child), please leave the
field blank

Country *

∨ Canada

City *

Fig. 15.2 "For Our Common Home" online pledge form. **Source**: Development and Peace, 2019b. "Intergenerational pledge for our common home." Available at https://www2.devp. org/en/campaign/forourcommonhome/pledge.

conversion. The advocacy component as initially planned was to involve strategic meetings with local members of Parliament to invite them to consider taking a specific stand on the destructive activities (mostly mining) of Canadian corporations in the Amazon; however, in-person advocacy was prevented by the COVID-19 pandemic situation and health restrictions.

Based on my observations, there was far more stress on expressing solidarity with Indigenous communities fighting in the Amazon than on taking individual and community actions to reduce carbon footprints at the local and parish levels in Canada. The part of the campaign that was geared toward individual and community pledges and actions to reduce carbon footprints at the local level seemed to be down-played and only minimally considered as the campaign went on. Figure 15.2 shows the pledge options that individuals and communities were called to select from. However, no data have been made available to demonstrate the pledges' trends or success rates.

Pledges, even though not much valued during the campaign, were in line with the ecological conversion Pope Francis called for, and the lifestyle dimension advocated by the *Laudato Si'* movement. In real life both dimensions cannot be strictly separated; a real conversion always leads to lifestyle changes. In this context, ecological conversion calls for a complete change of mindset to see the environment and nature differently, no longer simply as resources, but as a common home with which and in which we are all interconnected and have the shared responsibility to care for. The lifestyle dimension focuses on concrete actions that could be taken to reduce our carbon footprints and ensure the protection of our common home, both individually and as communities of faith. As communities, no church in Edmonton or Alberta, to the best of my knowledge, made pledges to reduce carbon footprints as a result of the campaign.

Although the COVID-19 pandemic caused difficulties that forced the campaign to put several of its activities on hold, Development and Peace Canada continued to work with local partners in the Amazon to maintain pressure on corporations involved in mining activities in the Amazonian region. Following Pope Francis's statement that "the colonizing interests that have continued to expand—legally and illegally—the timber and mining industries, and have expelled or marginalized the indigenous peoples, the river people and those of African descent are provoking a cry that rises up to heaven" (Pope Francis, 2020, p. 9), the official development organization of the Catholic church of Canada continued to stand by the Indigenous communities in the following terms, expressed in its letter of support prepared for the campaign and signed by 66,447 inhabitants of Canada:

Dear seringueiros of Machadinho d'Oeste and Mura people of Manaus,

Thank you for protecting the Amazon rainforest, your traditional home and humanity's common heritage.

We, the people of Canada, are pained to learn of your persecution, dispossession and criminalisation by those who would rob your lands, livelihoods, waters and way of life to exploit the gifts of the Amazon for profit.

We join you in urging your government to stop privileging corporate interests over your rights and the integrity of the forest.

We will impel our government to hold Canadian companies to account for what they do on your lands. Keeping you in our hearts, thoughts and prayers, we wish you more power in your fight for justice and dignity. (Development and Peace, 2019c)

However, it is important to mention that Catholic environmentalism, from a hierarchical perspective, is still quite limited in Alberta. Despite growing concerns for the protection of the environment and for environmental justice, there is limited engagement from the local church hierarchy. In Alberta, even though some church groups or movements such as Development and Peace are involved in the advancement of environmental justice, this is a movement pushed from below. The "For Our Common Home" campaign did not gather specific, strong, large-scale momentum, even though it was endorsed, at least in principle, by the hierarchy of the Catholic Church in Edmonton and the pastors of parishes and churches where it was implemented. But neither the Archdiocese of Edmonton nor the parishes where the campaign was implemented officially took any pledge, and there exists no structure to advance environmental education or advocacy at the archdiocesan or parish levels. This brings me to argue that, despite Pope Francis's calls, the efforts of Development and Peace, and the "For Our Common Home" campaign, issues related to integral ecology and ecological conversion are still seen as peripheral matters by the local church in Edmonton. Obviously, the deployment of the campaign, in a province largely dominated by a deeply rooted petro-culture, helped educate the faithful about concerns for the protection of the environment and for environmental justice, and helped enhance the culture of care through the development of a sense of solidarity with Indigenous environmental activists among Catholics in Alberta. However, it is important to mention that such solidarity is still very much oriented toward the Global South. The campaign, right from its conception, was externally oriented rather than focusing on internal situations of environmental persecutions of Indigenous peoples in Canada, taking responsibility for church-related environmentally harmful activities or investments, or taking a strong stand on provincial and national situations of environmental injustice.

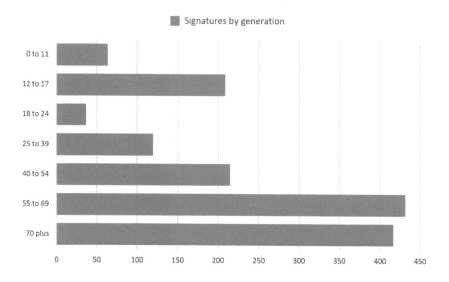

Fig. 15.3 Age distribution of "For Our Common Home" solidarity petition signers. **Source:** Presentation of solidarity signatures to Amazon Indigenous representatives on 4 October 2020.

As this campaign was being implemented, calling for solidarity with Indigenous Peoples of the Amazon who are persecuted and even killed for asking that their lands, environment, and livelihoods be preserved, the Wet'suwet'en blockades and struggles to stop the Coastal GasLink pipeline from crossing their unceded territory were intensifying amidst violent repression from the Royal Canadian Mountain Police (RCMP) in British Columbia, here in Canada. The "For Our Common Home" campaign did not issue any solidarity statement or action to support those Indigenous activists and Peoples struggling locally for the protection of their unceded lands, environment, and livelihoods. Proposals from some campaigners at the Sacred Heart Church of the First Peoples and Saint Thomas French church in Edmonton called for such local solidarity, but those calls were not really considered. I argue that this was an important shortcoming of the campaign because, even though the culture of care needs to be global in scope, it needs to start with our closest neighbors here in Canada. Truths spoken in support of environmental leaders in the Amazon apply equally to environmental leaders in Canada. Solidarity with Indigenous Peoples abroad is great, but it should be accompanied by solidarity with Indigenous Peoples of Canada.

Figure 15.3 shows the distribution of the internationally oriented Indigenous solidarity signatures collected by the campaign, at the national level, in Canada, by age groups. From a first-level observation, it appears that those above fifty-five years of age are more concerned than younger people with environmental justice issues, and ready to show their solidarity with Amazonian Indigenous communities and Peoples fighting for climate justice and for the protection of our common home.

A deeper observation and analysis of the situation brings me to the hypothesis that the dominant support from people aged fifty-five and above may not necessarily be as a result of their higher environmental sensitivity, but perhaps simply reflects their greater involvement (far more than younger generations) in churches' activities. Such a hypothesis seems plausible when we observe other environmental and climate justice mobilizations locally, nationally, and internationally. It would have made sense for more solidarity signatures to be from young people, perhaps in the eighteen- to twenty-four-age range, since social mobilizations in general, and climate justice mobilizations in particular, gather more momentum and are usually led by young people more than older generations. The Fridays for Future movement and the boldness of prominent, young, globally known climate-justice activists such as Autumn Peltier, Greta Thunberg, Xiuhtezcatl Martinez, Licypriya Kangujam, Lesein Mutunkei, Nyombi Morris, and Vanessa Nakate, among others, speak eloquently in this regard. In the specific case of religious climate justice activism, the global youth-dominated Catholic "*Laudato Si'* Generation" movement clearly demonstrates the vigour of youth involvement through faith-inspired climate justice activism (*Laudato Si'* Generation, 2019). The overall *Laudato Si'* work advances the teachings of Pope Francis through a variety of activities such as prayers and retreats, training of communities and animators, advocacy for the respect of people and nature, circles where people meet in small groups and deepen their relationship to God, to nature, and to ways leading to individual and community ecological conversion among others. The overall mission of the movement, based on the environmental teachings of Pope Francis developed in his Encyclical Letter *Laudato Si'*, is "to inspire and mobilize the Catholic community to care for our common home and achieve climate and ecological justice" (*Laudato Si'* Movement, n.d.). The force of such global movements, which include education, personal commitment, faith, and solidarity, lies in their ability to link these action steps and encourage people to begin where they can, hopefully expanding their engagement and learning from the situations and actions of others in the larger movement, both far and near. From this viewpoint, "For Our Common Home" is hopefully an initial step in a much wider and longer-term climate justice trajectory.

Conclusion

The implementation of the "For Our Common Home" campaign, at least in Alberta, did demonstrate a growing consciousness for environmental issues in general and, in particular, environmental justice concerns. Throughout my participation in the preparation and implementation of the campaign at the Saint Thomas of Aquinas Catholic Francophone parish and the Sacred Heart Church of the First Peoples in Edmonton, I witnessed strong involvement and desire to learn more from members of the two congregations. It is important to mention that both parishes had no prior history of environmental actions, and there are no known or self-declared climate justice/environmental activists' movements in either of the congregations. However, even though in both congregations, climate justice and environmental issues in general were not particularly familiar, their respective pastors took advantage of the campaign and created a space for members of their congregations to be educated on environmental justice issues and to engage, from a Christian perspective. Nevertheless, the limited interest in taking personal and community pledges to reduce their individual and collective carbon footprints, and the considerable level of participation in solidarity with persecuted Indigenous activists and communities fighting for the environment in the Amazon, were indications of a belief that fighting against climate change, and climate justice, are still largely other people's business; there seemed to be a conviction that the planet can be saved, but it has to be done largely through other people's actions, and that the expressed solidarity of communities from the Global North is enough. There is still a deeply rooted sense that Catholics in Alberta can continue maintaining their petro-culture, silently witnessing the persecution of local Indigenous climate justice activists and communities fighting to defend their unceded territories, environment, and livelihoods, while supporting Indigenous people leading similar struggles in the Global South. It made me wonder: How can we deepen our ecological conversion to face and include local climate justice?

Reference List

Alfred, T. (2010). "What is radical imagination? Indigenous struggles in Canada." *Affinities: A Journal of Radical Theory, Culture, and Action, 4*(2), 5–8.

Braden, J. (2021, June 24). "What is an ecological conversion?" *Laudato Si'* Movement. https://laudatosimovement.org/news/what-is-an-ecological-conversion-en-news/#:~:text=Laudato%20Si'%20Movement%20defines%20ecological,and%20renewing%20our%20common%20home .%E2%80%9D

Busby, J. (2016). "After Paris: Good Enough Climate Governance." *Current History, 19*, 3–9.

Butkus, R.A. (2002). "The Stewardship of Creation." https://www.baylor.edu/ifl/christianreflection/CreationarticleButkus.pdf

Cardinal, L. (2001). What is an Indigenous perspective? *Canadian Journal of Native Education, 25*(2), 180–182.

Development and Peace. (2019a). A future for the Amazon, a future for all. https://www.devp.org/wp-content/uploads/2021/11/future_for_the_amazon-backgrounder-en-v02.pdf

Development and Peace. (2019b). "Intergenerational pledge for our common home." https://www2.devp.org/en/campaign/forourcommonhome/pledge

Development and Peace. (2019c). "Solidarity Letter for the defenders of the Amazon." https://www.devp.org/wp-content/uploads/2021/11/devpeace_commonhome_solidarity_letter_en.pdf

Echeverria, E. (2014, July 26). "Ressourcement," "Aggiornamento," and Vatican II in Ecumenical Perspective. *Homiletic & Pastoral Review.* http://www.hprweb.com/2014/07/ressourcement-aggiornamento-and-vatican-ii-in-ecumenical-perspective/

Falkner, R. (2016). The Paris Agreement and the new logic of international climate politics. *International Affairs, 92*(5), 1107–1125.

Gatlif, K. (2012). Canticle of creation. *Publication of the Sisters of Saint Francis, 17*(1), 1–20.

Goodman, A. (2002). Transformative learning and cultures of peace. In E. O'Sullivan, A. Morrell, & M.A. O'Connor (Eds.), *Expanding the boundaries of transformative learning: Essays on theory and praxis* (pp. 185–198). Palgrave.

Hall, D.J. (1994). *The steward.* William B. Eerdmans Publishing.

Hrynkow, C. (2014). The new story, transformative learning and socio-ecological flourishing: Education at crucial juncture in planetary history. In F. Deer, T. Falkenberg, B. McMillan, & L. Sims, *Sustainable well-being: Concepts, issues, and educational practices* (pp. 105–120). Education for Sustainable Well-Being (ESWB) Press.

Hrynkow C., & Creamer D. (2015). Transformative praxis at work in Loreto Day School Sealdah: A remarkable fostering of positive peace. *Journal of Peace Education and Social Justice, 9*(1), 1–26.

Hrynkow, C. (2016a). *Laudato Si'*, transformative learning, and the healing of human-earth-divine relationships. *The Ecumenist: A Journal of Theology, Culture, and Society, 53*(4), 10–15.

Hrynkow, C. (2016b). No to war and yes to so much more: Pope Francis, principled nonviolence, and positive peace. In H. Eaton & L.M. Levesque (Eds.), *Advancing nonviolence and social transformation: New perspectives on nonviolence theories* (pp. 135–152). Equinox.

Interfaith Statement on Climate Change. (2016, November 10). Statement by religious, spiritual and faith-based leaders for the first meeting of the parties to the Paris Agreement (CMA1) during the twenty-second session of the Conference of the Parties (COP 22). https://d3n8a8pro7vhmx.cloudfront.net/bhumipledge/pages/192/attachments/original/1479807344/English_Statement_21–11–16.pdf?1479807344

Ivanova, M. (2016). Good COP, bad COP: Climate reality after Paris. *Global Policy, 7*(3), 411–419.

Jenkins, W., Berry, E., & Beck Kreider, L. (2018). Religion and climate change. *Annual Review of Environment and Resources, 43*, 85–108.

Kapyrka, J., & Dockstator, M. (2012). Indigenous knowledges and western knowledges in environmental education: Acknowledging the tensions for the benefits of a "two-worlds" approach. *Canadian Journal of Environmental Education, 17*, 97–112.

Kenfack, C.E. (2018). *Changing environment, just transition and job creation: Perspectives from the South*. CLACSO.

Kenfack, C.E. (2020). Social cohesion environmentalism: Theorizing the new wave of social alliances among climate, faith-inspired and social justice movements. *London Journal of Research in Humanities and Social Sciences (LJRHSS), 20*(5.1), 23–42.

Kovach, M. (2013). Treaties, truths, and transgressive pedagogies: Re-imagining Indigenous presence in the classroom. *Socialist Studies: The Journal of the Society for Socialist Studies, 9*(1), 109–127.

Kuhn, R.G., & Duerden F., (1996). A review of traditional environmental knowledge: An interdisciplinary Canadian Perspective. *Culture, 16*(1), 71–84.

Laudato Si' Movement. (2022). Mission of *laudato si'* movement. https://prev.laudatosimovement.org/learn/mission/

Laudato Si' Generation. (2019). Who we are. https://laudatosigeneration.org/

Laudato Si' Movement. (2021). Training of *laudato si'* animators. Available at https://drive.google.com/file/d/1ep7tIWtf1KR5suvoicv3P7e4A6M0dzX6/view

McGregor, D. (2004). Coming full circle: Indigenous knowledge, environment and our future. *American Indian Quarterly, 385*, 388.

Morgan, J. (2016). Paris COP 21: Power that speaks the truth? *Globalizations, 13*, 943–951.

Müller, T., & Ozuyürek, E. (2021, December 1). Religious communities can make the difference in winning the fight against climate change. *The Conversation.* https://theconversation.com/religious-communities-can-make-the-difference-in-winning-the-fight-against-climate-change-172192

Muthien, B. (2021). Rematriation: Reclaiming Indigenous matricentric egalitarianism. In B. Muthien and J. Bam (Eds.), *Rethinking Africa: Indigenous women re-interpret Southern Africa's pasts*. Jacana Media, chapter 2. (An earlier version of this chapter is available at http://www.gift-economy.com/articlesAndEssays/rematriation.pdf)

O'Malley, J.W. (2008). *What happened at Vatican II?* Belknap Press of Harvard University Press.

O'Sullivan, E. (1999). *Transformative learning: Educational vision for the 21st century*. University of Toronto Press.

Pope Francis. (2015). *Laudato Si'. [Encyclical letter on care for our common home.]* Libreria Editrice Vaticana.https://www.vatican.va/content/francesco/en/encyclicals/documents/papa-francesco_20150524_enciclica-laudato-si.html

Pope Francis. (2016). Pope's message for the 22nd meeting of the Conference of Parties to the United Nations Framework Agreement on Climate Change (COP 22): The fights against climate change and poverty are linked. Holy See Press Office. http://press.vatican.va/content/salastampa/en/bollettino/pubblico/2016/11/15/161115b.html

Pope Francis. (2020). Post-Synodal apostolic exhortation Querida Amazonia of the Holy Father Francis to the people of God and to all persons of good will. Libreria Editrice Vaticana. http://www.vatican.va/content/francesco/en/apost_exhortations/documents/papa-francesco_esortazione-ap_20200202_querida-amazonia.html

Richardson, B.J. (2008). The ties that bind: Indigenous Peoples and environmental governance. *Comparative Research in Law and Political Economy. Research Paper No. 26/2008*. http://digitalcommons.osgoode.yorku.ca/clpe/197

Sacred Heart. (2022). Sacred Heart Church of the First Peoples website. https://sacredpeoples.com/

Steinhauer, E. (2002). Thoughts on an Indigenous research methodology. *Canadian Journal of Native Education, 26*(2), 69–81.

Stonechild, B. (2005). Indigenous Peoples of Saskatchewan. *Indigenous Saskatchewan Encyclopedia*. University of Regina Press, Canadian Plains Research Centre. Retrieved 8 March 8 2022, from https://teaching.usask.ca/indigenoussk/import/indigenous_peoplesof_saskatchewan.php

United Nations Framework Convention on Climate Change (UNFCCC). (2015). *Adoption of the Paris Agreement*. http://unfccc.int/resource/docs/2015/cop21/eng/l09r01.pdf

United Nations Permanent Forum on Indigenous Issues. (2008). *Climate change and Indigenous Peoples*. http://www.un.org/en/events/indigenousday/pdf/Backgrounder_ClimateChange_FINAL.pdf

University of Alberta. (2018). *Panel on research ethics, informed consent template and guidelines*. Retrieved 17 October 2021, from https://www.ualberta.ca/research/research-support/research-ethics-office/forms-cabinet/forms-human.html

Verney, M.N. (2004). On authenticity. In A. Waters (Ed.), *American Indian thought* (pp. 133–139). Blackwell Publishing.

Action Research for Climate Justice: Challenging the Carbon Market and False Climate Solutions in Mozambique

Natacha Bruna and Boaventura Monjane[1]

Introduction: From Mining Extractivism to the Advent of the Carbon Market

The economy of Mozambique is a typical resource-based system. In general, the country's economic policy has focused on transforming the nation into a recipient of Foreign Direct Investment (FDI), adopting an extraction-transport-export scheme that, while resulting in high economic growth rates, fails to improve the welfare of its population (Mosca, Abbas, & Bruna, 2016; Castel-Branco, 2010).

Extractivism is prevalent in numerous other sectors, even beyond those traditionally associated with extractive activities. This is seen in agriculture, where agricultural commodities are harvested and exported with no or only minimal processing. These extractive schemes may generate environmental and social costs, which cause social marginalization and poverty in rural areas. Some studies show that rural livelihoods have been negatively affected in areas where extractive practices are carried out (Feijó, 2016; Mosca & Selemane, 2011). Negative effects on rural livelihoods and increasing poverty

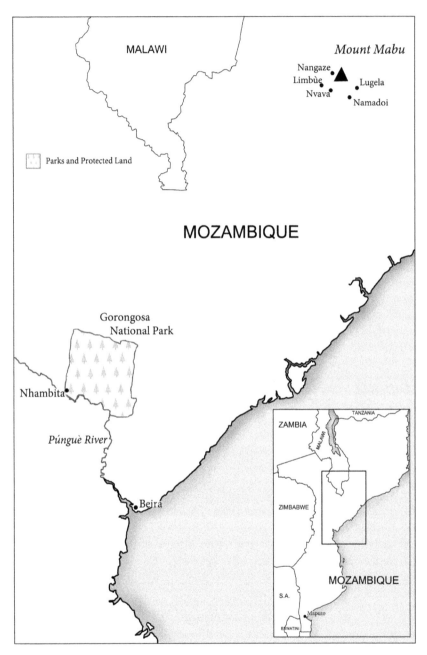

Map 7 Mozambique—Sofala and Zambézia Provinces

have been particularly identified—for instance, in coal mining areas in Tete province, in the natural gas extraction area in Inhambane, and in the cultivation of eucalyptus and rubber trees in various parts of the country.

Moreover, there are visible socio-economic costs as a result of climate change. Mozambique is considered one of the countries that is most vulnerable to climate change, largely because of its weak and fragile socio-economic and human development characteristics (Brito & Holman, 2012; World Bank, 2010). In the past few years, the country has experienced severe droughts, floods, and cyclones (Idai, in Central Mozambique, and Kenneth in the north of the country). Thousands of hectares of crops were destroyed, including cash crops, and the cyclones caused catastrophic health impacts, damaged infrastructure, and shut down numerous businesses (see for example Charrua et al, 2021; Feijó & Aiuba, 2019). In addition to the impact of extreme weather, many regions are silently impacted by climate variability (such as changes in rainfall patterns). This directly and negatively impacts the subsistence agriculture and other income-generating occupations of rural populations.

With the intention of mitigating the effects of the global environmental crisis through the limitation and reduction of greenhouse gas (GHG) emissions, the carbon market emerged from the Kyoto Protocol (UNFCCC 2008). The carbon market consists of buying and selling carbon credits (which are generated by the certification of carbon offsets) and this, in turn, allows buyers to continue to pollute a certain amount, measured in tons of carbon. Carbon is captured through different climate change mitigation actions, including reforestation (replanting in deforested areas) or through the preservation of environmental protection areas. After proper measurement and verification of carbon capture in tons of carbon, the credits are sold at a market price. The price of carbon credits on the international market fluctuated between 5 and 56 US dollars per ton between 2017 and 2022 (IHS Markit, n.d.). The implementation of a carbon capture project aims at capturing the maximum amount of carbon dioxide and selling the carbon credits to industrialized countries, polluting industries, or any other company or individual looking to offset carbon dioxide emissions.

Such mitigation and adaptation projects normally target less industrialized countries with high biodiversity potential, such as Mozambique, where about 25 per cent[2] of the national territory is a potentially protected area. With the global need for adaptation and mitigation, and potential biodiversity, Mozambique is a recipient of climate funds and a strategic destination for projects working on climate change adaptation and mitigation. However, studies have shown that these projects have a negative impact on rural livelihoods and, as we will see further below, they also stimulate new forms and dynamics of community resistance, given the overexploitation of natural resources (Fairhead et al., 2012; Bruna, 2019).

In researching these issues, we used an action-research methodology in collaboration with the communities affected by these projects—not about them but with them. This chapter seeks to highlight a new element in the rush for natural resources in countries like Mozambique: carbon. This new commodity, which is sold on international markets in the form of carbon credits, is the result of the implementation of "green" projects aimed at conservation and the reduction of GHG emissions.

The chapter looks at two cases. First, it analyzes the implementation of a REDD+ (reducing emissions from deforestation and forest degradation) project in Nhambita, Sofala province, as an example of the mainstream traditional solution to climate change (Map 7). Secondly, we describe an example of a type of climate change mitigation and adaptation initiative that does not follow the top-down and market-driven strategies of mainstream policies. The case we describe, in Mabu, Zambézia province, is an alternative to mainstream solutions—a joint project between community farmers and an environmental organization named JA! (*Justiça Ambiental*—Environmental Justice), where they work collaboratively on sustainable small-scale agriculture and livestock practices using local methods oriented towards the conservation of forests in the community. As these practices differ from those envisaged in mainstream climate change adaptation and mitigation policies, they constitute, we believe, first steps toward reducing the climate injustice that is evident in Mozambique and other extraction-driven countries.

Action Research and the Scholar-Activist Approach

The design of this study adopted an action research and scholar-activist approach. Action research arose from the need to bridge the gap between theory and practice. It is participatory, engaged, and committed research, as opposed to traditional research that is independent, non-reactive and objective. Action research seeks to bring research and action/practice together (Engel, 2000). The scholar-activist approach involves "rigorous academic work that aims to change the world or engaged activist research that is described by

Fig. 16.1 Mabu farmer.

Fig. 16.2 Dona Francisca.

detailed academic research, which is explicitly and unapologetically linked to political projects or movements" (Borras, 2016, p. 1). In practice, engagement with communities must be sensitive, socially friendly, and politically committed to the affected populations (Shivji, 2019, p. 15). Therefore, it consists of conducting rigorous—but not neutral—research (Santos, 2014).

As noted, our study was carried out in collaboration with civil society organizations and local social movements, namely JA! (https://justica-ambiental.org/), UNAC—*União Nacional de Camponeses* (National Peasant Movement Organization—https://www.unac.org.mz/), and the *Alternactiva* collective (Democratic Debate for Social Emancipation Platform—http://alternactiva.co.mz/).

These organizations work directly with rural communities in Nhambita and Mabu. JA! and UNAC (administration) were involved in the development of the project proposal and the research objectives. During the field study in Mabu and Nhambita, we worked in collaboration with local representatives and members of these organizations at the community level. Additionally, and always as a transversal factor, we focused on the inclusion of the research participants in knowledge-building, dissemination, and production of socially relevant and emancipatory content (Figures 16.1 and 16.2).

Community members from Mabu and Nhambita participated in this research not as objects of study in the classic and traditional sense of extractive research, but as active subjects. First, these communities were selected to some extent because of the relationship previously built between the authors, the organizations, and the individuals who participated in the study. Trust shared among authors, the organizations, and the leaders and members of the two communities contributed to this epistemic relationship (Monjane, 2021). This allowed the authors to ask certain questions and get more honest answers from the individuals interviewed. Meetings with all the participants, which may also be called focus groups, were opportunities for community discussions about the purpose of climate change mitigation and adaptation projects implemented in each community (Figures 16.3 and 16.4). These community discussions had not previously taken place in Nhambita and very few had been held in Mabu. Our workshops on climate justice were also an opportunity for mutual learning among the participants, research partners, and researchers. From these workshops emerged new understandings of climate change, greenhouse gas emissions, carbon markets, climate justice, and other topics. The sessions also allowed us to present and check the

Fig. 16.3 Climate justice workshop.

Fig. 16.4 Community group meeting outside.

preliminary results of the study, first in Nhambita and later in Lugela,[3] with members of both communities.

Exchange visits between the communities were a crucial part of this study. Besides the researchers and partner organizations, members of the Nhambita community visited the Mabu community. Both communities are affected by the same phenomena (climate-related changes and crises), but they have radically different response strategies. This made possible a deep understanding of resistance processes, and also the beginning of a cooperative relationship between the communities. It is well known in action-research that facilitating informal intercommunity exchange visits is essential for sharing new experiences among partners (Buti, 2021).

Another important element in this research was the production of audiovisual materials, namely a documentary film summarizing the experiences, struggles, and life alternatives in the two communities. Besides being produced with members of both communities telling their stories as active proponents, it is also a useful popular education tool on climate justice struggles and material for use in climate justice advocacy. This documentary will be returned to the communities in a coming phase of our action research.

Carbon Capture and Emissions Reduction in Nhambita (Gorongosa) and Socio-Economic Implications

The Global Rush to Natural Resources

Due to the dynamics mentioned above, a gradual change is evident in the plans of big multinational companies (e.g., Sasol,[4] which has increased its interest in natural gas in the name of climate change mitigation) and the redirection of global capital to so-called green investments—renewable energy, biofuels, forestry, and others (World Bank, 2010). This means that the rush to natural resources has been shaped to respond to the emerging need to capture carbon and/or reduce emissions. One of the main strategies promoted internationally and acclaimed (and funded in some cases) by organizations such as the United Nations, the Intergovernmental Panel on Climate Change (IPCC), the World Bank, and environmental organizations and foundations, is the REDD+ framework. This strategy is often associated with the carbon market through the implementation of carbon capture projects, through either reforestation or the (re)establishment of protected areas, so that emission reduction credits can be sold in the carbon market.

Mozambique already has numerous projects for climate change mitigation and adaptation set out in national strategies—the National Strategy for Reducing Emissions from Deforestation and Forest Degradation, Conservation of Forests and Increase of Carbon Reserves through Forests (Mozambique, 2016), and the National Strategy for Adaptation and Mitigation of Climate Change (Mozambique, 2012). From among the many such "green" projects operating in Mozambique, this chapter focuses on the experience of the Nhambita community, which is part of the buffer zone of Gorongosa National Park, one of the largest protected areas in Mozambique. A PES (payments for environmental services) project was implemented in this area under the country's REDD+ strategy. These new dynamics have not only triggered an increase in the land rush in Mozambique (some studies already confirm this; see for example Borras et al., 2011 and Bruna, 2019), but they have also promoted a rush to areas of high biodiversity in order to capture carbon and then sell carbon credits: the carbon rush. The Nhambita case near Gorongosa, which was a REDD+ project implemented by the company Envirotrade, shows how this carbon rush takes shape, the implications of these types of projects for the rural population, and the potential gains for the implementing stakeholders (usually foreigners).

Brief Descriptions of Nhambita and the Envirotrade Project

The community of Nhambita is located in the district of Gorongosa, in the province of Sofala. Gorongosa National Park (PNG) and its buffer zone cover an area of approximately 10,000 km² (Moçambique, 2016). Nhambita is located in the Púnguè region and is near one of the major rivers in the area, the Púnguè River (see Map 7, page 318).

Local families live mostly from dry-land farming (dependent on rainfall or using the low levels of the Púnguè River for cultivation) on small pieces of land (normally from 1.2 to 5 ha per household), which in Mozambique are called *machambas*. These families do not use fertilizers or pesticides and thus rely on shifting cultivation (crop rotation to allow the land to "rest" and restore the quality of the soil, called fallow periods). Besides farming, family members normally work in off-farm activities, such as casual, informal jobs (in the nearest towns or villages), in forestry-based jobs (carpentry, charcoal selling, sale of alcoholic beverages, traditional medicine), or handicrafts, among others.

It was in this area that the now-defunct British company Envirotrade started a REDD+ project in 2003, named the Sofala Community Carbon Project. According to the company, it was an operation to develop sustainable use of the land to achieve rural development in the region. It was a for-profit project, in which the carbon was captured from agroforestry (of native and non-native plants), forests protected, and deforestation avoided, with the carbon then traded on the open carbon market. The project comprised, in addition to agroforestry, the opening of a local carpentry workshop and a sawmill using local materials in a sustainable way, and a plant nursery for fruit and other species. The nursery supported tree planting and employed mostly women. In addition to the farmer-producers, the company also contracted carpenters, nursery workers, extension agents, and forest rangers who patrolled the forest against deforestation and wildfires.

According to the project's impact assessment, about 1,510 producers were involved in the project (Marzoli & Del Lungo, 2009) with the planting of numerous tree species, under a seven-year payment agreement, although the producers were supposed to protect the trees for longer than seven years. The project basically consisted of capturing carbon by planting trees of different species, and reducing emissions by not deforesting new areas for subsistence agriculture for food and other benefits. Marzoli and Del Lungo (2009) state that, between 2003 and 2008, the project made a total of USD $900,000 on the carbon market, generated mainly through agroforestry activities. The job of the producers was to plant the trees and provide all the necessary care during their growth period. The number of trees planted per producer varied, depending on the land available to each producer. In return, the producers received decreasing annual payments (represented as the equivalent of their labour needed to take care of the plants) according to the number of trees planted per producer—on the condition that no new areas were cleared for *machambas*. However, after fifteen years of operation in Mozambique, the company left the country, and the producers claim that the company stopped paying for the trees or for the investment the farmers had made. According to a former Envirotrade producer and technician,

> I received lemon, cashew and mango plants to be planted. I plant-
> ed them in the machamba of my house. So, we received tokens to

be exchanged for salary; in the first year, I got paid—it was about 300 (meticais)[5]; others received a lot of money. I was told the salary was based on the number of plants I received, but I did not have enough plants. I had about a hundred of them in my backyard. It got worse—they paid 290 meticais in the second year. In the third year, I was paid 90 meticais. With that money, I could buy only salt. In a meeting, they told us that the money would come, but it hasn't come [until now] and the project has ended. We were left with only plants. (Interview in July 2021, farmer, former Envirotrade producer)

According to another former producer, "[if] some plants died, the company would reduce our pay" (Interview, farmer, former Envirotrade producer, July 2021).

These statements suggest that the employment relations under the terms of the contract were hostile to farmers to the point that failure to comply with the provisions of the contract resulted in severe penalties enforced by the company, including the termination of the contract (Monjane, 2012). According to a former manager of the company, the payments were supported by the carbon price on the market. Revenues from the sales of carbon credits had three main purposes: (1) paying the producers; (2) covering operating costs of the project; and (3) covering costs related to the measurement and verification of carbon credits. The measurement and verification of carbon credits were done by third parties and not by the company itself. Meanwhile, due to the fall in the global price of carbon and the consequent financial infeasibility of the project, the company had to stop its operations and end the project.

Socioeconomic Implications

a) Disrupted Livelihoods and Social Reproduction Strategies

Although producers argue that planting trees has some benefits for farmers (they provide shade and fruit and protection from strong winds), negative socio-economic implications are evident. In addition to their debts and the drop-off in income after the company left, the planting of trees has affected the way the producers use the land, making them substitute agroforestry for food crops, thus jeopardizing food availability and access, besides the condition that no new areas can be cleared for other activities. Moreover, it was observed that agroforestry absorbed available local labour, which means that less work was spent on the *machambas*. This also triggered contradictions and conflicts among producers concerning their way of life and agricultural production, as fallowing land was no longer allowed (due to the prohibition on deforesting new areas). These conflicts were not managed with compensatory strategies or policies for the losses in access to land, constraints on traditional agricultural practices, or labour exploitation.

Fifteen years after the beginning of the project, it ended in 2018 and left behind unfinished work and hundreds of perplexed families. According to former producers, the company left the region without saying goodbye to the communities, and it failed to pay for plantation labour or ongoing care for the trees.

> Envirotrade did not leave properly. Envirotrade owes money to many people. First, they owe the producers for three years of planting. Second, they owe three years of salary for the work done by the nursery workers who raised seedlings. Third, they also owe three years of salary to the men who were protecting the areas and making firebreaks. Fourth, they owe three years of salary to the people who lived in the individual [forest] areas designated for the carbon project. Finally, they should indemnify the workers. (Interview in July 2021, former technician at Envirotrade)

According to the former carbon manager and coordinator at Envirotrade, who disagrees with the statements above, the deal ended due to the fall in global carbon prices and the project's resulting financial infeasibility, as carbon revenues funded the project. Also, the company claims to have been the victim of an "anti-REDD+ campaign" that supposedly discredited the work done by Envirotrade for all of those years. As mentioned earlier, the initial plan of the Nhambita carbon project was to pay producers for seven years after planting the trees, which the company says were advance payments, as the producer should take responsibility for caring for and protecting the trees for a much longer period, up to one hundred years (Kill, 2013). One hundred years is presumably the period in which a tree reaches its maximum capacity for carbon sequestration.

Opinions as to how effective the project was for the development of the region are mixed in Nhambita. Some former Envirotrade producers and technicians regret the discontinuation of the project, especially the loss of the monetary benefits from the annual payments they received.

Although there are no visible prospects of the company's returning to Nhambita, the question still remains within the community as to whether the Envirotrade project will be resumed by the company or any other interested parties.[6] Between uncertainty and hope, some producers continue to protect the planted trees—although without the obligation or pay to care for them—and, at the same time, to clear new areas for agriculture. While Envirotrade had operations in the region, the producers were not allowed—under the terms of the agreements—to clear new areas for other activities, including agriculture, since Envirotrade was interested in documenting greater amounts of vegetation and biomes for the purpose of capturing as much carbon as possible. Households in Nhambita seem to have plenty of fruit trees, mostly mango trees and cashew trees, planted through the project. Some of the producers had signed numerous contracts, adopting different methods (border strips, intercropping, yards), which was possible mainly for producers who had greater land availability. One of the concerns raised by the producers interviewed was that they did not know how to make effective use of the trees, which raises another question about how aware producers were of the objectives and specificities of the project.

According to a producer,

> We were left with only the plants.… In one area some farmers were cutting down trees in revolt, because they were not getting paid. They even started cutting plants in the machamba. I asked why they were cutting everything; they said [because they had] not been allowed to do that for many years [and then we] ended up not getting paid. The machamba is full of plants and they [say] we are going to cut them down. (Interview in July 2021, farmer, former Envirotrade producer)

Apart from the asymmetry in information between the company and the producers, also notable is that the company broke its promises to improve living conditions in the communities as a result of environmental projects. Instead, it is evident that the company created a significant level of economic dependency within the communities, which resulted in a disruption in income and living conditions shortly after the company left. Strategies promoting sovereignty and independence were not created—quite the contrary.

b) Compromised Food Sovereignty

One of the notable critiques from Nhambita carbon project researchers and activists involved the potential risks the project implied for food security in the region, since the hired producers (several hundred) tended to neglect food crop cultivation in favour of tree planting and maintenance. Although some of the trees planted provide fruit, this does not compensate for the other kinds of produce that are no longer grown in the *machambas*.

This was, in fact, the understanding of a teacher from the local elementary school after observing the dynamics of the project's implementation for about ten years. She stated that with the project, the Nhambita community developed a characteristic distinct from the other communities where she taught: farmers worked fewer hours in the *machambas* to dedicate more time to agroforestry.

> [The farmers] lost themselves a little, since they became more involved with the company and food production became their second priority. By abandoning food production they ended up with a loss. (Interview in July 2021, teacher from Nhambita)

It is premature, without a detailed study, to evaluate the changes that occurred in Nhambita regarding the reduction in local productivity and diets. The phenomenon that seems to have emerged with the closure of the project is a process of re-agrarianization, shown by the readoption of agriculture as the main household occupation.

As mentioned earlier, there are divergent opinions in the community about the economic impacts of the project. Those reminiscing about the project's benefits claimed that the project helped the producers purchase certain construction materials and consumer goods, such as cement bricks and metal roofing for home improvements, and some appliances (such as radios and solar panels), although few houses are built with imported materials, as observed by our research team.

Among the more skeptical voices is that of the leader of the Nhambita community, whose view was that Envirotrade simply *"exploited people"* (Interview in July 2021). This community leader refused to become a producer for Envirotrade, because he considered the salary offered too low for the hard work required to keep the trees alive and healthy. Moreover, he claimed that the terms of the contract benefitted only Envirotrade. The leader and his family decided to continue farming for food. Many other families also opted not to get involved with the project.

The experience of some women was different from that of other producers in the project. In an interview, one woman producer told us that she was contracted to work in the Envirotrade plant nursery from 6 a.m. to 4 p.m., and she also worked in her *machamba* before and after her shift; to this she also added household social reproduction activities. When questioned about the heavy workload burden that she carried and the low salary she received, she stated that it was necessary for her survival and, in particular, for the health-care and education of her children. After the company abandoned the project, women like her—still owed back pay by the company—lost their source of income from work in the plant nursery, and also the income from the planted trees; they went back to relying on their *machambas* for subsistence.

The Emergence of the Carbon Market and Its Social Costs: The Materialization of Climate Injustice in Mozambique

In Mozambique, mitigation and adaptation strategies envision the implementation of various land-based projects: increasing and consolidating conservation areas, increasing forest plantations such as eucalyptus or pine trees, creating biofuel monocultures (including for export), changing rural land-use methods (e.g., adopting climate-smart agriculture techniques), among others. Thus, some green projects involve large economic interests hiding behind environmental projects.

Generally, households are not adequately informed about these projects beforehand, as in the Nhambita case—and there are others, such as the Gilé National Reserve example and the implementation of conservation REDD+ (Bruna 2021). However, there are numerous players who profit from carbon trading, from measurement and verification companies to carbon offset purchasers (who are generally the biggest global polluters). Therefore, countries with a low ecological footprint,[7] as is the case in Mozambique, are encouraged to conserve and protect their biodiversity for the sake of fighting climate change, while other countries and industries buy these carbon credits and continue to industrialize, pollute, and generate wealth based on the extraction and expropriation of emissions rights and other ecological resources.

Made up of an asymmetrical relationship in which some win and others lose in the name of the environment, those who lose are precisely the ones who historically polluted the least. This green extractivism is at the heart of the materialization of climate injustice in Mozambique.

The Nhambita project and other carbon-capture environmental projects show the fragilities and contradictions of what we consider to be top-down climate solutions and policies (climate action from above). Although the narrative of the Nhambita project proponents presented it as a plan that would promote sustainable use of land, protect local biodiversity, and help develop rural areas, while paying for environmental services provided by the contractors, this project failed in the following ways:

a) Environmental Condescension

In addition to this project having been designed from the top down, its proponents ignored the opinions, expertise, experiences, and real interests of the beneficiaries. Although the farmers were informed of the environmental impacts and benefits of the project, the producers were not aware of the profit objectives of the project. For example, they did not know that carbon is a tradable commodity and that it could be sold on the international market, or who it would be sold to and for how much, what it was for, etc. In other words, there was considerable information asymmetry regarding the real financial interests and drivers of the project: carbon capture and subsequent sale of credits on the international market. Also, producers were not informed that the carbon credits were ultimately used to accommodate polluting activities in other parts of the world.

The fact that the project was designed without considering the aspirations and priorities of the producers worsened the impact of the drop in producers' incomes from the company's departure; producers had invested work and land in the project to gain economic benefits from the trees, instead of concentrating their efforts on activities that would provide long-term benefits for them without financial dependence on the company.

Although they had been named the beneficiaries of the forest inventory, currently the producers find themselves with areas filled with fruit trees and other species of little economic utility. For lack of markets and processing facilities, the fruit ends up rotting. Today's scenario in Nhambita is the result of policies that were inappropriate for local realities or priorities and which were designed through an asymmetrical top-down process to suit foreign economic interests.

b) The Failure of REDD+ and the Carbon Market

There are no known REDD+ projects that have been successful in their objective of stopping deforestation, but they are very successful in compensating for polluting activities. Several studies have provided evidence that these types of projects, in addition to adverse social effects, are not even effective in achieving their environmental objectives. That is to say, the studies question the effectiveness of such policies in mitigating and combating climate change (Casse et al., 2019). Moreover, we should reflect on and question the basis of policies such as REDD+ that depend on international market stimulus for their implementation. For instance, one of the reasons for the failure of the Envirotrade project in Mozambique, as mentioned earlier, was the fall in the international market price of carbon. Without the sale of carbon credits, the project became financially unviable, revealing its dependence on price variability and international market stimuli.

In the last five years, the price of carbon credits, as is the case for this type of REDD+ project, has ranged from 5 to 36 US dollar per ton (IHS Markit, n.d.). This variability poses risks for the implementation and sustainability of REDD+ projects that depend on the sale of carbon credits. In addition to the economic risk, this factor presents social risks, as a low carbon price can mean even fewer benefits for households affected by the project. It can also mean failure of the project, as happened with Envirotrade in Mozambique. Also, there is a risk in this scheme that arises from the volatility of the exchange rate between the US dollar and meticais (Mozambique's currency). The higher the dollar's value against the metical, the higher the income in the local currency and the more resources available for social projects. However, the opposite scenario poses a risk. Therefore, apart from the dependence on carbon prices, the success of these programs is also subject to exchange rate volatility. In other words, the livelihood of the producers involved depends on international-market and exchange-rate dynamics and will be subject to all the risks that this scheme entails.

Therefore, the way REDD+ was designed not only presents social risks and rural poverty intensification risks, but also promotes a scheme that continues to damage the environment insofar as it allows polluters to continue their polluting activities. In other words, the logic of the market in which

REDD+ functions makes its economic component more dominant than its environmental and social objectives.

What Alternatives? Emancipatory Actions for Climate Change Mitigation and Adaptation in Mabu

This section aims to explore, however tentatively, alternatives to the international top-down, market-based approach to climate action. The objective is not to deeply analyze the dynamics and implications of the alternatives, but to contribute to a debate that highlights fairer approaches which do not sustain existing climate injustices, but instead supply elements to support the construction of climate justice mainstays.

To this end, we investigated the case of Mabu, a community in the district of Lugela, Zambézia province, where the community and environmental organization JA![8] have been working together for over ten years. Over this time they have implemented sustainable small-scale agriculture and livestock activities based on local traditional practices and community forest conservation (forests of approximately 7,880 ha). These practices differ from mainstream climate change mitigation and adaptation policies, and they constitute, in our view, first steps towards minimizing the existing climate injustice in countries like Mozambique.

JA! started working with the population of Mabu (divided among four communities: Limbuè, Nvava, Namadoi, and Nangaze) in 2009 to understand and deal with the dynamics of illegal logging in the region and to promote the conservation of nearby Mount Mabu and its surrounding forest, which is considered a biodiversity hotspot.

> We did the work of raising awareness, which was a job that took us a long time, and we came to the conclusion that we needed to develop some activities with the communities, based on various discussions we had with them. And we tried in our discussions to find potential opportunities and challenges in the area, and understand what they wanted at the same time. So, we tried to create a convergence of all of these factors. (Interview in July 2021, René Machoco, JA!, Mabu)

The activities at the time involved raising awareness of the protection and conservation of the environment and the empowerment of communities in biodiversity protection and conservation. From 2009 to 2021, JA! assisted the community with (1) environmental awareness, (2) community DUAT[9] registrations, (3) registrations of community environmental licenses for forest and hill protection, (4) creation of farmer and poultry farmer associations for the purpose of developing income-generating occupations such as demonstration fields, aviaries, beehives for honey production, and others. According to JA! and the farmers interviewed during the field research, these activities were the result of a collaborative effort between the community and JA! (from planning to implementation) and are continuously (re)adapted and (re)negotiated to respond to the aspirations and priorities of the community itself.

From the interviews with the members of the associations, it was evident that the associations function as income-generators in the community and also as a mechanism for information sharing both within each community and among the communities. Also, they promote learning about techniques and strategies for making local and forestry products (as mentioned by the vice president of the Nangaze association in an interview). In every community, there are also women's associations, such as *Associação de Mulheres do Limbue*, which, among other activities, focus on poultry farming:

> As a single woman, I would think, how would I manage today? I don't have a husband, but the association helps me. By selling chickens, I help my life, by selling chickens, I help my children in school. Or if I get sick, I sell chickens and I can go to the hospital. That is how they have helped my life. (Interview in July 2021, Filomena, farmer in Limbue Community).

Conclusion

Although the implications of the carbon rush for rural subsistence are different from the implications of mining and agrarian extractivism, there are points of convergence between the two processes. The growing demand for land for the implementation of carbon sequestration projects, whether or not it involves the eviction of producers, ruptures locals' plans for survival (during and after the departure of the company from the land), leaving no compensation and posing high risks for food security. Along with these impacts comes the intensification of social inequality within the community. Our field visits showed us that it was the families who owned more land who planted more trees and made more money; this allowed them to invest more in agriculture by hiring local labour (more precisely, labour from less-favoured households with less land, including producers whose agreements with Envirotrade were cancelled as punishment for breaching contract clauses, mostly because they had opened up new areas for food crops).

In the case of Nhambita, resource usurpation did not expel farmers, as happens in cases of "traditional" mining and agrarian extractivism. However, the encroachment involved appropriation of control and management of the land so that the hired farmers no longer had decision-making power over the use and benefits from their own land. We call this practice "expropriation without expulsion." On the other hand, this also involved the usurpation of ecological resources, particularly the right to use biodiversity for their own subsistence; that is, the farmers lost their right to emit carbon by allowing carbon credit buyers to obtain it instead. This process of extracting emissions rights, legitimated by climate change mitigation and adaptation policies, is what we call "green extractivism" (Bruna, 2021).

Top-down climate crisis solution projects may seem appealing to rural farmers because of the monetary promises made and the better living conditions offered. However, this model has not been sustainable, as Nhambita's experience shows. In addition to the adverse effects of these policies, Mozambique has been the stage for extreme climate events with devastating impact. This shows how the countries that have contributed the least to the environmental crisis are often those that suffer the most from its impacts, and also those that host "false solutions" to climate change.

In Nhambita, evidence suggests that farmers seem to have joined the project only because they would be paid for it. While the value of trees planted in the community cannot be minimized—for example because they provide shade, fruit, and protection from high winds and cyclones—from a broader perspective, planted trees do not seem to be of much use to the producers. While some choose to clear new spaces, others cut down some of the trees planted (on a small scale), which indicates that the project will likely end up producing the opposite effect to the one desired by the stakeholders, namely Envirotrade, the funders and carbon offset purchasers. What really determined the acceptability of the project by the community was the structural issue of rural unemployment and low wages in Mozambique.

It is this context that emphasizes the need to deepen our notion and concept of climate justice, envisioning policies and solutions to environmental crises that are economically sustainable and socially just, and holding in mind the history of ecological footprints and the varying priorities of countries at different levels of industrialization and economic development. In other words, the conception and design of climate change mitigation and adaptation policies should not stray far from the principles that guide climate justice. The Mabu case study gives clues in this regard. Mainstream climate change mitigation and adaptation policies are usually implemented without the equitable participation of local actors and rural communities, and cause adverse implications for rural livelihoods, as discussed above. Mabu's experience is completely different insofar as there is an absence of information asymmetry among the actors involved, greater participation and engagement of the community in decision-making processes, and ownership of the project by the community participants, who share and support its objectives.

The various actions developed to mitigate and adapt to climate change in Mabu—namely agroecology, conservation of community forests, and promotion of environmentally correct livelihood strategies—are designed in a participatory manner and implemented according to the community's wishes. They are also greener than extractivist, export-oriented agriculture. Agroecology does not cause carbon emissions and is even claimed to cool the planet (LVC, 2007); honey production and small animal farming are also considered environmentally friendly because they are practiced in a sustainable way, unlike the mass animal and meat production of industrial agriculture. Related to this are the consumption habits and patterns practiced in the community, especially regarding agricultural products, which largely involve local produce for local consumption.

JA! and the community collaborate systematically and horizontally to ensure that the aspirations and needs of the community are met and that community members themselves assume leadership in planning and implementation. Aware that such initiatives are not a complete and integrated demonstration of climate justice, we highlight the importance of these actions for building climate justice in countries in the southern hemisphere.

This study is the result of action research with an academic-activist approach—an approach that is still emerging in the context of Mozambique and that, as shown in this study, has great potential to provide information for climate change and climate justice studies, as well as in other areas of

Fig. 16.5 Climate justice workshop participants, July 2021, Nhambita, Gorongosa, Mozambique.

knowledge. This study makes an important contribution by starting a relatively new debate linking the field of extractivism, green extractivism, and climate justice, oriented towards the efforts of anti-extractivism activists in the world, especially in the Global South.

NOTES

1 This study was supported by a Queen Elizabeth Scholarship (QES—York University) in collaboration with *Observatório do Meio Rural* (The Rural Observatory). Translated from the Portuguese by Evandro Rodriquez and P.E. Perkins.

2 https://www.mta.gov.mz/conservacao/potencial-da-biodiversidade/.

3 Lugela is the capital of the District of Lugela, in the Province of Zambézia, about 30 km away from Mabu. The workshops were held in Lugela because of the central location of the town and better infrastructure.

4 Sasol Limited is an integrated energy and chemical company based in South Africa.

5 US$ 5.00 = 65 Meticais at the time, so 300 meticais was about 23 US dollars.

6 Interview 1, former producer.

7 The ecological footprint is a way to measure the pollution levels and GHG emissions in every country.

8   *Justiça Ambiental* (also known by its acronym JA!, which means 'now!' in Portuguese, is a volunteer) is a volunteer non-governmental organization operating in Mozambique. "JA! members see the environment as a holistic concept, and environmental justice as a vehicle to assure equity and equality in society as a whole by means of the environment. In support of sustainable development, we try to view the concept of equality on a large scale, and thus value and assure the rights of future generations to a healthy and safe environment, in the same way that we value this right for ourselves," posted on 14 December 2021. https://justica-ambiental.org/sobre/.

9   DUAT is the Portuguese acronym for rights to use land, which is a kind of land-use grant provided to applicants (individuals and entities).

Reference List

Borras, S.M. Jr. (2016, April 14). *Land politics, agrarian movements and scholar-activism.* Inaugural Lecture of the International Institute of Social Studies.

Borras, S.M., Fig, D., & Suárez, S.M. (2011). The politics of agrofuels and mega-land and water deals: Insights from the ProCana case, Mozambique. *Review of African Political Economy, 38*(128), 215–234. https://doi.org/10.1080/03056244.2011.582758

Brito, R., & Holman, E. (2012). *Responding to climate change in Mozambique: Theme 6— Agriculture.* National Institute for Disaster Management (INGC), Phase II.

Bruna, N. (2019). Land of plenty, land of misery: Synergetic resource grabbing in Mozambique. *Land, 8*(8), 113.

Bruna, N. (2021). *From a threat to an opportunity: Climate change as the new frontier of accumulation.* YARA Working Paper 8, PLAAS Institute for Poverty, Land and Agrarian Studies, University of the Western Cape. http://www.yara.org.za/wp-content/uploads/2021/10/YARA-Natacha-Bruna-Oct-2021.pdf

Buti, R.P. (2021). História quilombola no chão: No caminho para o ensino de uma antropologia imersa na vida. *Pedagógicas, 7*(1), 2–24.

Casse, T., Milhøj, A., Neilsen, M.R., Meilby, H., & Rochmayanto, Y. (2019). Lost in implementation? REDD+ and country readiness experiences in Indonesia and Vietnam. *Climate and Development, 11*(9), 799–811. https://doi.org/10.1080/175655 29.2018.1562870

Castel-Branco, C. (2010). Economia extractiva e desafios de industrialização em Moçambique. *Cadernos IESE (Instituto de Estudos Sociais e Económicos), 1.*

Charrua, A.B., Padmanaban, R., Cabral, P., Bandeira, S., & Romeiras, M.M. (2021). Impacts of the tropical cyclone idai in mozambique: A multi-temporal landsat satellite imagery analysis. *Remote Sensing, 13*(2), 201.

Engel, G.I. (2000). Pesquisa-ação. *Educar Em Revista, 16*(16), 181–191. https://doi.org/10.1590/0104-4060.214

Fairhead, J., Leach, M., & Scoones, I. (2012). Green grabbing: A new appropriation of nature? *Journal of Peasant Studies*, *39*(2), 237–261. https://doi.org/10.1080/0306615 0.2012.671770

Feijó, J. (2016, September). *Investimentos, assimetrias e movimentos de protesto na Província de Tete* [Working paper]. Observador Rural No 44. Observatório do Meio Rural.

Feijó, J., & Aiuba, R. (2019, April). *Recuperando do ciclone Idai na província de Sofala— alguns desafios próximos*. Destaque Rural No 56. Observatório do Meio Rural.

IHS Markit. (n.d.). Global carbon index. Retrieved 15 December 2022, from https:// indices.ihsmarkit.com/#/Carbonindex

Kill, J. (2013). *Carbon discredited: Why the EU should steer clear of forest carbon offsets*. FERN and Les Amis de la Terre. https://www.fern.org/fileadmin/uploads/fern/ Documents/Nhambita_internet.pdf

LVC (La Via Campesina). (2007, November 9). *Small scale sustainable farmers are cooling down the earth*. La Via Campesina Policy Documents. Retrieved 15 March 2022, from https://viacampesina.org/en/small-scale-sustainable-farmers-are-cooling-down-the-earth/

Marzoli, A., & Del Lungo, P. (2009, November). *Evaluation of N'hambita pilot project*. AGRECO Consortium.

Moçambique, República de (2016). *Boletim da Republica No 88, 2016*. https://www. portaldogoverno.gov.mz/por/Governo/Legislacao/Boletins-da-Republica/Boletins-da-Republica-2016/BR-N.1-88-III-SERIE-2016

Monjane, B. (2012, August 16). Escravatura de carbono e REDD+ em Moçambique: camponeses "cultivam" carbono ao serviço de poluidores. *@Verdade*. https:// verdade.co.mz/escravatura-de-carbono-e-redd-em-mocambique-camponeses-cultivam-carbono-ao-servico-de-poluidores/

Monjane, B. (2021). *Rural struggles and emancipation in Southern Africa: Agrarian neoliberalism, peasants' movements and rural politics in Mozambique, South Africa and Zimbabwe* [Unpublished doctoral dissertation]. University of Coimbra. https:// eg.uc.pt/handle/10316/95250

Mosca, J., Abbas, M., & Bruna, N. (2016). *Governação 2004–2014: Poder, estado, economia e sociedade*. Alcance Editores.

Mosca, J., & Selemane, T. (2011). *El dorado Tete: Os mega projectos de mineração*. Centro de Integridade Pública.

Mozambique, Republic of (2012, November 13). *National climate change adaptation and mitigation strategy*. https://www.ctc-n.org/sites/www.ctc-n.org/files/resources/ mozambique_national_climate_change_strategy.pdf

Mozambique, Republic of (2016, November 2). *Estratégia nacional para a redução de emissões de desmatamento e degradação florestal, conservação de florestas e aumento de reservas de carbono através de florestas (REDD+) 2016–2030*.

https://www.biofund.org.mz/wp-content/uploads/2019/01/1548332243-ESTRAT%C3%89GIA%20NACIONAL%20DO%20REDD+.pdf

Santos, B. de S. (2014). *O- direito dos oprimidos*. Edições Almedina.

Shivji, I.G. (2019). Sam Moyo and Samir Amin on the peasant question. *Agrarian South: Journal of Political Economy*, 1–16. https://doi.org/10.1177/2277976019845737

UNFCCC (United Nations Framework Convention on Climate Change). (2008). *Kyoto protocol reference manual on accounting of emissions and assigned amount*. https://unfccc.int/sites/default/files/08_unfccc_kp_ref_manual.pdf

World Bank. (2010). *Economics of adaptation to climate change*.

Youth Climate Activism: Mobilizing for a Common Future

Patricia Figueiredo Walker

Introduction

Historically, young people globally, especially marginalized, disenfranchised children and youth—those who are disadvantaged, Indigenous, racialized, immigrants, refugees, and disabled—have been largely excluded from consideration as a group in global climate change mitigation and adaptation decision-making processes. This is perhaps because they are below voting age, are not seen as important consumption decision-makers, and/or are assumed to fall under family categories in relation to food, housing, transportation, recycling, leisure activities, and other climate mitigation and adaptation factors. The traditional climate change narrative often represents children and young people as "victims," because of their young age and longer lifetime exposure to climate change impacts, rather than as capable agents of change. However, engaging children and young people in climate change research, policy, and practice and supporting their participation at the highest levels of climate change decision-making is important for several reasons. Many scholars have argued that young people's climate change engagement is a *moral imperative*—that young people have a right to be informed and consulted regarding issues that will affect their future (Chawla & Heft, 2002; Hicks & Holden, 2007; Page, 2007; Trott, 2020)—while others have argued that children's engagement is necessary in preparing them to face and address future climatic changes (Ballantyne et al., 1998; see also Ojala, 2012; Koger, 2013; Schreiner et al., 2005).

Of course, all young people carry gender, ethnic, racial, national, sexual, ability, and other identity characteristics that are part of their relationships to both climate change and political action. Children and youth also differ from each other by age, maturity, physical ability, and many other factors. Rather than understanding children and youth as a single group, our focus here is on how age is an additional intersectional category that differentiates how people are impacted by climate change, as well as their ability to influence their own future.

As noted by Haynes and Tanner (2015), young people's "capacities to inform decision-making processes, communicate risks to their communities and take direct action to reduce risks" have been largely neglected to date (p. 357). However, when properly informed, empowered, and enabled, young people have the capacity to engage in constructive climate change action, influence adults, parents/caregivers, peers, and the public, and inform climate change research, policy, and practice. Participatory youth-centred research studies in the Global South and the Canadian Arctic have demonstrated the importance of directly involving young people in climate change research and including their voices in policy discussions, as important pathways for enhancing their agency and adaptive capacity and facilitating their adaptation to climate change (Haynes & Tanner, 2015; MacDonald et al., 2015). Youth Participatory Action Research (YPAR), and other participatory methodologies, are emerging as promising ways to explore young peoples' perspectives and elicit their participation in public forums on climate change.

Lawson et al. (2018, p. 204) argue that "children have unique perspectives on climate change, represent an audience that is easily reached through schools and are arguably best equipped to navigate the ideologically fraught topic of climate change with older generations in ways that inspire action." They further state that "children may be able to overcome anti-reflexive tendencies of adults through intergenerational learning (IGL) in the context of climate change" (p. 205). According to Lawson et al. (2018), anti-reflexive "forces" or "tendencies" include "individuals' political ideologies and worldviews" and "politically driven climate change skepticism" (pp. 204–205). For instance, "the bond between parent and child helps facilitate conversations around uncomfortable topics" and parents, in general, tend to perceive their children as being more trustworthy and "ideologically neutral" sources of climate change information (Lawson et al., 2018, p. 205). Given their effectiveness as climate change communicators, children and young people are ideally suited to communicate climate-related risks to their communities and raise awareness of these issues with government officials and other stakeholders. In addition, given their situated knowledge, observations, and lived experiences, children and young people are important stakeholders in climate change processes and can offer unique perspectives and policy ideas. Thus, supporting their engagement and developing their sense of agency can be very important politically—as seen, for example, in the impact of the Fridays for Future (FFF) movement.

This chapter focuses on intersectional climate justice for children and youth, and their engagement as subjects in climate action. Focusing mainly on Canada, we outline several ways that adults in home, school, or community settings, and young people themselves, can work to include children and youth in climate discussions—as well as the importance and potential of doing so.

Agency

Children and young people need agentic experiences to develop their sense of agency and become more resilient and adaptive. Agency refers to one's "ability to imagine and effect change" (O'Brien et al., 2018) as well as the belief in one's own capacity or competence. Agency enables children and young people, especially those who are disadvantaged and marginalized, to play active roles in shaping, improving, and preserving their communities (and cultural identities) and advocating for equitable adaptation. Furthermore, agency is one of the factors that contribute "to shaping patterns of public (dis) engagement with climate change" (Bieler et al., 2017, p. 65).

Research to date suggests that environmental, climate change, and social engagement as well as the opportunity to become involved and actively participate in climate change research and decision-making processes can enhance young people's agency, adaptive capacity, resilience, and adaptation (MacDonald et al., 2015; Trott, 2019). For example, "the opportunity to be meaningfully involved in their community, whether through research projects or community programs, is one of the many protective factors known to enhance circumpolar Indigenous youth resilience to a variety of stresses, including climate change challenges" (MacDonald et al., 2015, p. 487).

Fostering Hope, Engagement, and Action

A growing body of research has examined the association between hope and environmental engagement. Ojala (2008; 2012) found a positive association between hope and pro-environmental behaviour. According to Ojala (2012) "hope about a better, alternative, future could play an important part in motivating people to take action concerning global problems" (p. 626). Thus, children and young people who experience a high degree of (constructive) hope concerning climate change are more likely to act and seek out solutions (Ojala, 2008; Ojala, 2012; Li & Monroe, 2019).

How climate change is presented or framed may influence the response and engagement of children and youth with climate change. For example, "framing of climate change as an impending environmental disaster may contribute to a sense of despair and feelings of helplessness, which can lead to disillusion, apathy, and inactivity, or a perceived lack of potential to influence sustainability outcomes" (Hayes et al., 2018, p. 2). In addition, as Ojala (2012) points out, "education about global issues sometimes increases" pessimism (p. 626). Therefore, climate change education should go beyond enhancing scientific literacy to foster hope and facilitate action.

Research suggests that scientific literacy alone is insufficient to spur and sustain young people's political engagement and action on climate change (Trott, 2020; Hargis & McKenzie, 2021). As Groulx et al. (2017) explain, "rigorous science is an integral part of defining and promoting action in the face of climate change, but so are legitimate opportunities for citizens to engage with the climate change discourse, define local priorities, and meaningfully influence decisions" (p. 69). Thus, experiential opportunities for children and youth to become directly involved and engaged with local climate issues can foster meaningful action on climate-related public policies as well as personal decisions.

Additionally, engagement is thought to contribute to one's mental health and can be used as a strategy to address *climate anxiety*—an ever-growing problem among children and youth (Cunsolo Willox & Ellis, 2013; Ojala, 2018; Clayton, 2020; Wray, 2022). Trott (2020) argues that "children's constructive engagement enables [them] to envision alternatives and to believe they can be agents of transformative change" (p. 535). As Corner et al. (2015) note, when young people's "perceived self-efficacy is limited, personal engagement with climate change is likely to be lower" (p. 530). In addition to directly benefitting children and young people in the short- and long-term, their sustained, constructive engagement is beneficial to society, as it can lead to "societal transformation to sustainability" (Trott, 2020, p. 535).

Participatory Methods to Enhance the Agency and Promote the Voices of Young People

Emerging research indicates that participatory research methods, like digital photography (e.g., photovoice) and participatory video, are effective ways to enhance children's and youth's agency, adaptive capacity, climate change awareness, knowledge, engagement, visibility, and influence (MacDonald et al., 2015; Trott, 2019; Trott, 2020). Participatory research projects that employ these methodologies offer young people opportunities to be in control of the research process and share their unique perspectives on climate change, local problems, and solutions. In addition, they allow young people to "shape the outcome according to their own interests, ideas, skills, and values and [...] contribute rich, unanticipated, and meaningful understandings of [the] research questions" (MacDonald et al., 2015, p. 490). This approach to research challenges the narrative that children and young people are *victims* of climate change who require the protection and assistance of adults and caregivers to speak and make decisions on their behalf; rather, children and young people are positioned as "negotiators who are powerful experts" (Marr & Malone, 2007, p. 4).

As Trott (2019) explains, participatory methods stand out for their potential to empower young people's agency and facilitate their constructive climate change engagement. For example, YPAR "provides young people with opportunities to study social problems affecting their lives and then determine actions to rectify these problems" (Cammarota & Fine, 2010, p. 2). Most importantly, "YPAR teaches young people that conditions of injustice are produced, not natural; are designed to privilege and oppress; but are ultimately challengeable and thus changeable" (Cammarota & Fine, 2010, p. 2). As such, YPAR may contribute to young people's sense of empowerment and agency, or belief in their capacity to affect change, thereby addressing feelings of hopelessness and helplessness—known factors that contribute to apathy and disengagement. However, as Trott (2019) points out, participatory action research (PAR)-based studies involving "children as social actors, change agents, collaborators, or co-researchers" remain rare (p. 46).

The following sections explore participatory methods, including photovoice, participatory video, and citizen science, to empower young people's agency and facilitate their constructive climate change engagement. Furthermore, they provide a brief overview of climate change activism in Canada, including young Canadians' active involvement in the fossil fuel divestment (FFD) movement and climate change litigation in this country.

Photovoice

Photovoice, a PAR method based on feminist theory and Paulo Freire's educational approach for critical consciousness, "is a process by which people can identify, represent, and enhance their community through a specific photographic technique" (Wang & Burris, 1997, p. 369). This strategy enables children and youth to "record and vivify their community's strengths and concerns; promote critical dialogue and knowledge about community issues through group discussion of photographs; and reach policy makers" (Wang, 2006, p. 147). Furthermore, photovoice is emancipatory and agentic (Derr & Simons, 2020), as "it entrusts cameras to the hands of people to enable them to act as recorders, and potential catalysts for change, in their own communities" (Wang & Burris, 1997, p. 369).

To date, photovoice remains underutilized as a strategy for facilitating young people's engagement with climate change (Trott, 2019). However, Trott demonstrated the potential of this strategy in supporting children's constructive climate change engagement through a collaborative PAR project with children ages ten to twelve. *Science, Camera, Action!* combined hands-on climate change educational activities with photovoice, integrating "transformative pedagogy with arts-based and participatory methodology to empower children's agency through personally relevant and locally meaningful action projects addressing climate change" (Trott, 2019, p. 58). Project participants "acquired new knowledge about climate change and its local impacts" and "developed stronger beliefs in their agentic capabilities, while taking tangible steps towards the sustainable transformation of their communities" (Trott, 2019, p. 58).

Participatory Video

As MacDonald et al. (2015) explain, "participatory video (PV) is a digital media research method with roots in community activism and social development that aims to shift power dynamics by having participants direct and control the creation of a film on a topic of research and community" (p. 488). Although youth-centred PV research examining climate change is still uncommon, this research approach has become more popular in recent years (Haynes & Tanner, 2015). As Kindon (2003) explains, PV offers "a *feminist* practice of looking, which actively works to engage with and challenge conventional relationships of power associated with the gaze in geographic research, and results in more equitable outcomes and/or transformation for research participants" (p. 143). This makes PV appealing as a strategy for child- and youth-centred climate change organizing.

Empirical research using PV methods with groups of young people in the Philippines has demonstrated the potential and efficacy of this method in increasing young people's awareness and knowledge of local disaster and climate-related risks and empowering them to engage with community members and decision-makers around climate change. In addition, young people, ages 13 to 21, were able to "document and raise awareness of disaster risk and use screening events to mobilise and advocate for risk reduction measures in their communities" (Haynes & Tanner, 2015, p. 357). Similarly, MacDonald et al. (2015) examined "the potential of youth-led PV as a strategy to foster known protective factors that underpin the resilience of youth and their capacity to adapt to various stresses, including impacts of climate change" (p. 486). This research, which focused on Inuit youth in Labrador, Canada, showed "that PV may be a pathway to greater adaptive capacities because the process connects to known protective factors that enhance resilience of circumpolar [I]ndigenous youth. PV also shows promise as a strategy to engage youth in sharing insights and knowledge, connect generations, and involve young Inuit in planning decision making in general" (MacDonald et al., 2015, p. 486).

In general, participatory research methods such as PV and photovoice "bring awareness and respect of the culture and context of the partner community, facilitate capacity development, and highlight local knowledge, voices, and experience that advance research in a way Western science cannot do alone" (MacDonald et al., 2015, p. 487). When employed by young people, these participatory research methods can facilitate and highlight the importance of collective engagement, which "can promote children's hope and well-being—by creating conditions that allow children to feel part of a collaborative effort rather than acting in isolation" (Trott, 2019, pp. 57–58, citing Kelsey & Armstrong, 2012). In addition, participatory research methods can be effective strategies for children and youth to share knowledge, information, and local perspectives with diverse stakeholders, including policy makers. They also offer opportunities for political advocacy and youth climate activism.

Citizen Science

Like the above-mentioned participatory research methods, citizen science can enhance youth participation, representation, and climate change engagement and action. "Co-created and collaborative citizen science designs, for instance, can offer citizens some control over what research questions are asked, and how data is collected" (Groulx et al., 2017, p. 47). Additionally, citizen science "can promote the co-creation of scientific and environmental knowledge" and "individual changes in environmental attitudes," specifically "by fostering experiential learning" and "(re) connecting participants to the natural world" (Groulx et al., 2017, p. 50). Hargis and McKenzie (2020) highlight the "critical" role of place-based pedagogies "in moving beyond climate and environmental awareness to empowerment and action" (p. 2). A growing body of research supports place-based approaches to climate change education (Field, 2017; Hargis & McKenzie, 2020) as they serve to contextualize climate change and help young people understand that climate change is not a far-away/distant problem with complex and inaccessible solutions, but rather a process with very concrete local impacts.

Youth Climate Change Activism

In recent years, children and youth globally have—in unprecedented ways and numbers—engaged in climate change activism to express opposition to the business-as-usual status quo that is causing global warming, and to demand systemic change, climate justice, and political action on this issue. O'Brien et al. (2018) argue that when young people engage with climate change, they are "implicitly or explicitly entering into debates that involve dissenting from prevailing norms, beliefs, and practices, including economic and social norms like consumption, fossil energy use, and the unjust use of power in decision making" (p. 42). This is evidenced by the FFF (Fridays for the Future) Global Climate Strike, a youth-led grassroots movement initiated and led by Swedish youth climate activist Greta Thunberg beginning in 2018.

The FFF movement, which continues to organize local and global, physical and digital school strikes, rallies, and marches has adopted an *intersectional* approach to climate justice, outlining "collective demands, that include [I]ndigenous rights and sovereignty; defending land, water, and life; zero-carbon economy; separation of oil and state, universal public services and infrastructure; justice for migrants and refugees and a sustainable future for all" (March 2019). The FFF movement gained international attention in 2019, leading to the largest climate demonstration in human history, which took place on 20 September 2019—with over four million people worldwide, including hundreds of thousands of Canadians from at least eighty-five Canadian cities and towns, joining the Global Climate Strike. Montreal held the largest single climate march yet (estimated at five hundred thousand people). The FFF movement has been a catalyst for youth climate activism in Canada and globally, demonstrating the potential of young people to exert political and intergenerational influence at national and global scales.

Fossil Fuel Divestment Movement

The FFD movement aims to eliminate public and private investment in fossil fuel companies. The first FFD campaign took place in 2010 in Philadelphia, where a Swarthmore College student group called upon their institution to stop investing in fossil fuel companies after learning about the environmental impacts of mountaintop removal (Maina et al., 2020). In 2012, environmentalist and 350.org co-founder Bill McKibben (2012) wrote a radical essay for *Rolling Stone* magazine urging the public to "view the fossil-fuel industry in a new light" (p. 6). In his essay, McKibben (2012) boldly declared: "[The fossil fuel industry] has become a rogue industry, reckless like no other force on Earth. It is Public Enemy Number One to the survival of our planetary civilization" (p. 6). The essay sparked a global FFD movement, with higher education institutions and students in particular playing a key role.

Canada has the third-largest proven oil reserves in the world—most of which are found in Alberta's oil sands—and is the fourth largest global producer and exporter of oil (Natural Resources Canada, 2019). According to a recent report by Environmental Defence (2021), "In 2020, the federal government either announced or provided a minimum of nearly $18 billion to the oil and gas sector" (p. 1). According to the Canadian Association of Petroleum Producers (2021), the oil sands are responsible for 11 per cent of total national greenhouse gas emissions (). However, recent aircraft measurements over the Canadian oil sands "indicate that CO_2 emission intensities for OS [oil sands] facilities are 13–123% larger than those estimated using publicly available data. This leads to [...] 30% higher overall OS GHG [greenhouse gas] emissions (17 Mt) compared to that reported by industry" (Liggio et al., 2019, p. 1).

Divesting from fossil fuel companies is seen by many as a vital step in eroding the fossil fuel industry's social license to operate, or the public's perception and acceptance of the industry's legitimacy. Furthermore, it is seen as crucial to addressing climate change and the environmental and social impacts of fossil fuel extraction in Canada. Given the fact that "post-secondary institutions have a significant amount of their endowment funds invested in fossil fuel companies" (Maina, 2016, p. 1), they can play an important role in the divestment movement and influence similar action by other investors. FFD campaigns across higher education institutions, led primarily by students, have increased steadily over the last decade (Maina et al., 2020). According to Maine et al. (2020), students in the Canadian higher education institution FFD movement are responsible for initiating thirty-one of the existing thirty-seven FFD campaigns in Canada.

In January 2021, FFF Toronto, Sustainabiliteens Vancouver, FFF Calgary, School Strike for Climate Halifax, and Climate Justice Guelph initiated a series of "bank switch" actions to bring attention to the fact that "Canada's big five banks—TD, RBC, Scotiabank, BMO and CIBC—are among the biggest financiers of fossil fuels in the world" (Speers-Roesch, 2021). Through these actions, youth intended to pressure the banks "by threatening to remove [their] money from [these] banks unless they made stringent, concise plans to fully divest from fossil fuels" (FFF Toronto, 2021). Similarly, students across Ontario released a video urging teachers "to demand that the Ontario Teachers' Pension Plan (OTPP) stop investing their retirement savings in oil, gas, coal, and pipeline companies that fuel the climate crisis" (Shift: Action for Pension Wealth & Planet Health, 2021).

These groups have used a number of tactics including "signing of petitions, sit-ins, rallies, and protests, facilitated through face-to-face and online platforms" to mobilize FFD campaigns and promote climate and ecological justice more broadly (Maina et al., 2020, p. 1). In Canada, the FFD movement has given young people opportunities to engage in constructive climate change activism and express opposition to neo-liberal capitalism, continuing legacies of colonialism, and the status quo that is contributing to climate change (Saad, 2019).

Through their collaborations, innovations, and resilience, young people across Canada are not only leading the FFD divestment movement, but also the fight against climate inaction and social, intergenerational, and ecological injustice. In addition, they are turning to the Canadian legal system in unprecedented efforts to demand action on climate change.

Climate Change Litigation in Canada

In the year 2019, children and youth across Canada filed climate justice lawsuits against federal and provincial governments alleging violations to their rights and freedoms. In June 2019, ENvironnement JEUnesse (ENJEU, n.d.), an environmental non-profit, filed a climate lawsuit against the Canadian government at the Superior Court of Quebec, on behalf of young Quebeckers thirty-five years old and under. In July 2019, the court refused "to grant ENvironnement JEUnesse the authorization to institute a class action" (ENJEU, n.d.) on behalf of the plaintiffs because it "found the age 35 cutoff to be arbitrary and inappropriate, since it did not consider the rationale for choosing it to be adequately justified" (Amnesty International, 2020). In August 2019, ENJEU appealed the court's decision and in February 2021 it "presented its application for authorization to institute its class action to the Quebec Court of Appeal" (ENJEU, n.d.). Unfortunately, the court dismissed the appeal, but "given the importance of the matter, ENvironnement JEUnesse filed an application for leave to appeal to the Supreme Court of Canada on February 11, 2022" (ENJEU, n.d.).

In October 2019, fifteen young people—ages ten to nineteen years—from seven provinces and one territory filed a lawsuit (*La Rose et al. v. Her Majesty the Queen*) against the Canadian Government alleging Canada's actions on climate change violate their rights to life, liberty, and security of the person under Section 7 of the Canadian Charter of Rights and Freedoms and their right to equality under Section 15, given the disproportionate impacts of climate change on young people. The federal government responded with a motion to strike the plaintiffs' claim in order to stop the case from proceeding to trial. On 27 October 2020 the Federal Court of Canada granted the government's motion, despite acknowledging that "the negative impact of climate change to the Plaintiffs and all Canadians is significant, both now and looking forward into the future" (Our Children's Trust). On 24 November 2020 the attorneys for the plaintiffs filed a Notice of Appeal with the Federal Court of Appeals. "The youth plaintiffs are currently awaiting a date for oral argument in the Federal Court of Appeals" (Our Children's Trust).

In November 2019, seven young Ontario climate activists, between the ages of thirteen and twenty-four, filed a similar lawsuit (*Mathur et al. v. Her Majesty in Right of Ontario*) arguing that Ontario's new greenhouse gas reduction target and the repeal of the old Climate Change Act "violate the rights of Ontario youth and future generations under ss. 7 and 15 of the *Charter*" (Chen, n.d.). Like the federal government, the Ontario government responded with a motion to strike the lawsuit. However, in July 2020, the youth plaintiffs countered the motion and the court ruled in their favour. For the first time in Canadian history, a court "ruled that fundamental rights protected under the Charter can be threatened by climate change and citizens have the ability to challenge a Canadian government's action on the climate crisis under the highest law in the land" (Thomson, 2021). In response, the Ontario Government applied for leave to appeal the ruling, which the court dismissed. According to Ecojustice, *Mathur et al.*, which proceeded to a full hearing before the Ontario Superior Court in September 2022, is "the first case of its kind to clear key procedural hurdles" (n.d.). As noted by Ecojustice, the "case is already changing the Canadian legal landscape" (Page & Thomson, 2021). More specifically, the case, 1) "established the courts are a viable avenue for citizens to challenge government actions that threaten their Charter rights and the climate," 2) "established that the harms from climate change are not speculative nor impossible to prove," and 3) "established that climate change can impact Canadians' rights to life, liberty and security of the person" (Page & Thomson, 2021).

These lawsuits represent a potential turning point in climate change litigation and youth climate activism in Canada. They illustrate how children and youth are expressing their agency by taking legal action to challenge government actions and contributions to climate change—a trend also seen in other parts of the world, including Colombia and the Netherlands (Savaresi & Auz, 2019). Despite lack of educational opportunities to engage with climate change in schools and contribute to solutions inside and outside of formal institutions of education (see Bieler et al., 2017; Wynes & Nicholas, 2019), young Canadians are demonstrating remarkable agency, resilience, and resourcefulness, successfully engaging in organizing and climate activism and mobilizing other youth and actors for climate action, fossil fuel divestment, and social and ecological justice.

Conclusion

As the Earth continues to heat up, young people will continue to share their perceptions, inform climate research and policy, and contribute actively to local solutions. For both ethical and pragmatic reasons, it is crucial to respect and enhance young people's rights to express their views and become active participants in research and decisions on issues that may affect their lives and futures, especially climate change, due to its huge long-term impacts. From a procedural justice perspective, children and youth must be represented in climate policy and decision-making processes, considering they have a stake in the outcomes and will be directly or indirectly affected long into the future by decisions made today.

According to the United Nations Convention on the Rights of the Child, children have the right to express their views freely (Article 12), the right to freedom of expression (Article 13), and the right to grow up in a healthy environment (Article 24). It is, therefore, every child's *right* to participate in decision-making processes concerning their present and future adaptation and to grow up in a healthy and climate just world. Promoting the participatory rights of *all* children and youth is essential in addressing their vulnerability to climate change and ensuring their effective adaptation.

Consulting with and including the voices, experiences, concerns, and perceptions of young people in climate change research, as well as encouraging and facilitating their meaningful participation in decision-making processes, is essential to ensure that the data collected, and measures developed, are complete and representative of young people and the challenges they face. Although this chapter has focused specifically on children and youth in Canada, the concepts discussed are applicable and relevant to young people everywhere.

Reference List

Amnesty International's Legal Team. (2020, February 26). *Environnement jeunesse (ENJEU) v. attorney general of Canada.* https://www.amnesty.ca/legal-brief/environnement-jeunesse-enjeu-v-attorney-general-canada

Ballantyne, R., Connell, S., & Fien, J. (1998). Students as catalysts of environmental change: A framework for researching intergenerational influence through environmental education. *Environmental Education Research, 4*(3), 285–298.

Bieler, A., Haluza-Delay, R., Dale, A., & Mckenzie, M. (2017). A national overview of climate change education policy: Policy coherence between subnational climate and education policies in Canada (K–12). *Journal of Education for Sustainable Development, 11*(2), 63–85.

Cammarota, J., & Fine, M. (2010). Youth participatory action research: A pedagogy for transformational resistance. In *Revolutionizing education* (pp. 9–20). Routledge.

Canadian Association of Petroleum Producers (2021). *Greenhouse gas emissions: Oil sands GHG emissions.* https://www.capp.ca/explore/greenhouse-gas-emissions/

Chawla, L., & Heft, H. (2002). Children's competence and the ecology of communities: A functional approach to the evaluation of participation. *Journal of Environmental Psychology, 22*(1–2), 201–216.

Chen, A. (n.d.). *Young climate activists attempt to hold province accountable for inadequate emissions target.* David Asper Centre for Constitutional Rights, University of Toronto. https://aspercentre.ca/young-climate-activists-attempt-to-hold-province-accountable-for-inadequate-emissions-target/

Clayton, S. (2020). Climate anxiety: Psychological responses to climate change. *Journal of Anxiety Disorders, 74*, 102263.

Corner, A., Roberts, O., Chiari, S., Völler, S., Mayrhuber, E.S., Mandl, S., & Monson, K. (2015). How do young people engage with climate change? The role of knowledge, values, message framing, and trusted communicators. *Wiley Interdisciplinary Reviews: Climate Change, 6*(5), 523–534.

Cunsolo Willox, A., Harper, S.L., Ford, J.D., Edge, V.L., Landman, K., Houle, K., Blake, S., & Wolfrey, C. (2013). Climate change and mental health: an exploratory case study from Rigolet, Nunatsiavut, Canada. *Climatic Change, 121*(2), 255–270.

Derr, V., & Simons, J. (2020). A review of photovoice applications in environment, sustainability, and conservation contexts: is the method maintaining its emancipatory intents? *Environmental Education Research, 26*(3), 359–380.

Ecojustice. (n.d.). *#GenClimateAction: Mathur et. al. v. Her Majesty in Right of Ontario.* Retrieved 12 January 2023, from https://ecojustice.ca/case/genclimateaction-mathur-et-al-v-her-majesty-in-right-of-ontario/

Ecojustice. (2020, November 13). *Victory! Young Ontarians prevail over Ford government's attempt to shut down climate case.* https://ecojustice.ca/pressrelease/victory-young-ontarians-prevail-over-ford-governments/

ENJEU (ENvironnement JEUnesse). (n.d.). *ENvironnement JEUnesse to the Québec Court of Appeal.* https://enjeu.qc.ca/justice-eng/

Environmental Defence. (2021, April). *Paying polluters: Federal financial support to oil and gas in 2020.* https://environmentaldefence.ca/report/federal_fossil_fuel_subsidies_2020/

FFF (Fridays for Future) Toronto. (2021, February 11). *FFFTO said fossil banks? No thanks!* February Love from FFFTO Newsletter.

Field, E. (2017). Climate change: Imagining, negotiating, and co-creating future(s) with children and youth. *Curriculum Perspectives, 37*(1), 83–89.

Groulx, M., Brisbois, M.C., Lemieux, C.J., Winegardner, A., & Fishback, L. (2017). A role for nature-based citizen science in promoting individual and collective climate change action? A systematic review of learning outcomes. *Science Communication, 39*(1), 45–76.

Hargis, K., & McKenzie, M. (2020). *Responding to climate change education: A primer for K-12 education.* The Sustainability and Education Policy Network, Saskatoon, Canada.

Hayes, K., Blashki, G., Wiseman, J., Burke, S., & Reifels, L. (2018). Climate change and mental health: Risks, impacts and priority actions. *International Journal of Mental Health Systems, 12*(1), 1–12.

Haynes, K., & Tanner, T.M. (2015). Empowering young people and strengthening resilience: Youth-centred participatory video as a tool for climate change adaptation and disaster risk reduction. *Children's Geographies, 13*(3), 357–371.

Hicks, D., & Holden, C. (2007). Remembering the future: What do children think? *Environmental Education Research, 13*(4), 501–512.

Kindon, S. (2003). Participatory video in geographic research: A feminist practice of looking? *Area, 35*(2), 142–153.

Koger, S.M. (2013). Psychological and behavioral aspects of sustainability. *Sustainability, 5*(7), 3006–3008.

Lawson, D.F., Stevenson, K.T., Peterson, M.N., Carrier, S.J., Strnad, R., & Seekamp, E. (2018). Intergenerational learning: Are children key in spurring climate action? *Global Environmental Change, 53,* 204–208.

Li, C.J., & Monroe, M.C. (2019). Exploring the essential psychological factors in fostering hope concerning climate change. *Environmental Education Research, 25*(6), 936–954.

Liggio, J., Li, S.M., Staebler, R.M., Hayden, K., Darlington, A., Mittermeier, R.L., O'Brien, J., McLaren, R., Wolde, M. Worthy, D., & Vogel, F. (2019). Measured Canadian oil sands CO_2 emissions are higher than estimates made using internationally recommended methods. *Nature Communications, 10*(1), 1–9.

MacDonald, J., Ford, J., Willox, A., & Mitchell, C. (2015). Youth-led participatory video as a strategy to enhance Inuit youth adaptive capacities for dealing with climate change. *Arctic, 68*(4), 486–499.

Maina, N. (2016). *The state of fossil fuel divestment in Canadian post-secondary institutions.* Sustainability and Education Policy Network, University of Saskatchewan, Saskatoon, Canada.

Maina, N.M., Murray, J., & McKenzie, M. (2020). Climate change and the fossil fuel divestment movement in Canadian higher education: The mobilities of actions, actors, and tactics. *Journal of Cleaner Production, 253,* 119874.

Mar, K. (2019, November 29). *Fridays for Future Toronto: An intersectional approach to climate justice.* Greenpeace Canada. https://www.greenpeace.org/canada/en/story/27927/fridays-for-future-toronto-an-intersectional-approach-to-climate-justice/

Marr, P., & Malone, K. (2007, December). What about me? Children as co-researchers. In *AARE 2007 International Education Research Conference* (p. 27).

McKibben, B. (2012). Global warming's terrifying new math. *Rolling Stone, 19*(7), 2012.

Natural Resources Canada. (2019, December 16). Oil resources. https://www.nrcan.gc.ca/energy/energy-sources-distribution/crude-oil/oil-resources/18085

O'Brien, K., Selboe, E., & Hayward, B.M. (2018). Exploring youth activism on climate change. *Ecology and Society, 23*(3).

Ojala, M. (2008). Recycling and ambivalence: Quantitative and qualitative analyses of household recycling among young adults. *Environment and Behavior, 40*(6), 777–797.

Ojala, M. (2012). Hope and climate change: The importance of hope for environmental engagement among young people. *Environmental Education Research, 18*(5), 625–642.

Ojala, M. (2018). Eco-anxiety. *RSA Journal, 164*(4 (5576), 10–15.

Our Children's Trust. *Youth v. Gov.* https://www.ourchildrenstrust.org/canada

Page, D., & Thomson, F. (2021, April 22). *Earth Day special: Four ways the Mathur et. al. youth climate case is making history.* Ecojustice. https://ecojustice.ca/earth-day-special-four-ways-the-mathur-et-al-youth-climate-case-is-making-history/

Page, E.A. (2007). *Climate change, justice and future generations.* Edward Elgar Publishing.

Saad, A. (2019). The fossil fuel divestment movement: A view from Toronto. In P.E. Perkins (Ed.), *Local activism for global climate justice: The Great Lakes watershed.* Routledge.

Savaresi, A., & Auz, J. (2019). Climate change litigation and human rights: Pushing the boundaries. *Climate Law, 9*(3), 244–262.

Schreiner, C., Henriksen, E.K., & Kirkeby Hansen, P.J. (2005). Climate education: Empowering today's youth to meet tomorrow's challenges. *Studies in Science Education, 41*(1), 3–49. https://doi.org/10.1080/03057260508560213

Shift: Action for Pension Wealth & Planet Health. (2021, January 7). *A message from the students of Ontario to their teachers* [Video]. YouTube. https://www.youtube.com/watch?v=JPOiHGJqCmM

Speers-Roesch, A. (2021, March 1). *Canada's big banks are bankrolling the climate crisis.* Greenpeace Canada. https://www.greenpeace.org/canada/en/story/46186/canadas-big-banks-are-bankrolling-the-climate-crisis/

Thomson, F. (2021, December 8). *Young people could make Canadian legal history in 2022.* Ecojustice. https://ecojustice.ca/young-people-could-make-canadian-legal-history-in-2022/

Trott, C.D. (2019). Reshaping our world: Collaborating with children for community-based climate change action. *Action Research, 17*(1), 42–62.

Trott, C.D. (2020). Children's constructive climate change engagement: Empowering awareness, agency, and action. *Environmental Education Research, 26*(4), 532–554.

United Nations Convention on the Rights of the Child, 20 November 1989, https://www.ohchr.org/en/instruments-mechanisms/instruments/convention-rights-child

Wang, C., & Burris, M.A. (1997). Photovoice: Concept, methodology, and use for participatory needs assessment. *Health Education & Behavior, 24*(3), 369–387.

Wang, C.C. (2006). Youth participation in photovoice as a strategy for community change. *Journal of Community Practice, 14*(1–2), 147–161.

Wray, B. (2022). *Generation dread: Finding purpose in an age of climate crisis.* Knopf Canada.

Wynes, S., & Nicholas, K.A. (2019). Climate science curricula in Canadian secondary schools focus on human warming, not scientific consensus, impacts or solutions. *PloS one, 14*(7), e0218305.

List of Contributors

GUY DONALD ABASSOMBE, a QES Scholar, is a doctoral candidate in Geography at the University of Yaoundé I, and a researcher at the Global Mapping and Environmental Monitoring (GMEM) research office in Yaoundé, Cameroon. abassguydonald@gmail.com

FERRIAL ADAM, a QES Scholar, completed her PhD in Development Studies at the University of Johannesburg and is the manager of WaterCAN (Community Action Network) focusing on activist citizen science to monitor water resources in South Africa. feradam@gmail.com

SARITA ALBAGLI is a Professor/Researcher at the Brazilian Science and Technology Information Institute (IBICT), Federal University of Rio de Janeiro, Brazil. sarita.albagli@gmail.com

JOAQUIN ALMONACID is a graduate student at the Universidad de Los Lagos, Chile. joaco.almon@gmail.com

FRANCISCO ARAOS, a QES Scholar, is a Professor at the University of Los Lagos, Chile. francisco.araos@ulagos.cl

AYANSINA AYANLADE, a QES Scholar, is an Associate Professor at Obafemi Awolowo University in Ile-Ife, Nigeria and University Assistant at the University of Vienna, Austria. aayanlade@oauife.edu.ng

OLUWATOYIN SEUN AYANLADE is a Senior research fellow at Obafemi Awolowo University in Ile-Ife, Nigeria. osayanlade@oauife.edu.ng

ADEFUNKE F. O. AYINDE is an Associate Professor at Federal University of Agriculture in Abeokuta, Nigeria. ayindeafo@funaab.edu.ng

CAMILA BAÑALES-SEGUEL, a QES Scholar and agronomist, is completing a PhD in Environmental Science at the Universidad de Concepción, Chile. caspicamila@gmail.com

PETRA BENYEI is a Juan de la Cierva Postdoctoral Fellow at the Institute of Economy, Geography and Demography (Spanish National Research Council - CSIC), Madrid, Spain. petra.benyei@gmail.com

FRANCISCO BRAÑAS is a professional in Marine Protected Areas, Ministry of Environment, Punta Areas, Chile. fbranasloayza@gmail.com

NATACHA BRUNA, a QES Scholar, is a Researcher at the Rural Observatory (Observatório do Meio Rural) in Maputo, Mozambique. natachabruna89@gmail.com

DANIELA CAMPOLINA, a QES Scholar and high school teacher in Rio Acima, Brazil, holds a PhD in Education from the Federal University of Minas Gerais and is the Co-Coordinator of the Education, Mining and Territory Research Group (EduMiTe), based in Belo Horizonte, Brazil. danicampolina@gmail.com

KÁTIA CAROLINO, a QES Scholar and lawyer, holds a PhD in Environmental Science and is a professor at the State University of Minas Gerais (UEMG) in Frutal, Minas Gerais, Brazil. carolino.ea@gmail.com

MARCONDES G. COELHO JUNIOR, a QES Scholar, is a PhD student in Environmental and Forest Sciences at the Federal Rural University of Rio de Janeiro, Seropédica, Brazil and a Socio-Environmental Analyst at the Instituto Centro de Vida, Cuiabá, Brazil. marcondescoelho22@gmail.com

DANIELA COLLAO is an independent investigator, Valdivia, Chile. danielacollaonavia@gmail.com

JAIME CURSACH is a Postdoctoral Researcher at the Universidad de Los Lagos, Chile. jcurval@gmail.com

ELIANE M. R. DA SILVA is a Researcher at Brazilian Agricultural Research Corporation (Embrapa Agrobiologia), Seropédica, Brazil, and a Professor in the Graduate Program in Environmental and Forest Sciences at the Federal Rural University of Rio de Janeiro, Seropédica, Brazil. eliane.silva@embrapa.br

EDUARDO C. DA SILVA NETO is an Assistant Professor at Federal Rural University of Rio de Janeiro, Seropédica, Brazil. eduardoneto@ufrrj.br

RHYS DAVIES is a custom cartographer and illustrator who has worked with authors and publishing companies large and small. https://www.rhysspieces.com

FLORENCIA DIESTRE is a graduate student at the University of Los Lagos, Chile. f.diestre@gmail.com

RONALDO DOS SANTOS is a Community Leader at Associação dos Moradores do Quilombo do Campinho da Independência (AMOQC), Angra dos Reis, Brazil, and a Coordinator at the National Coordination for the Articulation of Black, Rural, and Quilombola Communities, Brasília, Brazil. rscampinho@gmail.com

VADEL ENECKDEM TSOPGNI is a PhD student in Geography at the University of Yaoundé I, and a researcher at the Global Mapping and Environmental Monitoring (GMEM) research office in Yaoundé, Cameroon. vteneckdem@yahoo.com

CHRISLAIN ERIC KENFACK, a QES Scholar, holds a PhD in Political Science and Sociology from the University of Coimbra, Portugal and a Master's degree in International Relations from the University of Yaounde II, Cameroon. His postdoctoral research at the Department of Political Science, University of Alberta, focused on Indigenous climate justice. chrislaineric@yahoo.fr

MARY GALVIN is an Associate Professor in the Department of Anthropology and Development Studies at the University of Johannesburg, South Africa. MGalvin@uj.ac.za

LUSSANDRA MARTINS GIANASI is an Associate Professor, vice-head of the Geography Department, and Coordinator of the Geosciences Institute's academic and administrative extension activities organization (CENEX-IGC) at the Federal University of Minas Gerais, Belo Horizonte, Brazil. lussandrams@gmail.com

ALLAN YU IWAMA, a QES Scholar, is a Visiting Professor at the Federal University of Paraiba, João Pessoa, Brazil. allan.iwama@dse.ufpb.br

MARGARET O. JEGEDE is a Senior research fellow at Obafemi Awolowo University in Ile- Ife, Nigeria. mojegede@oauife.edu.ng

CARLA LANYON is a researcher at Stockholm Resilience Centre, Sweden. carla.lanyon@gmail.com

MANUEL LEMUS is a member of the Mapuche-Huilliche Communities Association of Carelmapu (Asociación de Comunidades Mapuche-Huilliche de Carelmapu), Chile.

JOSÉ MOLINA-HUEICHÁN is the Coordinator of the Mapuche-Huilliche Communities Association of Carelmapu (Asociación de Comunidades Mapuche-Huilliche de Carelmapu), Chile. josemolina1971@gmail.com

BOAVENTURA MONJANE, a QES Scholar, is an Associate Researcher at Eduardo Mondlane University's African Studies Center (Centro de Estudos Africanos, Universidade Eduardo Mondlane) in Maputo, Mozambique, and a postdoctoral researcher at the Institute for Poverty, Land and Agrarian Studies, University of the Western Cape (PLAAS, UWC) in Cape Town, South Africa. boa.monjane@gmail.com

ANDRIES MOTAU, a QES Scholar, is a doctoral candidate with the African Climate and Development Institute and the Minerals to Metals group at the University of Cape Town, South Africa. He organized and facilitated a Climate Justice webinar series in 2020-2021 at the Centre for Civil Society, University of KwaZulu-Natal, Durban South Africa https://sobeds.ukzn.ac.za/events/category/special-webinar-series-climate-justice/list/?eventDisplay=past. latumotau@gmail.com

PATIENCE MUKUYU, a QES Scholar, is a doctoral candidate in Development Studies at the University of Johannesburg, South Africa. pmukuyu@gmail.com

SOLOMON NJENGA, a QES Scholar, holds a PhD in Earth and Climate Sciences from the Institute for Climate Change and Adaptation (ICCA), University of Nairobi, Kenya. He is a Climate Research Fellow at the US International University - Africa (USIU-Africa) in Nairobi, Kenya. njengasolomon@gmail.com

AICO NOGUEIRA, a QES Scholar, is a Research Collaborator at the Luiz de Queiroz Higher School of Agriculture at the University of São Paulo, and a Research Fellow at the Maria Sibylla Merian Center for Advanced Latin American Studies in the Humanities and Social Sciences (CALAS), Guadalajara, México. aico.nogueira@gmail.com

DAVID NÚÑEZ is an independent investigator and member of the ICCA consortium - Territories and areas conserved by Indigenous peoples and local communities, Chile. tokoiwe@gmail.com

MOSES O. OLAWOLE is an Associate Professor at Obafemi Awolowo University in Ile-Ife, Nigeria. molawole@oauife.edu.ng

ADEWALE M. OLAYIWOLA is an Associate Professor at Obafemi Awolowo University in Ile-Ife, Nigeria. amolayiwola@oauife.edu.ng

ABIMBOLA OLUWARANTI is an Associate Professor at Obafemi Awolowo University in Ile-Ife, Nigeria. boluwaranti@oauife.edu.ng

CLAUDIO OYARZÚN is a member of the Mapuche-Huilliche Communities Association of Carelmapu (Asociación de Comunidades Mapuche-Huilliche de Carelmapu), Chile.

MARCOS GERVASIO PEREIRA is a Professor at the Federal Rural University of Rio de Janeiro, Seropédica, Brazil. mgervasiopereira01@gmail.com

RAFAEL PEREIRA is a researcher at the National Centre for Early Warning and Monitoring of Natural Disasters, Brazil. rafael.pereira@cemaden.gov.br

PATRICIA E. PERKINS is a Professor in the Faculty of Environmental and Urban Change, York University, Toronto, Canada. esperk@yorku.ca

EMERSON RAMOS is a Community Leader at Associação dos Remanescentes de Quilombo de Santa Rita do Bracuí (Arquisabra), Angra dos Reis, Brazil. emerson-mec@hotmail.com

WLADIMIR RIQUELME is a graduate student at the Pontificia Universidad Católica de Chile, Santiago, Chile. wladiriquelme@gmail.com

DANIELLA RUIZ is a member of the Mapuche-Huilliche Communities Association of Carelmapu (Asociación de Comunidades Mapuche-Huilliche de Carelmapu), Chile.

LARA DA SILVA is a MSc student in Environmental Sciences at UNISUL, Tubarão, Brazil. lara934166@gmail.com

RAMIN SOLEYMANI-FARD is an investigator at the Universitat Autònoma de Barcelona, Barcelona, Spain. Ramin.Soleymani@uab.cat

MARCOS SORRENTINO is a senior professor at the Department of Forest Sciences and coordinator of the Laboratory of Education and Environmental Policy (Oca) at the Luiz de Queiroz Higher School of Agriculture at the University of São Paulo, and a visiting professor in the Graduate Program in Education at the Federal University of Bahia, Brazil. sorrentino.ea@gmail.com

MESMIN TCHINDJANG holds a PhD in Geomorphology from the University of Paris 7, France and is a Professor in the Department of Geography, University of Yaoundé I, Cameroon. Tchindjang.mesmin@gmail.com

FRANCISCO THER-RÍOS is an Associate Professor at the University of Los Lagos, Osorno, Chile. fther@ulagos.cl

ADRIEN TOFIGHI-NIAKI is an investigator at the Universitat Autònoma de Barcelona, Spain. adrien.tofighi@uab.cat

ANA P. D. TURETTA is a Researcher at Brazilian Agricultural Research Corporation (Embrapa Solos), Rio de Janeiro, Brazil, and a Professor in the Graduate Program in Territorial Development and Public Policy at the Federal Rural University of Rio de Janeiro, Seropédica, Brazil. ana.turetta@embrapa.br

DARLYS VARGAS is a member of the Mapuche-Huilliche Communities Association of Carelmapu (Asociación de Comunidades Mapuche-Huilliche de Carelmapu), Chile.

PATRICIA FIGUEIREDO WALKER holds a Master's in Environmental Studies from York University, Toronto, Canada. pattyf@yorku.ca

LEMLEM F. WELDEMARIAM is a PhD student at the University of Vienna, Austria. lemlem.weldemariam@univie.ac.at

KATHRYN WELLS holds a Master of Arts in Sociology from York University, Toronto, Canada. kathrynwells@live.ca

GONZALO ZAMORANO is an Independent Researcher in Chile. gonzalozamoranoh@gmail.com

Index

Page numbers with an *f* refer to figures.
Page numbers with a *t* refer to tables.

A

Abenaki (First Nations, Canada), 1
academic research: challenges with, 12, 17; responsibilities of, 20
ACCRA (Africa Climate Change Resilience Alliance) initiative, 57
ACS. *See* activist citizen science (ACS)
Acselrad, H., 158, 163
action-oriented research: benefits of, 6, 7*f*, 320, 322; and participatory geographic information systems (GIS), 26–27
action research. *See* action-oriented research
activist citizen science (ACS): about, 272; benefits of, 272; science, basic training, 265, 267, 275n6; Vaal Environmental Justice Alliance (VEJA), 273; water justice, 259, 260. *See also* citizen science (CS)
activities: as data collected, participatory community monitoring, 229
adaptation strategies: farmers, smallholder, 60–62; oil palm planters, 86–90, 87*f*, 89*f*
Africa: apartheid regime, 239, 241, 244–245, 247, 253–254, 259, 282, 287; challenges worsened by climate change, 49, 58–59; extreme weather events, 48–49. *See also* Cameroon; Kenya; Mozambique; Nigeria; South Africa
Africa Climate Change Resilience Alliance (ACCRA) initiative, 57
Agbeloba Farmers Society of Ilu-Aje (Nigeria), 52
agency: community, 6, 7*f*, 33, 48, 140, 146–147; and resilience, 106, 110; and youth, 345–346
Agenda 21 pilot project (Guapiruvu community, Brazil), 143–144

aggiornamento, 302–303
"A Green New Eskom" (South Africa), 283
agricultural inputs, 75, 76–78, 77*f*, 88–89, 89*f*, 93n8
agriculture: as cause of climate change, 69; impacts of climate change on, 69–70
agroecology, 129, 337
agroforestry: benefits of, 126–127, 127*f*, 129; and food insecurity, 330–331; as income source, 120, 326; socioeconomic implications of, 327–329
AGUA. *See* Solidarity Economy and Sustainable Development Association of Guapiruvu (AGUA)
Aiê Eleteloju (*Quilombo do Bracuí*), 129
air pollution: and citizen science (CS), 228, 260, 269, 273; coal mining, 283, 285
Alberta: "For Our Common Home" campaign, 307, 308–310, 308*f*, 311–312, 311*f*; oil sands, 351–352
Almeida, A.W.B., 130–131
Al-Shabaab extremists, 180, 195
Alternactiva collective, 322
Alto Ribiero State Park (Brazil), 143, 144
Amazon: Francis (pope) on Indigenous Peoples of, 301–302; religious environmental campaigns in, 298. *See also* Brazil; "For Our Common Home" campaign
Amazon Forest, 161, 307
Amazon river, 155
AMOQC (*Associação de Moradores do Quilombo Campinho da Independencia*) (Brazil), 122
Aneja, V.P., 285
Angostura dam (Biobío River, Chile), 208
Angra dos Reis Nuclear Power Plants (Brazil), 122

climate activism: exclusion of youth from, 343; youth, opportunities for, 347–349, 350–353

climate anxiety, 346–347

climate change: as data collected, participatory community monitoring, 229; effects of, 25; environmental justice response to, 137–138; Francis (pope) on, 300; research, 25–26

climate change causes: agriculture, 69; coal mining, 286–287. *See also* climate change

climate change impacts: on Africa, 70, 90, 91; on agriculture, 69–70; on Cameroon, 70, 91; classification of, 34, 35*f*; on farmers, 49, 58–59, 62; on Global South, 137–138; on oil palm sector, 82–83, 82*t*, 84*f*, 85, 85*f*, 86*f*, 91. *See also* climate change

climate justice: and carbon credits, 331–334; coal industry, 284, 290; and commons governance, 7*f*; community engagement processes, 198–201; environmental justice, compared to, 14–15; explained, 2, 15, 207; and just transformations, 13–14; and Just Transition, 288; lawsuits, 353–355; mining, 162, 171, 172; top-down, 332–333, 336; water-related, 156; workshops on, 322, 323*f*, 324. *See also* participatory research

Climate Justice Guelph, 352

Climate Justice International (CJI), 180

climate variability, 319

climatological indicators (of climate change impacts), 34, 35*f*

cloud harvesting, 267

CoAdapta | Litoral project (Chile), 32, 33, 36

CoAdapta methodology, 31–32, 32*f*, 34–36, 35*f*, 39

coal-fired power plants: protest movements against, 179

coal industry: as cause of climate change, 286–287; disinvestment in, 282; environmental impacts of, 283, 285–286; Just Transition movement, 281–282, 283, 284; resistance to, 289–291; social impacts of, 286

Coastal GasLink pipeline (Canada), 311

Cobquecura earthquake (2010) (Chile), 31

Cock, J., 288

co-created citizen science, 268–269, 271. *See also* citizen science (CS)

Coelho, T.P., 165–166

cognitive justice, 17, 261–262, 274

collaborative citizen science, 268–269, 271. *See also* citizen science (CS)

colonial globalization, 11

colonialism: culture of silence, 166–167, 167*f*, 173n5; dismantling structures of, 16, 17, 20;

displacement of original peoples, 1, 21n1, 102, 141–142, 208–209; globalization of, 11; Global North (Western) impositions of, 12, 14–15; inequities caused by, 1, 2, 6, 15, 69, 309; Just Transitions, challenges with, 13, 284, 287–289, 290–292; and land access inequality, 129–131; oil palm exploitation, 93n3; *Quilombola* communities and, 116–117, 120, 121, 129–131; and water access inequality, 239–240, 241, 244–247, 248–249, 253–254. *See also* capitalism; decolonization

colonization. *See* colonialism

common-pool natural resources: common property, 105–106, 110–111; described, 99, 113n2; governance of, 100, 101, 102–106, 110–111, 112; open access, 102, 105, 109; private property, 103–104; public or state property, 104; resilience, concept of, 106, 110–111

common property regime: common-pool natural resources, 105–106; rules governing, 110

commons: described, 2, 8n1; tragedy of the, 2. *See also* common-pool natural resources; knowledge commons; marine commons; "The" Tragedy of the Commons (Hardin)

commons-building, 3

commons governance: benefits of, 6, 7*f*; challenges of, 141–143; commons spaces, protecting, 19; communal ownership, 105–106

commons justice, 15

commons-reclamation, 3

commons spaces: protecting, 19

Commune of Ngwéi (Cameroon): adaptation strategies, 86–90, 87*f*, 89*f*; agricultural inputs, access, 76–78, 77*f*, 88–89, 89*f*, 93n8; climate change impact, 82–83, 82*t*, 84*f*, 85, 85*f*, 86*f*; climate trends, 78–81, 79*f*, 80*f*, 82*f*, 90–91; described, 71, 72*f*, 73, 73*f*, 93n1; exploitation, oil palm, 74–76, 93n3; land-management system, 75–76, 76*f*; Oil Palm and Adaptive Landscape (OPAL) project, 74, 93n5; oil palm production, 70, 71, 72; participatory research, 70–71, 74

communities (local): engagement process, 198–201; initiatives in response to climate change, 138–139; ownership of natural resources, 105–106, 109–110; Protected Areas (PAs), challenges with management of, 141–143; resilience, factors that provide, 106, 110–111; social classification system criteria, 145

community gardens: benefits of, 19, 100–101; engagement, importance of local, 107;

Mother Earth (Mother Nature), 301, 302
Mount Mabu (Mozambique), 334
Mozambique: carbon capture project, 319–320, 324–325; climate change mitigation strategies, 325; economy, 317; extreme weather events, 319; green extractivism, 317, 319, 331–332; *machambas* (small pieces of land), 325, 326–327, 328, 329, 330, 331; map of, 318*f*; REDD+ carbon-sequestration projects, 3, 320. *See also* Mabu, Zambézia province (Mozambique); Nhambita, Sofala province (Mozambique)
Mpumalanga Province (South Africa): coal industry, resistance against, 284, 289–290; environmental effects of coal mining, 285–286; Just Transition (coal industry), 282; map of, 242*f*; participatory research, 291; social impacts of coal mining, 286
multiplying vulnerability index, 50, 52
Munnik, V., 284
Mura people of Manaus (Amazonian Indigenous community), 307, 309–310

N

Nakoda (First Nations, Canada), 1
Nalule, V. R., 283
Namadoi community (Mabu, Mozambique), 334
Nangaze community (Mabu, Mozambique), 334, 335
National Development Plan (South Africa), 281, 283
National Environmental Policy (PNMA) (Brazil), 159
National Information System on Dams (Brazil), 171, 174n7
National Institute of Cartography (INC) (Brazil), 93n6
National Institute of Colonization and Agrarian Reform (INCRA) (Brazil), 141, 144
National Mining Agency (Brazil), 160–161
National Policy on Dam Security (Política Nacional de Segurança de Barragens PNSB) (Brazil), 160
National Strategy for Adaptation and Mitigation of Climate Change (Mozambique), 325
National Strategy for Reducing Emissions from Deforestation and Forest Degradation (Mozambique), 325
National Water Act (NWA) (1998) (South Africa), 245–246, 247, 248–249, 264
National Water Policy (1997) (South Africa), 264

National Water Resources Policy (PNRH) (Water Law) (Brazil), 159–160
National Water Resources Strategy 2 (NWRS) (South Africa), 251–252, 263, 264
Natives Land Act (1913) (South Africa), 244
natural disasters: LICCI indicators, 34, 35*f*; risks, 31
Nature: as a commodity, 211; rights of, 211–212; spiritual relationship with, 299–300
Nature's Contributions to People (NCP): defined, 118; soils' role in delivering, 118–119, 118*t*
Nêhiyawêwin (First Nations, Canada), 1
New Mining Code (Brazil), 159, 161–162
ngütram (traditional format of extended conversation), 213
Ngwéi, Commune of (Cameroon). *See* Commune of Ngwéi (Cameroon)
Nhambita, Sofala province (Mozambique): about, 325–326; action-oriented research, 322, 323*f*, 324; carbon capture project, 319–320, 324–325; climate justice, workshops on, 322, 323*f*, 324; exchange visits with Mabu, 324; food insecurity, 330–331; implications of REDD+ project, 327–329, 330–331, 332; map of, 318*f*; REDD+ (Sofala Community Carbon) project, 320, 325; top-down climate solutions, 332–333, 336; tree planting for carbon credits, 326–327
Nigeria: farmers, effect of climate change on, 49, 58–59; focus group discussions (FGDs) with farmers, 50, 52, 53–54, 54*f*; National Adaptation Programmes of Action (NAPAs), 60; training workshops with farmers, 52, 53, 53*f*, 54*f*, 55, 56*f*. *See also* Africa; farmers, smallholder (small-scale)
Niitsitapi (First Nations, Canada), 1
nitrogen dioxide (NO_2) emissions, 285
Nvava community (Mabu, Mozambique), 334
NWRS. *See* National Water Resources Strategy 2 (NWRS) (South Africa)

O

Obafemi Awolowo University (Ile-Ife, Nigeria), 53
Oblo (LICCION digital platform), 32, 34, 36, 38*f*, 39, 40
OBM (Mining Dams Observatory) (Brazil), 172–173
O'Brien, K., 350–351
Observatory of Sustainable and Healthy Territories of Bocaina (OTSS), 125
oceans. *See* common-pool natural resources

private property regime: common-pool natural resources, 103–104

privatization: of water, 265–266

production: of environmental resources, 157–158

Programa Operaçao Trabalho (POT) (São Paulo), 108

Projecto de Desenvolvimento Sustentável (PDS) (Brazil), 113n4

property regimes: for common-pool resources, 102–106, 109, 110–111, 112

Prospectors & Development Association of Canada (PDAC), 166

Protected Areas (PAs): challenges with the management of, 141–143; restrictions on, 138

PSBM (Mining Dam Safety Plans/Plano de Segurança de Barragem) (Brazil), 160

public (state) property regime: common-pool natural resources, 104

Public SIGBM (Brazil), 171

public sphere dimension (*Laudato Si'* movement), 304, 304f, 305–306

Púnguè River (Mozambique), 318f, 325

PV. *See* participatory video (PV)

Q

Queer, 21n1

Quellón (Chile), 28, 29f, 32. *See also* Chile

Quepuca Ralco community. *See* Mapuche-Pewenche Indigenous People (Wallmapu)

Queuco River (Chile): about, 208, 212–213; participatory research, river monitoring strategy, 213–214, 215f, 216; social cartography (participatory mapping), 216–218, 217f; water rights, 210–211

Queuco River Defense Network. *See* Red por la Defensa del Río Queuco (RDRQ)

Quilombo do Bracuí (was *Santa Rita do Bracuí*): about, 121–122; map of, 30f; soils, social value of, 129

Quilombo do Campinho (was *do Campinho da Independência*): about, 120–121; agroforests, 126; map of, 30f

Quilombola communities (Brazil): agroforests, 126–127, 127f; challenges faced by, 120, 121–122; collective land ownership, 3, 19, 113n4, 130–131; history of, 116–117; land acquisition, inequality in, 129–130; land rights, 130–131; participatory research, 119–120, 122, 123f, 124f, 125; soils, local knowledge of, 126; transfer of traditional knowledge, 127, 128f, 129. *See also* Brazil;

specific Quilombos (e.g., Quilombo do Campinho)

Quilombos, 117. *See also Quilombola* communities (Brazil)

R

race, 14; Environmental Justice Movement (EJM) and, 14, 15; Just Transition programs, 290, 291; knowledge sharing across, 6; and poverty, 251–252, 271; and water access, 248–249, 250, 251–252, 253, 254; and youth, climate change activism, 343, 344. *See also* environmental racism; historically disadvantaged individuals (HDI) (Black people); institutional racism; socio-economic levels (class)

racism. *See* environmental racism; institutional racism

Rain Forest agro-climatic zone (Nigeria), 50, 51f, 61–62

rainwater harvesting, 267

Ralco dam (Biobío River, Chile), 208, 209

Ralco Lepoy community. *See* Mapuche-Pewenche Indigenous People (Wallmapu)

Ramos, E.L., 129

RBC (Canada bank), 352

RDRQ. *See* Red por la Defensa del Río Queuco (RDRQ)

RDS (*Reserva de Desenvolvimento Sustentável* Sustainable Development Reserves) (Brazil), 113n4

RECSOIL recarbonization program (Brazil), 116

REDD+ (reducing emissions from deforestation and forest degradation): carbon capture, 324–325; failure of, 3, 333–334. *See also* Nhambita

Red por la Defensa del Río Queuco (RDRQ) (Chile), 211, 212, 213

red tides (algal blooms), 31

reducing emissions from deforestation and forest degradation (REDD+). *See* REDD+

Reguemos Chile (private corporation), 209–210

religion: and climate action, 297–298

religious environmentalism, 303, 305–306

rematriation (Indigenous concept of), 299, 301, 303

renewable energies: concerns with, 207–208; hydropower, impacts of, 209

research: climate change, 25–26. *See also* academic research; action-oriented research; participatory research

researchers: responsibilities of, 20

Printed in the USA
CPSIA information can be obtained
at www.ICGtesting.com
LVHW061721141223
766477LV00009B/70